STATISTICAL METHODS FOR COMMUNICATION RESEARCHERS AND PROFESSIONALS

René Weber, Ph.D., M.D.
Ryan Fuller, M.B.A.

University of California, Santa Barbara

Kendall Hunt
publishing company

Kendall Hunt
publishing company

www.kendallhunt.com
Send all inquiries to:
4050 Westmark Drive
Dubuque, IA 52004-1840

ISBN 978-1-4652-1223-8

Printed in the United States of America
10 9 8 7 6 5 4 3 2 1

ABOUT THE AUTHORS

René Weber is an Associate Professor in the Department of Communication at the University of California, Santa Barbara. He holds a Ph.D. (Dr.rer.nat.) in Psychology (Berlin University of Technology, Germany) and an M.D. (Dr.rer.medic.) in Cognitive Neuroscience (RWTH University Aachen, Germany). He also holds Bachelor and Master's Degrees in Communication and Business Administration with an emphasis on econometrics and statistics. In his academic research he focuses on mass communication/audience behaviors and effects as a product and determinant of complex cognitions. He develops and applies both traditional social scientific and neuroscientific methodology (e.g., fMRI) to test media related theories. His research has been published in major communication and neuroscience journals, in numerous book chapters, and in three books.

Ryan Fuller is a Doctoral Candidate in the Department of Communication at the University of California, Santa Barbara, and Research Associate at the Carsey-Wolf Center. He holds a Bachelor Degree in Communication (UC Davis) and a Master Degree in Business Administration (San Francisco State University). His research interests include organizational communication, conflict and negotiation, labor unions, media industries, and communication research methods.

WEBSITE & SUGGESTIONS

The authors maintain a website with additional information about the book. The site can be accessed at http://statsbook.dr-rene-weber.com (click on "Statistics Textbook"). At the book's website, readers can find datasets, additional examples and problems, errata, links to other statistics sources, course materials, etc. The website also provides a link to a "suggestion database." The authors welcome any suggestion for improvement and appreciate any information regarding errors, poor wording or other issues. Via the suggestion database readers can easily and anonymously submit their suggestions and retrieve suggestions that other readers have submitted. Suggestions can also be submitted via email to René Weber (renew@comm.ucsb.edu) or Ryan Fuller (ryanfuller@gmail.com).

ACKNOWLEDGMENTS

There are many people we would both like to acknowledge and thank for helping to make this book a reality. First, we would like to thank Kendall Hunt's Senior Acquisitions Editor Joseph Wells for approaching us about this project. We also thank Kendall Hunt's Project Coordinators Wendy Pauly (for the preliminary edition) and Thalia Cutsforth (for the first edition), as well as our publisher's design and copyediting staff for all their efforts and good advice. Second, we would like to thank our research assistant, Audrey Abeyta, who helped us to find examples to include in this book. Not only did she help us acquire recent examples in flagship communication journals, she also evaluated whether examples were accessible and interesting for undergraduate students. Third, we would like to thank UC Santa Barbara graduate students Christopher Seaman and Alex Markov. They served as teaching assistants for the class "Statistical Analysis for Communication Researchers (COMM87)," and provided us insights about challenges students face with statistical methods, helped to steer students to create problem set examples, and helped to filter out better examples for inclusion in this book. We thank UC Santa Barbara graduate student James Michael Mangus for setting up and maintaining our "textbook suggestion database." This database is an important tool for the ongoing improvement of this book. We would also both like to thank our students for serving as test cases for this project, and helping us to improve the quality of this textbook for future students.

In addition to these shared acknowledgements, René would like to thank his former advisor, the German Psychologist and Statistician Dr. Jürgen Bortz (* March 8, 1943, † September 22, 2007). His guidance was invaluable for René's academic career. His lifelong mission to further the *understanding* of statistics and methods among social scientists lives on in this book. René also would like to thank Ryan for being such a patient co-author.

Ryan would like to thank René for helping him to expand on his knowledge of statistics, and for being an outstanding co-author and mentor throughout this process. He would also like to thank his wife, Serena, for her tremendous support during the writing of this book.

Several other people along the way helped to bring this book to fruition. We are tremendously grateful for all support we have experienced. We look forward to the feedback from our current and future students so that we may continue to improve the content of this book.

CONTENTS

INTRODUCTION

RATIONALE

The purpose of *Statistical Methods for Communication Researchers and Professionals* is to provide students with a beginning-to-intermediate overview of statistical methods in the context of communication research and practice. Communication researchers focus on how people use messages to create meaning within and across contexts, cultures, and media. Professional communicators' work revolves around creating messages. While there are many textbooks on statistical methods, and some that deal with their application in the social sciences, there are fewer still that deal specifically with communication processes and effects. Drawing on a range of examples from different arenas in communication research and practice, this text is designed to orient students toward the use of statistical methods and to ask appropriate and analytical questions about knowledge claims. This text will prepare students who go on to careers as research scientists or as professional communicators (e.g., advertisers, journalists, marketers, and public relations consultants), and who are responsible for describing, evaluating, and interpreting data they confront daily in their professions.

We are well aware of students' misgivings about statistical methods. Statistics anxiety is well documented in the education literature (Onwuegbuzie & Wilson, 2003). From our own experience, the statistical methods course requirement is sometimes met with groans or jeers from our students. We sometimes hear comments such as "Statistics is hard and sucks"; "Stats is boring"; and even "Statisticians are liars."

We know that students bring a number of assumptions or biases when they approach a new subject matter, and we are aware of the baggage associated with statistical methods. These biases are evident in some representative comments from famous personalities our students can quote when asked what they associate with statistics. Author Mark Twain is credited with saying, "There are lies, damned lies, and statistics." The former British Prime Minister Winston Churchill once said, "I only believe in statistics that I forged myself." While we acknowledge the misuse of statistics, we also believe these biases emerge and persist when the purposes and application of statistical methods are not well understood. Unfortunately, statisticians have neither done the best job with reputation management, nor had such a quotable proponent.

The principal goal of *Statistical Methods for Communication Researchers and Professionals* is to help students develop the statistical competency necessary to become good researchers and good "statistical citizens" in the information age. This includes promoting students' *understanding* of ideas and terms of statistical methods, and cultivating their abilities to evaluate data and the claims they confront, make decisions, and explain those choices to statisticians and non-statisticians alike. We draw on our experience teaching thousands of undergraduates and graduate students and extant literature on teaching statistical methods. Our approach emphasizes using real data and authentic applications. Further, our approach presents data in a context that is meaningful to students (a focus on communication phenomena for communication majors), and emphasizes the science behind the data.

Importance of Statistical Methods

Every day we are inundated with information in our professional as well as personal lives. We simply cannot attend to everything, and need help in separating out and evaluating claims. Understanding statistical methods provides us with an efficient way to explore, analyze and interpret data, as well as evaluate the strength of evidence presented in support of or opposition to claims. In this chapter, we offer the value of statistical methods to you, either as a communication researcher or as a communication professional.

Statistical Methods Enable Scholars and Professionals to Share Ideas

Although our aim in this text is to focus on statistical methods in communication studies and professions, we want to underscore that statistical methods are widely used, and knowing the language helps scholars and professionals share ideas. Statistical methods provide a common platform on which many disciplines with a quantitative orientation rely. Having solid training in statistical methods can lead to a broader understanding of research and its interdisciplinary nature. For example, in addition to our own conferences and journals, communication scholars have spread ideas into the fields of psychology, sociology, health, and business and management, to name a few. From a professional standpoint, knowledge of statistical methods has a similar benefit. For example, a science journalist with a solid understanding of statistical methods will easily understand general principles of a research project, be it in engineering and physics or communication and psychology. While engineers speak of signals and noise, and communication and psychology scholars may speak of treatment and error, the statistical concept behind both pairs of terms is the same. As well, statistical methods training helps the journalist understand the limitations of any claims generated from a particular project, and translate those findings for news audiences. In sum, solid training in statistical methods helps scholars or professionals understand and share ideas that advance their disciplines or have a broader public impact, as well as prevent the promulgation of bad ideas by filtering out studies with dubious findings and weak research designs.

Statistical Methods Provide Scholars and Professionals with Skills That Will Benefit Their Careers

Understanding and applying statistical methods can have immense value for your career. For example, recent news reports have underscored how today's job seekers gain a competitive advantage if they are able to explore, analyze, and interpret the massive amounts of data available in the digital age (Lohr, 2009; The Economist, 2010). Students whose undergraduate training is in a communication discipline go on to careers in advertising, journalism, marketing, and public relations, to name a few. These are careers in which the ability to apply statistical ideas to decision making can be important. Whether you are trying to make sense of polling data as a journalist or attempting to tease out consumer preferences as a marketer, your understanding of statistical methods will provide you with more power behind the evidence you use to support your claims. Additionally, your ability to read and translate academic research can help you bring together domains that have not talked to one another but benefit from such bridge-building. Ultimately, a strong education in statistical methods trains students in critical thinking, including generating and ruling out alternative explanations for an outcome or relationship and addressing contradictory findings, and leads to better decisions. Better decisions lead to better performance and success. There is an obvious relationship between success and job security and job satisfaction.

Understanding Statistical Methods will Help You to Become a Good "Statistical Citizen"

If you remain convinced that you will never come across concepts employed in statistical methods again, we hope to emphasize the practical importance of marshaling this knowledge. We are betting that you will come across these concepts, though. Understanding the ideas behind statistical methods will help you to become a better consumer of information, and enable you to critically analyze claims made in both academic and non-academic settings. No doubt you have heard, read, or watched a number of news stories that report academic findings (e.g., "New study finds . . ."). Or you have likely come across advertisements or other statements that make claims that seem "too good to be true" (e.g., think of weight-loss drugs or exercise equipment or regimens) and which may be based on less-than-solid evidence. Solid training in statistical methods will help you to cultivate a sense of healthy skepticism, separate sound from dubious claims, generate possible alternative explanations for observed outcomes or relationships, and hopefully settle or resolve contradictory findings.

OVERVIEW OF TEXT

Throughout *Statistical Methods for Communication Researchers and Professionals,* we will address three primary goals of statistical methods. First is the design of samples, including those based on probability and non-probability procedures. Second is describing, exploring, and summarizing data in samples or non-samples. Third is making predictions and generalizations from a sample to the population we are interested in, or testing assumptions in populations with data in samples. Our textbook is aligned with addressing each of these three goals of statistical methods.

Our examples and problem sets center on promoting students' statistical competency, including 1) data awareness, 2) understanding statistical concepts and terminology, 3) understanding data collection and statistics generation, 4) interpretation skills, and 5) communication skills (Rumsey, 2002). Data awareness focuses on how data are pervasive, how data are (mis)used, and how data-driven decisions can greatly influence our lives. Understanding of statistical concepts goes beyond calculations, use of formulas, and explaining in statistical language, or what we call the cook-book approach to statistical methods. Instead, evidence of statistical understanding centers on the student's ability to use a concept within a broader problem and to answer questions about it. By understanding data collection and generating statistics, we mean providing students opportunities to learn firsthand to create their own research questions/hypotheses, gather data, and generate statistics based on their data. By interpretation skills, we mean having students provide in their own words what the results mean in the context of their hypothesis or research question. And by communication skills, we mean promoting the student's ability to demonstrate and exchange statistical ideas for the benefit of different audiences. And after all, communication skills (spoken and written) are frequently mentioned by employers as highly desired skills of new graduates. When our students move beyond the classroom and into the workforce, our goal is that they are able to apply statistical reasoning to explain their decisions in the context of a particular problem, regardless of whether they are in a research or applied setting and whether their audience has statistical methods training. Even if students themselves do not apply statistical reasoning in their jobs directly, we hope that this text helps them to ask the right questions to statisticians and research professionals that provide information for their business decisions. Asking informed questions usually translates into higher quality information and, as a consequence, to better decisions.

Our textbook examples are drawn from recent work published in communication journals. We caution that our examples' data are not the authors' actual data, but instead are simulated. The results more or less correspond to the authors' findings on which these examples are based. In some cases, we have made the effect sizes larger for didactic purposes. We present these examples as if students are running replication studies (something which is often encouraged).

In the following section, we provide an overview of chapter content. These represent broad strokes for each chapter.

Chapters 1 through 6 provide the conceptual groundwork for students to understand statistical methods. **Chapter 1: Introduction to Statistical Methods and Measurement** provides the foundational ideas necessary to understanding and using the language of statistical methods, including distinguishing among populations and samples, parameters and statistics, and different levels of measurement. **Chapter 2: Descriptive Statistics** reviews graphical and numerical description of data, including measures of central tendency and measures of variation. **Chapter 3: Probability and Probability Distributions** overviews the core concepts of probability, an important prerequisite for the chapters to follow. **Chapter 4: Sampling Methods** reviews non-probability and probability sampling methods which helps to understand how researchers obtain a representative selection of study elements. **Chapter 5: Sampling Distribution & Central Limit Theorem** introduces the Central Limit Theorem, one of the core concepts in (parametric) statistics. Good knowledge of this chapter is essential for the understanding of subsequent chapters. **Chapter 6: Estimation (Confidence Intervals) & Testing** discusses point and interval estimation; i.e., how to predict conditions (usually parameters) in populations based on data in samples. This chapter also provides the conceptual basis of hypothesis testing, including specifying directional and non-directional research hypotheses, null hypothesis, decisions and types of errors in tests of hypotheses, statistical significance, practical significance of test results, test power, and effect size.

Chapters 7 and 8 demonstrate specific tests and the application of concepts learned from in Chapters 1 through 6. **Chapter 7: Specific Significance Tests for Differences** reviews specific tests for differences between groups, including *t*-test for two independent samples, *t*-test for dependent samples, Chi-square tests, and one-way Analysis of Variance (ANOVA). **Chapter 8: Specific Significance Tests for Dependencies** overviews concepts related to tests of relationships, including regression, covariance, and correlation.

Finally, **Chapter 9: Epilogue (A Critical Perspective on Statistics)** provides some reflection on the use and misuse of statistical reasoning by communication researchers and professionals. For writing a critical chapter we have invited our colleague, Timothy R. Levine, from Michigan State University, one of the most accomplished and published scholars in the field of communication. In his chapter he argues (among other things) that a statistical analysis can never make up for a poor research rationale or for a poor research design, and he criticizes the undue use of advanced statistical methods in communication research, which seems to be a trend in recent years. The chapter also discusses the "tyranny of *p*-values" and widespread "hypothetico-deductive extremism" in communication research.

Our textbook's appendices are also worth mentioning here. First, **Appendix A** provides tables that we refer to throughout the textbook, including the standard normal distribution, the *t*- and *F*-distribution, and the Chi-square distribution. Second, students will find brief solutions for the problem sets in **Appendix B. Appendix C** offers recommended course schedules for classes running on either semester or quarter systems. Finally, **Appendix D** is a how-to manual for a graphical user interface for R, a free, open-source statistical computing program. This includes twelve problems, and covers everything from downloading and installing R and R Commander to running the most important statistical tests in this textbook.

Recommendations for Class Schedules

Our textbook can be used for statistics classes at universities that operate with a quarter system (10 weeks) or a semester (14 weeks) system. In our experience, the semester system provides the necessary time to discuss the content of all chapters in sufficient detail. Instructors that need to teach the class in a quarter system will not be able to discuss all topics in detail—for those instructors it will be important to remind students that studying our textbook, including the appendices and problem sets, is crucial for the successful completion of a statistics course. Our recommendations for instructors on a quarter system are as follows: Content of Chapters 1 and 2 should be discussed in detail. Lectures that refer to Chapter 3, and especially to Chapter 4, can be shortened. We would not recommend shortening the lecture on Chapters 5 and 6 as those are the crucial for the understanding of statistical inference. Likewise, while Chapters 7 and 8 should be discussed in detail, Chapter 9 can be shortened or not discussed at all in classes on a quarter system. Appendix C contains examples of schedules for classes on a quarter and on a semester system.

Recommendations for Class Tutorials & Statistics Software

We recommend that introductory statistics classes for communication researchers and professionals are accompanied by class tutorials. In our experience, a bisectional organization of these tutorials proved to be effective. For example, the first hour of a two-hour tutorial should be spent with the discussion of lecture content, review of the homework, and problem sets, etc. The second hour should be spent with solving statistical problems, including the problem sets of this textbook, with statistics software. Most statistics textbooks for social science researchers recommend using SPSS© (Statistical Package for the Social Sciences) for this purpose. However, we do not follow this trend and suggest an alternative recommendation for two reasons: First, increasingly students struggle to buy student licenses of expensive, commercial statistics software for their computers. At the same time, we experienced that students who install statistics software on their own computers and learn to work with it, versus students who are forced to access statistics software at their university's computer labs, learn better, gain knowledge that is more sustainable, and are more motivated to do their own analyses and apply what they have learned in lecture and tutorial. Second, complicated copyright rules make it more difficult for instructors as necessary to use screenshots, software codes, etc. in their lecture materials and textbooks. At the same time, freely available statistics software applications have reached a high level of sophistication and user friendliness and are available for all computer platforms. Therefore, we recommend using "R" for class tutorials (http://www.r-project.org). Appendix D (Using R and R Commander for Statistical Analyses) contains a detailed and tailored introductory R manual for this textbook. It can be easily used in class tutorials or by communication professionals who are interested in conducting statistical analyses.

REFERENCES

Lohr, S. (2009, August 5). "For today's graduate, just one word: Statistics." *The New York Times.* Retrieved from http://www.nytimes.com/2009/08/06/technology/06stats.html?_r=2

"Data, data everywhere: A special report on managing information." (2010, February 27). *The Economist,* p. 1–16.

Onwuegbuzie, A. J., & Wilson, V. A. (2003). Statistics anxiety: Nature, etiology, antecedents, effects, and treatments—a comprehensive review of the literature. *Teaching in Higher Education,* 8(2), 195–209.

Rumsey, D. J. (2002). Statistical literacy as a goal for introductory statistics courses. *Journal of Statistics Education* Volume 10 (3), www.amstat.org/publications/jse/v10n3/rumsey2.html

Introduction to Statistical Methods and Measurement

CHAPTER 1

The primary focus of this chapter is to introduce important terms related to statistical methods and measurement.

In designing their studies, researchers have to make a number of decisions. These decisions are often guided by different sources, including current scholarly literature, the researcher's particular interests, and priorities of funding agencies, to name a few. These decisions drive who or what to collect information on, and how to gather data on the particular phenomenon of interest. In the following section, we define some important terms that briefly introduce statistical methods and measurement.

A study element is the unit about which information is collected and that provides the basis of analysis. In communication research, some examples of study elements include people (e.g., individuals, dyads, groups), newspaper articles, television news stories, television programs, television scenes, music lyrics, Web pages, feature films, short video clips, and video games, to name a few. Researchers collect information on variables, or characteristics for each study element that can vary in value among study elements. We will discuss variables in greater detail in the section Levels of Measurement.

Quantitative communication researchers mostly are not interested in case-studies, i.e., making claims about one study element, but instead investigating and recognizing patterns in aggregates. A population is defined as the theoretically specified aggregation of study elements. In other words, a population defines all elements to which you would like to generalize your findings in a particular study. The population of interest can be wide, such as all registered U.S. voters, or narrow, such as all registered student voters on a particular college campus. Wide populations are usually assumed to be infinite. Narrow populations are labeled as finite populations. Whenever the concept of a population is relevant in this book, we assume an infinite or wide population. In either case, your decision of the population to study is driven by what you are investigating, and determines the scope of your claims.

Often times, it is not practical to gather information on an entire population, due to time, money, or (even if we had time and money) the changing nature of a population of interest. A sample is a special subset of a population observed in order to make inferences about the nature of the population itself. If a researcher were interested in registered voters' knowledge of a particular campaign, her sample would need to represent the population of voters at large. Otherwise the researcher's findings would be biased. In a sense, a sample can be thought of as a "mini-population." Similarly, if a researcher were interested in students' likelihood to vote in an upcoming election, he would need to ensure that all registered student voters had a fair chance of being included in his study. We will discuss more on the strategies of sampling in Chapter 4.

Now that we have distinguished between population and sample, we have another important distinction to make: parameter and statistic.

A parameter is a numerical summary of a given variable *in a population.* Some examples of parameters of interest to communication scholars include average number of hours U.S. adults spend watching television or average number of hours U.S. adults spend surfing the Web.

A statistic is a numerical summary of *sample* data. A statistic is usually an estimator of a parameter. So called *inferential statistical methods* provide information on the accuracy of such estimations. (We will discuss this in greater detail in Chapter 5, which covers confidence intervals.) In the above examples, the statistic refers to the sample data of hours U.S. adults spend in front of a television or surfing the Web.

In statistical methods, a parameter, such as a population mean, is represented by a Greek and/or capital letter (μ), whereas a statistic, such as the sample mean, is represented by a Latin and/or a lowercase letter (y).

We will discuss more on types of parameters and statistics in subsequent chapters. Now we want only to sensitize you to the notion that one applies to populations (parameter) and the other refers to sample data (statistic). We now turn to the levels of measurement used in statistical methods.

LEVELS OF MEASUREMENT

By measurement, we are referring to assigning numbers to attributes of a variable. As mentioned above, a variable is a characteristic measured for each subject that can vary in value among study elements. Levels of measurement, also called a measurement scale, define rules for *how* numbers are assigned to attributes of a variable. We distinguish among different types of measurement scales (nominal, ordinal, interval, and ratio, all defined in the following section) because this often determines the statistical procedures that are applied.

Nominal

With a nominal measurement scale, we assign numbers to attributes of variables in a way that the numbers provide information on whether or not two subjects have the same attributes.

For example, if our study elements are people, we might be interested in recording their biological sex. We measure the attributes "male" and "female" by assigning the number "1" to male and "2" to female. The numbers have no particular meaning; they could also be "20" and "30" as long as they differentiate the biological sexes of the subjects. If we have two subjects with two different numbers on the variable sex, we know that they are different biological sexes; one is male and one is female. Figure 1.1 displays this relationship.

A variable that is measured on a nominal scale is referred to as a qualitative variable. In other words, the scale of measurement of a qualitative variable is a set of unordered categories.

Figure 1.1. Nominal variable scale

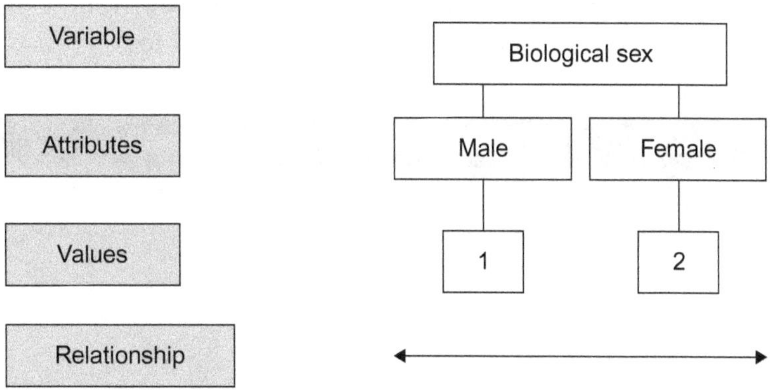

Biological sex is the variable on which we are collecting information on our study elements. Biological sex has the attributes "male" and "female," and the corresponding values "1" and "2." The level of measurement describes the relationship among these two values. In this case, sex is a nominal variable, so the numbers only differentiate two sexes.

It is important that nominal measures are exhaustive and mutually exclusive. Exhaustive means that all possibilities are accounted for by your categories. For example, if you were to ask your study subjects on your campus for whom they were most likely to vote, you would need to include all the candidates running for office. Considering that elections allow people to write in candidates as well, if you didn't include an "other" category, your list would not have been truly exhaustive. If you only asked one question, then you would also need to provide the option for your respondents to indicate that they are not planning to vote in the election. Table 1.1 provides a good example that is exhaustive because it includes all three candidates for a local election (Candidates A–C), an "other candidate" option, and an option for respondents to indicate they don't intend to vote.

Table 1.1. Exhaustive example

In the upcoming election, for whom are you most likely to vote?
A. Candidate A B. Candidate B C. Candidate C D. Other Candidate (Write-in) E. I don't plan to vote

Mutually exclusive means that study elements cannot fit in more than one category. We'll continue with our election example. Let's say that you were interested in whether voting intentions are related to where people grew up. For example, consider the following bad example in Table 1.2, which provides the respondent with an ambiguous choice.

Table 1.2. Mutual exclusivity example 1

Where did you grow up?[1]
A. Country
B. Farm
C. City
[1]http://www.statpac.com/surveys/question-qualities.htm

A person who grew up on a farm in the country might not know how to answer this question. The modification in Table 1.3 alleviates the ambiguity.

Table 1.3. Mutual exclusivity example 2

Where did you grow up?
A. Rural
B. Urban
C. Suburban

Ordinal

With an ordinal measurement scale, we assign numbers to attributes of variables such that the numbers provide information on the rank-order of the subjects. The numbers represent relatively more or less of a variable.

For example, again our study elements are people and we are interested in their level of education. Our ordinal measurement scale could represent "highest degree earned": 1 = doctorate, 2 = master's, 3 = bachelor's, 4 = associate's, 5 = high school.

The different numbers represent different academic degrees (like in nominal measurement), but we have even more information now. If we have two subjects with different numbers on the variable "highest degree earned" we know they have a different academic degree (like in a nominal measurement). We also know that the individual with the higher number (based on our scale) has a lower academic degree. However, we don't know how much lower the academic degree is, only that it is lower. To average the numbers has no meaning. The missing information on intensities of attributes is provided by interval or ratio measurements.

Interval

With interval scales, we assign numbers to attributes of variables in a way that the difference between the assigned numbers indicates the subjects' difference on a certain variable.

For example, our study elements are people, who we might ask to report on self-disclosure in their relationships as in Table 1.4.

Table 1.4. Self-disclosure question

When I am self-disclosing, I am consciously aware of what I am revealing (c.f., Wheeless, Nesser, & McCroskey, 1986).				
1 = Disagree	2 = Somewhat disagree	3 = Don't know	4 = Somewhat agree	5 = Agree

Statements on such rating scales are assumed to define an interval measurement, because we assume that two subjects that rate 1 and 2 feel the same difference in agreement as two subjects that rate 4 and 5, just at another level. An interval scale defines a continuum of a variable (in the above example, awareness of self-disclosures), and averages can be calculated and meaningfully interpreted. We cannot say that the person who marked 4 (Somewhat agree) compared with a person who marked 2 (Somewhat disagree) agrees twice as much. This is the case because in the above scale we do not have a true zero point. It is easy to imagine that we could have used a scale ranging from 1 (Completely disagree) to 7 (Completely agree) with "Somewhat agree" now represented by the scale value 5 and "Somewhat disagree" by the value 3. How we as researchers define the anchors of a scale is more or less arbitrary. Simply by adding more extreme anchors, the ratio between "Somewhat agree" to "Somewhat disagree" changed from 4/2 to 5/3; that is ratios of values are not constant along the continuum of the measurement scale. This is the reason why we cannot say that a study participant who checked 4 agreed twice as much to the statement as a participant who checked the value 2.

Ratio

With ratio scales, we assign numbers to attributes of variables in a way that the ratio between the assigned numbers indicates the subjects' ratio on a certain variable. This is typically the case if the scale has a true or naturally given zero point and, therefore, is not arbitrary or a result of a researcher's definition. Some examples of data that have zero points include time, distance, and income.

For example, again our study elements are people and we are interested in the number of hours they spend surfing the Web. We know that some people may not have access to the Web, or may deliberately not access it, so there is a true zero point. That is, they spend zero time surfing the Web. Two people who daily spend 3 hours and 4 hours on the Internet show the same time difference as two people who daily spend 5 and 6 hours on the Internet. The person who spends 6 hours surfing daily is online twice as long as the person who is on the Internet 3 hours daily.

To pull this all together, Figure 1.2 provides a summary of the levels of measurement and their relative strength.

Figure 1.2. Hierarchy of levels of measurement

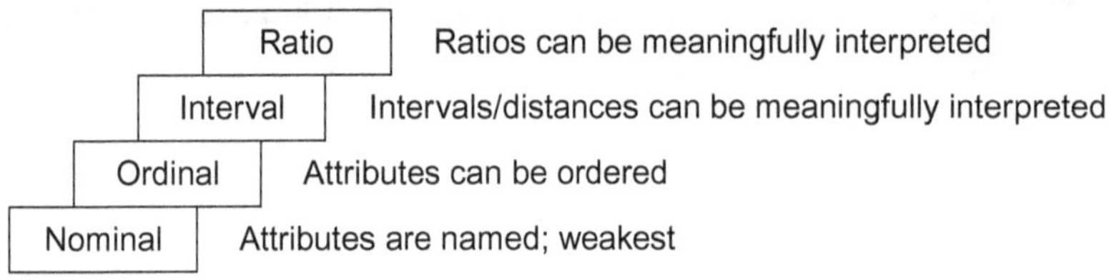

Table 1.5 provides an illustration of the different levels of measurement, using standard demographic variables, as well as some examples from interpersonal, group, organizational, and media contexts of communication research.

Recommendations

It is important to be aware of the level of measurement in order to apply the appropriate statistical procedure and to interpret findings correctly.

Nominal measurements provide the lowest amount of information, and ratio measurements the highest amount of information. You should always try to measure attributes on a high level of measurement if possible. For example, let respondents provide their age, income, time spent using certain media, rather than having them check pre-defined categories. You can always transfer interval and ratio level measurements (i.e., higher levels) into ordinal or nominal values (i.e., lower levels) by categorizing values (see below), but it is difficult to move from nominal and ordinal to interval or ratio levels.

SUMMARY

It is impossible to learn a new language (here statistical language) without studying the "vocabulary." This chapter introduced important terms related to statistics and measurement. We focused on distinguishing populations from samples, as well as the rationale for using samples. In addition, we defined the different levels of measurement that researchers use and provided examples. Based on the level of measurement researchers decide which statistical procedure is appropriate to analyze their data.

Table 1.5. Examples of different levels of measurement in communication studies

	Nominal	Ordinal	Interval	Ratio
Demographic	Biological sex[a]; political affiliation[a]; race/ethnicity[a]; religion[a]; sexual orientation[a]; marital status[a]; country[a]; zip code[a]; generational cohort[a]; occupation[a]; employment status[a]; academic major[a]	Highest academic degree[a]; academic grade[a] (e.g., K-12)	Grade point average[a]	Income[a]; age[a]; household size[b]; household income[b]; test scores[a]; crime rate[a]
Interpersonal communication	Disclosure partners[a]; types of disclosures[a]	Rank-order preference of disclosure partners[a]	Self-reported relational satisfaction[a]	Time in relationship (days, weeks, months)[a,b]; number of disclosures[a,b]
Intergroup communication	Membership in a group (yes/no)[a]	Rank-order preference of group memberships[a]; rank-order preference of group members	Self-reported identification with the group[a]	Length of group membership[a]; frequency of intragroup communication[a,b]
Organizational communication	Membership in an organization (yes/no)[a]	Rank-order preference of organizational members[a,b]	Self-reported identification with the organization[a]	Length of organizational membership[a]
Media	Self-reported use of particular media (yes/no)[a]	Rank-order preference of media[a]	Self-reported enjoyment of particular media[a]	Hours per day spent using particular media[a]

a = individual is the study element; b = study element is unit other than the individual, such as a household, dyad (e.g., friends or romantic partners), group, organization, or some geographical entity (e.g., city, county, state, country)

TERMS

- A **study element** is the unit about which information is collected and that provides the basis of analysis.
- A **population** is defined as the theoretically specified aggregation of study elements.
- A **sample** is a special subset of a population observed in order to make inferences about the nature of the population itself.
- A **parameter** is a numerical summary of a given variable in a population.
- A **statistic** is a numerical summary of sample data. A statistic is usually an estimator of a parameter.
- A **variable** is a characteristic measured for each subject that can vary in value among study elements.
- A **measurement scale** is a defined set of rules for how numbers are assigned to attributes of a variable.
- A **qualitative variable** is one for which the scale of measurement is a set of unordered categories. This applies to nominal-level measurement.
- A **quantitative variable** is one for which the possible values of variables differ in magnitude. Variables with attributes measured at the ordinal, interval, and ratio levels are quantitative.
- **Nominal measurement** refers to assigning numbers to attributes of variables such that the numbers provide information on whether or not two subjects have the same attributes.
- **Ordinal measurement** refers to assigning numbers to attributes of variables such that the numbers provide information on the rank-order of the subjects.
- **Interval measurement** refers to assigning numbers to attributes of variables such that the difference between the assigned numbers indicates the subjects' difference on a certain variable.
- **Ratio measurement** refers to assigning numbers attributes of variables such that the ratio between the assigned numbers indicates the subjects' ratios on a certain variable. To calculate ratios meaningfully, there needs to be a true or natural given zero point, not one arbitrarily defined by the researcher.

CHAPTER 1 PROBLEM SET

1. What is the difference between a population and a sample?
2. What is the difference between a parameter and a statistic?
3. What are the different levels of measurement used in statistical methods? Why do you need to be aware of which level of measurement is applied?

4. Find a study in a communication-oriented scholarly journal, such as *Human Communication Research, Communication Monographs, Journal of International and Intercultural Communication, Journal of Communication,* or *Journal of Computer-Mediated Communication.* As you read the study, answer the following questions.
   a. What is/are the study element(s) in the paper?
   b. What is the sample in the study? What is the population about which the authors are trying to draw conclusions?
   c. What are the variables that the authors include in the study?
   d. What are the levels of measurement for the variables of interest in the study?
5. Find a study reported by a news agency (television, print, or Web). In your search, it might help to use a phrase such as "study finds" or "new study finds."
   a. What is the study about?
   b. Who conducted the study? Is it conducted by industry researcher or is it conducted by an academic researcher?
   c. What are the variables included in the study? From the study, can you identify the levels of measurement for the variables? What are they? If not, what do you think they are?

REFERENCES

Wheeless, L. R., Nesser, K., & McCroskey, J. C. (1986). The relationships of self-disclosure and disclosiveness to high and low communication apprehension. *Communication Research Reports, 3,* 129–134.

CHAPTER 2

Descriptive Statistics

As a communication professional or researcher, you will be interested in making statements to characterize your sample or population of interest. In other words, you are interested in knowing what a typical study element is like in your population, as well as the variation that exists within your population of interest. For example, you may want to describe typical media consumption patterns of your population of interest, or their attitudes toward new technologies. In this chapter, we will provide an overview of ways to describe data with tables and graphs, and numerical summaries to describe the center and spread of data.

TABULAR AND GRAPHICAL DESCRIPTIVE METHODS

Many analyses begin with creating tables and graphs to get an idea for the distribution of data.

Frequencies: Nominal-Level Measurement

A frequency distribution shows all possible values for a variable, and includes the number of observations at each value. A relative frequency distribution lists all the possible values together with the proportions or percentages. The example in Table 2.1 is a frequency distribution and a relative frequency distribution of the categorical variable "Willingness to pay for online news content."

Table 2.1. Are online news users willing to pay for their favorite news sites?

Answers/Values	Frequency	Proportion	Percentage
No	1,852	.82	82
Yes	339	.15	15
Don't know	68	.03	3
Total	2,259	1.00	100

Source: PRC-Project for Excellence in Journalism and PRC-Internet & American Life Project Online News Survey: December 28, 2009–January 19, 2010.

Another way to visualize categorical data is to use a bar chart. A bar is drawn over each value, with the height of the bar representing the proportion or percentage of that value. Figure 2.1 is a bar graph of the data in Table 2.1.

Figure 2.1. U.S. online users' willingness to pay for access to favorite news sites

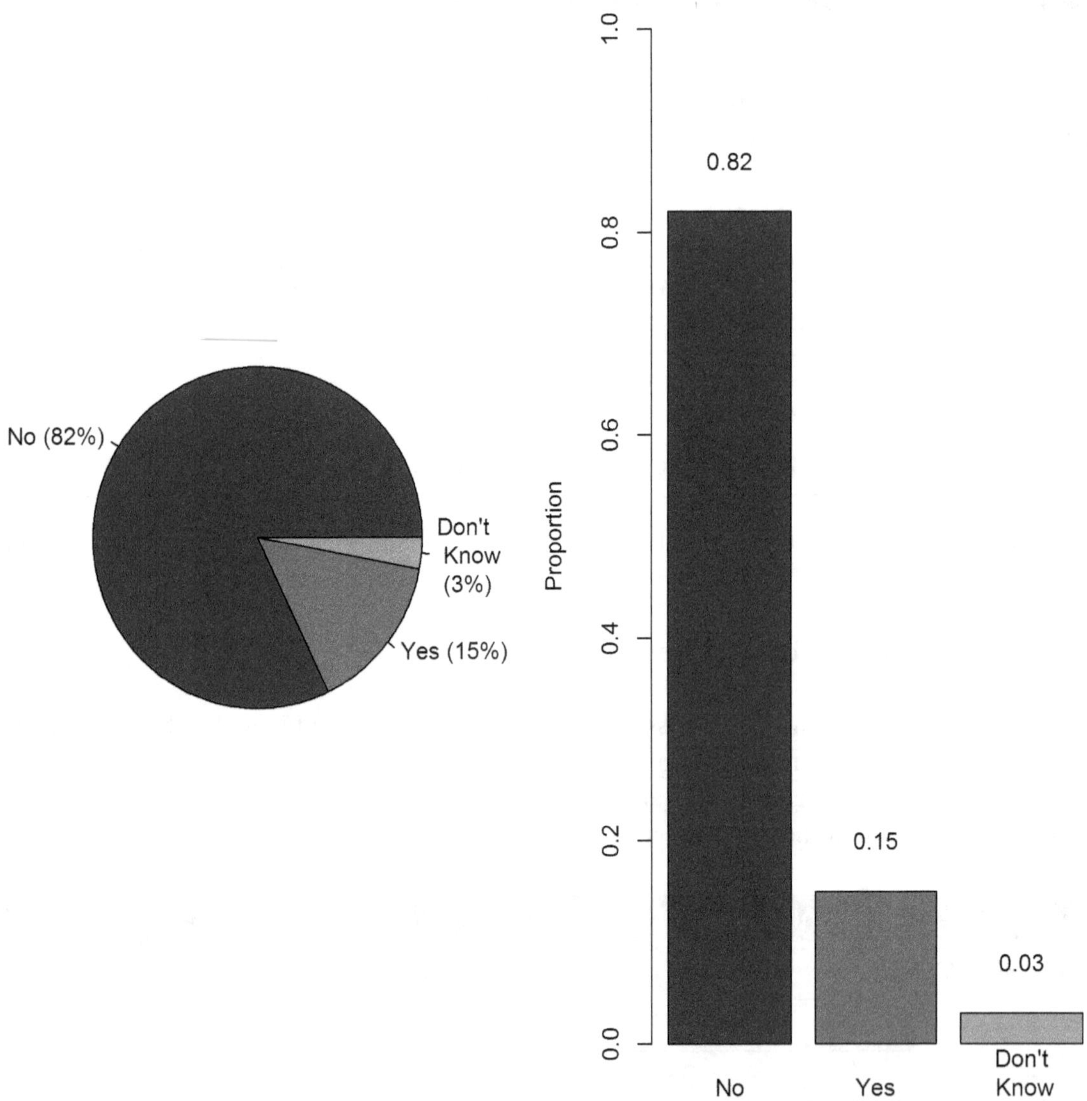

Two ways to display the same information of the variable "Willingness to pay for online content." The left panel shows the percentage of each level in a pie chart. The right panel shows the same information in a bar chart. It is always a good idea to label your charts with the summary values represented by the area.

In a bar chart, the order of the bars is not determined. However, by convention, they are ordered in descending frequency (most frequent to least frequent). Bar charts can also be used for variables that are measured on an ordinal-level. In this case, the order follows the natural ordering of the variables' values or levels. The bars in a bar chart are separated in order to emphasize that the variable is categorical rather than interval or ratio.

Another common way to visualize categorical data is a pie chart. In a pie chart the segments represent the proportions or percentages of the values or categories. Using a pie chart is not recommended if there are many categories with a low frequency. Also, the eye is bad at judging relative areas; so if you do choose to use a pie chart, always include the percentage of the area for each category in the figure. The left panel of Figure 2.1 is a pie chart of the data in Table 2.1.

Frequencies: Interval/Ratio-Level Measurement

Tables and graphs are used as well to show trends for variables using interval- and ratio-level measurement (sometimes called quantitative level measurement).

Table 2.2 is an unordered print out of self-reported hours per week of television watched by fifty communication undergraduate students who were asked to keep track of their viewing habits. There are several ways to explore the structure of the data.

Table 2.2. Self-reported hours per week of television watched
0, 25, 44, 38, 25, 46, 22, 54, 46, 12, 37, 33, 19, 10, 47, 29, 32, 56, 12, 13, 6, 23, 27, 25, 28, 25, 38, 20, 24, 5, 7, 20, 43, 42, 33, 24, 33, 41, 38, 4, 9, 41, 42, 27, 24, 20, 13, 8, 26, 45

To begin, we construct a frequency table, illustrated in Table 2.3. This requires us to categorize the data in Table 2.2.

A common question is how to choose the intervals for this categorization. With intervals that are too large it is difficult to see differences and too much information may be lost. With intervals that are too small we cannot get a feeling for the distribution behind the data because too many irregularities might obscure the overall distribution of the data. There are no hard and fast rules for the optimal width of intervals, just some "rules of thumb:"

- The greater the total number of observations (sample size), the more categories are used and the smaller the width of the intervals.
- The higher the range (maximum value—minimum value) of a variable the larger the width of intervals.
- The interval should not be so wide that two measurements included in it have a difference between them that is considered major. If categories cannot distinguish relevant differences between values, the width of the intervals is too large.
- Categories have to be mutually exclusive. Any possible value has to fit in one and only one interval; i.e., intervals must not overlap.

- The number of intervals should not exceed 20 (usually between 5 and 15; otherwise they are too difficult to read).
- Important: Intervals should all have the same width (otherwise one might get a wrong impression of the data's underlying distribution).

In the given example, the range is 56; we choose 6 intervals, each with equal width.

Table 2.3. Frequency table of hours of TV watched

Interval	Frequency	Relative Frequency	Percentage	Cumulative Frequency
0 under 10	7	0.14	14	14
10 under 20	6	0.12	12	26
20 under 30	17	0.34	34	60
30 under 40	8	0.16	16	76
40 under 50	10	0.2	20	96
50 under 60	2	0.04	4	100
Total	50	1	100	

The frequency table in Table 2.3 contains an additional column labeled "cumulative frequency." In this column we simply add up the percentages from the previous column. Cumulative frequencies inform about the percentage of observations that either do not or do exceed a value. For instance, the cumulative frequency of 60% in Table 2.3 (row three) tells us that 60% of all surveyed students watched less than 30 hours TV; i.e., 0 to less than 30 hours. This number also tells us that 100% − 60% = 40% of all surveyed students watched at least 30 hours per week; i.e., 30 hours and more.

Because the data are continuous, i.e., measured on an interval/ratio-level, we are able to visualize the data's distribution with a histogram. A bar is drawn over each interval of numbers, with the height of the bar representing the number of observations in that interval. The bars are not separated in order to visualize that the histogram represents the distribution of an interval/ratio variable. Also intervals that have no frequency do occur. Due to the interval/ratio-level measurement, the order of the bars is fixed and cannot be changed. Figure 2.2 shows a histogram of self-reported hours per week of television watched.

Another way to get a picture of the distribution of the data is by constructing a stem-and-leaf plot. In a stem-and-leaf plot, each observation is represented by its leading digit, which de-

Figure 2.2. Histogram of self-reported hours per week of television watched

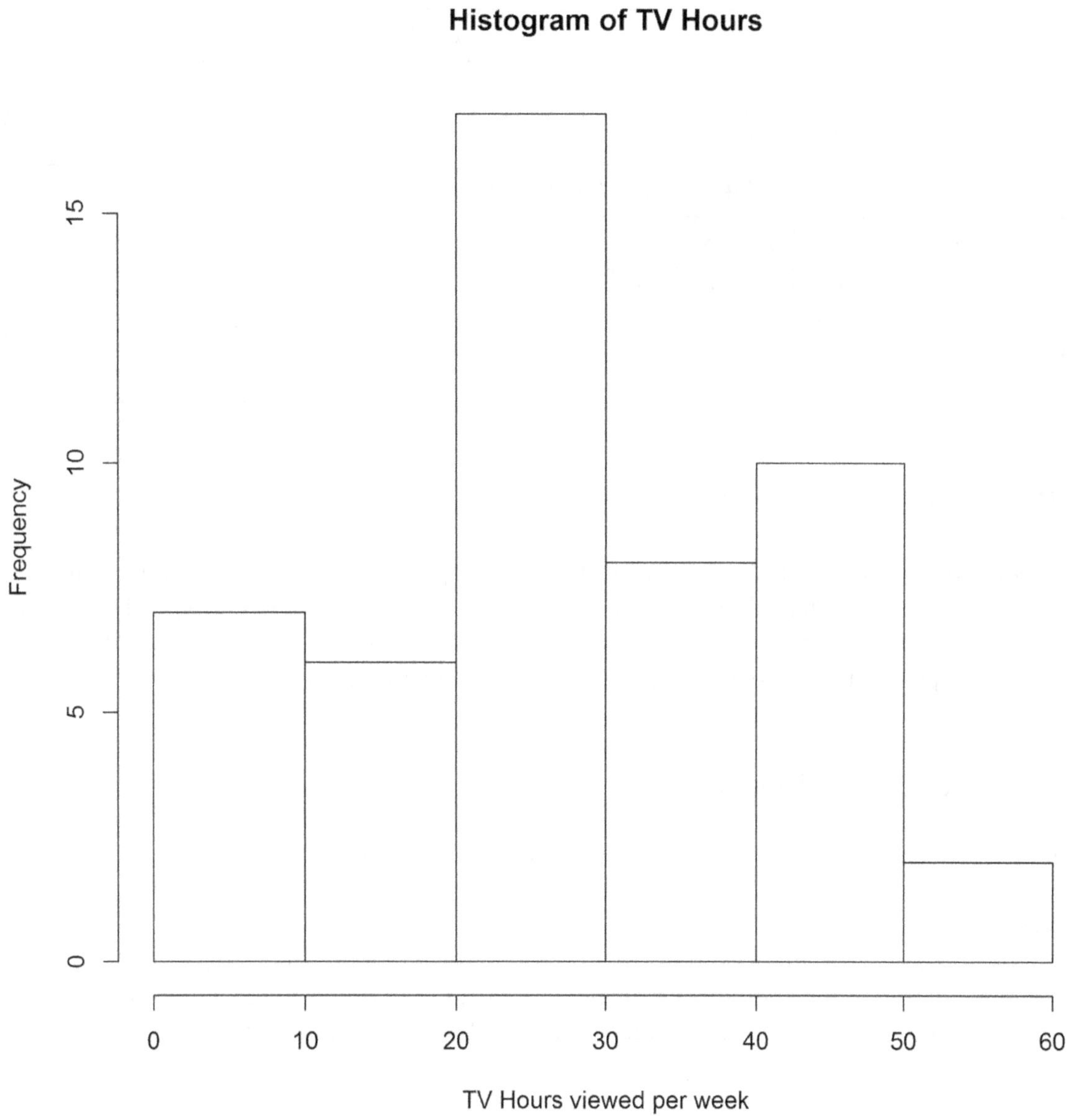

The bars touch in a histogram because the data are continuous.

fines the stem, and by its final digit, which defines the leaf. Stem-and-leaf plots arrange the leaves in order on each line, from smallest to largest. The displayed information is essentially the same as in a histogram. It even provides information on the exact values of the variable under consideration—information which is lost in a histogram. Figure 2.3 shows a stem-and-leaf plot of the data on hours of TV watched.

Figure 2.3. Stem-and-leaf plot of hours of TV watched

Stem	Leaf
0	0 4 5 6 7 8 9
1	0 2 2 3 3 9
2	0 0 0 2 3 4 4 4 5 5 5 5 6 7 7 8 9
3	2 3 3 3 7 8 8 8
4	1 1 2 2 3 4 5 6 6 7
5	4 6

NUMERICAL DESCRIPTIVE METHODS

Numerical description includes measures of central tendency, measures of variation, and measures of position. Measures of central tendency provide information about a typical or representative observation. Measures of variation describe the spread of observations around the center. Measures of position provide more detailed information about the observations' distribution.

Measures of Central Tendency

Measures of central tendency include mean, median, and mode.

Mean

The mean is the sum of all measurements divided by the number of measurements. The mean is represented in the formula below. The formula is read as the mean is equal to the sum of all measurements from 1 to *n*, divided by *n*. Table 2.4 provides the formula for the mean.

Table 2.4. Mean

$$\bar{y} = \frac{\sum_{i=1}^{n} y_i}{n}$$

n represents the sample size; Σ represents the sum of all measurements from 1 to *n*; $\bar{Y}$ represents the mean or average for the sample.

Consider the example in Table 2.5.

Table 2.5. Mean of ad click data for 1 hour

The following is ad click data for the last 1 hour from 9 web sites your company advertises on . . . 3, 5, 6, 7, 9, 11, 15, 16, 19

$$\bar{y} = \frac{3 + 5 + 6 + 7 + 9 + 11 + 15 + 16 + 19}{9} = 10.1$$

There are some important characteristics of the mean to consider. First, the mean is only appropriate for variables that are measured on at least an interval level (See Chapter 1 for levels of measurement). For instance, the mean of the variable "preferred news medium" with the values 1 (television), 2 (radio), 3 (internet), 4 (newspaper), 5 (others), cannot meaningfully be interpreted. A theoretical mean of 2.3 would not represent any of the categories. In addition, the assignment of values to the attributes of this variable is arbitrary, and as a consequence, results for mean calculations are arbitrary. Second, the mean can be highly influenced by outliers or data points that are far away (above or below) from the rest. In the case of outliers, the mean is not a measure that represents the distribution well. For instance, if we added to the ad click data in Table 2.5 a measurement of 90, the mean would change to 18.1 and as such would no longer represent the majority of the data well. The mean is usually pulled in the direction of the longer tail of a skewed distribution. Outliers above the mean lead to skewness to the right (the longer tail is on the right). Outliers below the mean lead to skewness to the left (the longer tail is to the left). The more highly skewed the distribution, the less representative the mean is for that distribution.

Median

The median splits a distribution into two equal parts, when observations are ordered from lowest to highest. The median is the middle value in an ordered sequence of values. A more technical definition of the median is provided in Table 2.6.

Table 2.6. Median

The median is determined as the value with the index $\frac{n+1}{2}$ in an ordered sequence of n values

If n is an uneven number, this definition points to exactly one value. If n is an even number, however, this definition indicates a value between two values as the median. In this case, the median is defined as the mean of the two indicated values. Consider the example in Table 2.7.

Table 2.7. Median for ad click data

The following is an ordered sequence (smallest to largest) of ad click data for the last 1 hour from 9 web sites your company advertises on. Sample size is uneven ($n = 9$).

3, 5, 6, 7, 9, 11, 15, 16, 19

The median is the value with index $\frac{9 + 1}{2} = 5$, or the 5th value, which is Md = 9.

The following is an ordered sequence (smallest to largest) of ad click data for the last 1 hour from 10 web sites your company advertises on. Sample size is even ($n = 10$)

3, 5, 6, 7, 9, 11, 15, 16, 17, 19

The median is the value with index $\frac{10 + 1}{2} = 5.5$; i.e., the median lies between the 5th and 6th observations. In this case, the median is the mean of these two observations. Md = (9 + 11) / 2 = 10.

There are important properties of the median to consider when using this as your measure of central tendency. First, with regard to level of measurement, the median is appropriate for ordinal-, interval-, and ratio-level data. However, the median is not appropriate for nominal data, because calculations require an order of the data, which nominal measurement does not provide. Second, with regard to the distribution of data, the median represents the central tendency of your data appropriately when the distribution is highly skewed. That is, when your data set contains measurements that are far above or below the rest. In other words, the median is far less sensitive to outliers than the mean. For instance, if you replaced the value 19 in Table 2.5 with the value 90, or even 900, the median would remain the same.

Mode

The mode is the most frequent value that occurs in a distribution. The mode is usually used to represent the central tendency of nominal-level (categorical) data but can also be used for ordinal-, interval- and ratio-level data. The mode should only be used for distributions that indicate only one most frequent value (unimodal distributions). For bimodal distributions (i.e., two values occur with the same highest frequency) or multimodal distributions (i.e., many values occur with the same highest frequency) the mode is—in a strict sense—undefined. However, sometimes it makes sense to report two modes for bimodal distributions and multiple modes for multimodal distributions. Consider the examples in Table 2.8.

Table 2.8. Example of Mode

The following is ad click data for the last 1 hour from 9 web sites your company advertises on.

3, 5, 6, 7, 9, 11, 11, 16, 19

The mode is 11, which is the most frequent value in the data, occurring 2 times.

Or, consider the example from earlier in Table 2.1. The value that occurs most often for the categorical variable "Willingness to pay for online news" is no, with 82%.

Similar to the median, the mode is insensitive to outliers. Assuming that outliers occur infrequently (otherwise those values wouldn't be called outliers), adding extreme values to the data in Table 2.6 does not change the distribution's mode.

Mean, median, and mode provide information on the central tendency of data. In addition, a simple comparison of mean, median, and mode of the same data provides valuable information on the shape of the data's distribution. If mean, median, and mode are all equal, then the shape of the distribution is symmetric (see Figure 2.4). As mentioned, the mean is pulled to the longer tail of a skewed distribution. That is, if the mean is greater than the median, and the median is greater than the mode, then the shape of the distribution can be characterized as skewed to the right (the longer tail falls to the right). Accordingly, a distribution would be described as skewed to the left if the mean is smaller than the median, and the median is smaller than the mode.

Figure 2.4. Mode, median, mean, and the shape of a distribution

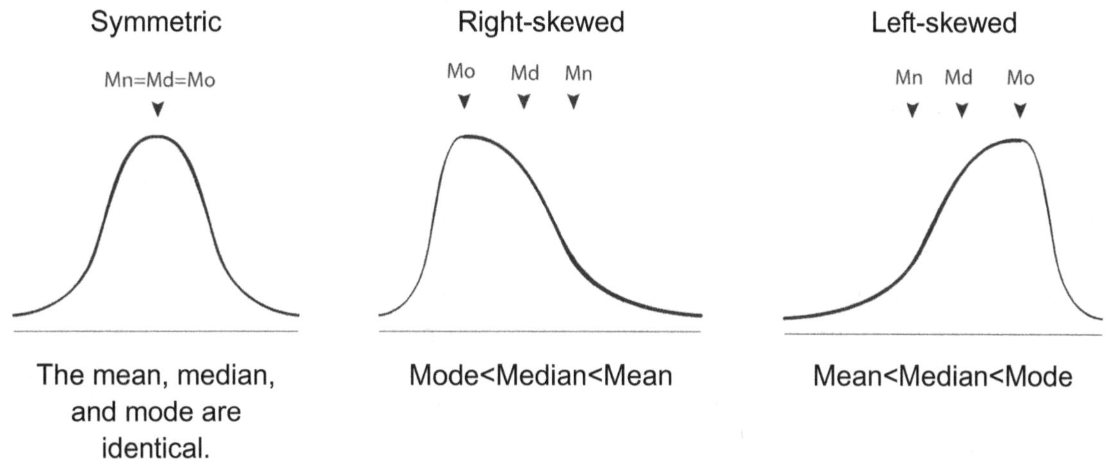

Recommendations

In choosing a measure of central tendency, the goal is to optimally represent the distribution in one value. The measure we choose to express central tendency depends on what we mean with optimal representation. If we are interested in the value of a randomly selected subject, it makes sense to use the most frequent value in the sample, which defines the mode. If we are interested in a measure of which all values deviate, on average, the least we would select the median. If we are interested in a measure that represents a distribution while avoiding large errors when estimating single values of this distribution with this measure, we would select the mean, as the squared deviation of all values from their mean is minimal. You won't find another measure other than the mean that minimizes squared deviations. Table 2.9 summarizes the characteristics of the measures of central tendency.

Table 2.9. Measures of central tendency—Summary of characteristics

	Mean	Median	Mode
Level of measurement	Interval and ratio	Ordinal, interval, and ratio	Nominal, interval, and ratio
Limitations	Highly influenced by outliers	Not appropriate for nominal data	Appropriate only for unimodal distributions
When does it optimally represent central tendency?	When you are interested in avoiding large errors when estimating single values of this distribution with this measure.	When you are interested in estimating a value with this measure of which all values deviate on average the least.	When you are interested in estimating the value of a randomly selected study element.

Measures of Variation

Measures of variation are extremely important in social sciences and communication research because researchers are primarily interested in explaining the differences or variation among study elements. Moreover, measures of central tendency alone are not sufficient to describe a distribution. Consider Figure 2.5, which shows the distributions of "willingness to express opinion" for individuals from different countries. Both distributions have the same mean, but the variation is quite different. Distribution A (u-shaped) suggests an average willingness to express opinion in public with many people willing to speak their minds, but also many people unwilling to speak their minds. There is almost no middle group of individuals willing to express their opinions. Distribution B (bell-shaped) suggests the same average willingness to speak their minds with a few individuals willing to take expressive action, but most people are characterized by an average willingness to speak their minds. In which country would you prefer to live? Which is healthier for democracy and free speech?

Figure 2.5. Two distributions with the same mean, but different variability

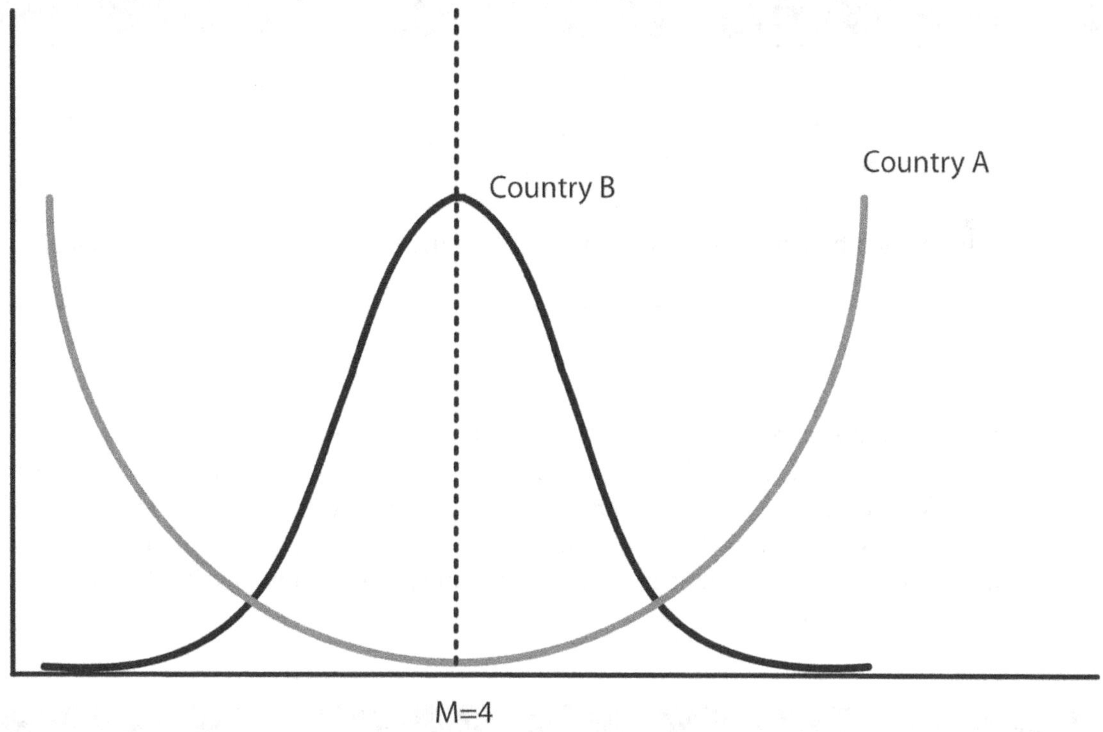

Range

The range is a very simple measure of variation. It is the difference between the largest and the smallest observation (maximum—minimum). The range provides a first impression on the data's variability. However, the range has a serious downfall in that it does not consider the center of a distribution and how the values deviate from the center. For instance, it is impossible to distinguish the different qualities of the two distributions in Figure 2.5 above as both distributions have identical maximum and minimum values.

Variance and Standard Deviation

To consider the center of a distribution we use the two most common (and most important) measures of variation, the variance and the standard deviation. These measures of variation are based on the deviations of values from their mean. Each value in the sample of values has a deviation, which is the difference between the value and the mean of all values $(y_i - \bar{y})$. When an observation falls above the mean, that deviation is positive. The deviation is negative when an observation falls below the mean.

The variance can be understood as an average of all squared deviations (see Table 2.10).

Table 2.10. Variance
$$s^2 = \frac{\sum_{i=1}^{n} (y_i - \bar{y})^2}{n}$$
Variance is computed by squaring each deviation from the mean, summing those values, and dividing by *n*.

Because deviations are squared, the unit of the variance is the squared unit of the original observations, e.g., income squared or clicks squared. This makes the variance difficult to interpret. What does "clicks squared" mean? The square root of the variance is better for the purpose of interpretations, because it converts the variance to a measure with the original unit of the observations. The square root of the variance is defined as standard deviation (see Table 2.11).

Table 2.11. Standard deviation
$$s = \sqrt{s^2} = \sqrt{\frac{\sum_{i=1}^{n} (y_i - \bar{y})^2}{n}}$$
Standard deviation is simply the square root of the variance.

Table 2.12 illustrates the computation for both variance and standard deviation step by step.

Properties of Variance and Standard Deviation

Variance and standard deviation share the following properties: (1) Both measures can only be positive and are zero when all observations have the same value (i.e., s and $s^2 \geq 0$). (2) The greater

Table 2.12. Computation of variance and standard deviation

Value/Observation	$(y_i - \bar{y})$	$(y_i - \bar{y})^2$
10	$(10 - 5) = 5$	25
4	$(4 - 5) = -1$	1
2	$(2 - 5) = -3$	9
8	$(8 - 5) = 3$	9
8	$(8 - 5) = 3$	9
1	$(1 - 5) = -4$	16
9	$(9 - 5) = 4$	16
5	$(5 - 5) = 0$	0
2	$(2 - 5) = -3$	9
1	$(1 - 5) = -4$	16
Sums		
50	0	110

Step 1. Sum up the values/observations and compute the mean. $\bar{y} = \dfrac{\sum_{i=1}^{n} y_i}{n} = \dfrac{50}{10} = 5$

Step 2. Subtract the mean from each observation and sum $(y_i - \bar{y})$.

Step 3. Square the difference of the observation and the mean $(y_i - \bar{y})^2$ and compute sum of squares.

Step 4. Compute the variance. $s^2 = \dfrac{\sum_{i=1}^{n} (y_i - \bar{y})^2}{n} = \dfrac{110}{10} = 11$

Step 5. Compute the standard deviation. $s = \sqrt{s^2} = \sqrt{11} = 3.32$

the deviations of values from their mean, the larger are s and s^2. (3) If s or s^2 is used to compare the variation of values across groups (e.g., how diverse political attitudes of newspaper A's readers are compared to newspaper B's readers), then the group with the higher value for s and s^2 shows higher variation/diversity. Finally, (4) if the data are multiplied by a constant (e.g., multiplied by 2) then the standard deviation can be multiplied with the same constant and the variance can be multiplied with the squared constant. For instance, if all data values are doubled the standard deviation doubles as well and the variance increases with the factor $2^2 = 4$.

Interpreting Standard Deviations

It may be difficult to interpret the meaning of a standard deviation. As an empirical rule, if the values' distribution is bell-shaped and approximately normally distributed, about 68% of all observations fall within ± one standard deviation (i.e., $\pm 1s$) and about 95% of all observations fall within ± two standard deviations (i.e., $\pm 2s$). All or nearly all observations fall within $\pm 3s$. Figure 2.6 illustrates this empirical rule.

Figure 2.6. Normal distribution and standard deviations

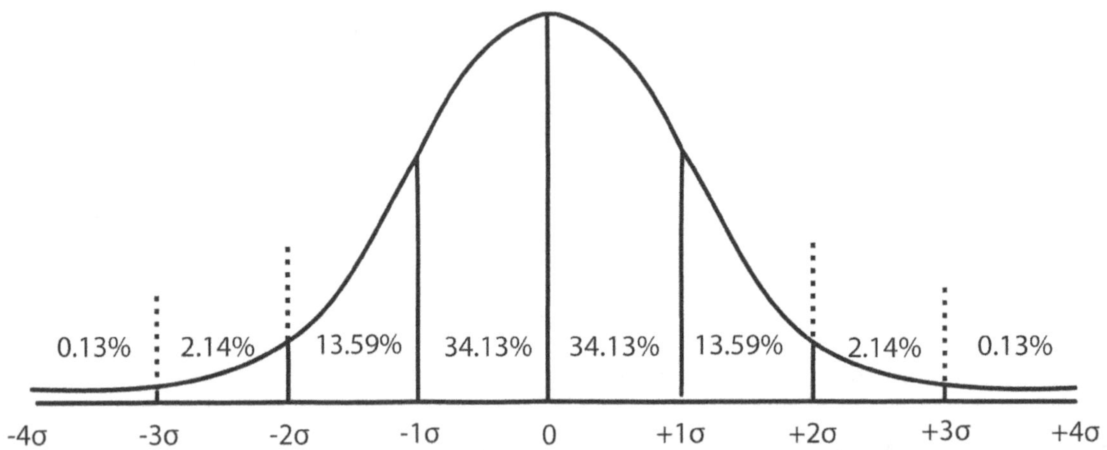

The empirical rule does not work well for distributions that are highly skewed or for those that take on only a few values. We will discuss normal distributions in Chapter Five in more detail.

z-Scores

Above, we introduced standard deviation, and the empirical rule. As researchers and practitioners, we will often be interested in comparing scores from different distributions, et al., distributions that differ in mean and standard deviation. For example, we want to compare communication competence scores (CC scores) of two students on two different scales. Let us assume that we know from prior research that the Wiemann (1977) CC scale has a mean of 60 and a standard deviation of 10, and the Monge, Backman, Dillard, & Eisenberg (1982) scale has a mean of 90 and a standard deviation of 8. A higher score on each scale indicates greater communication competence. Marcia scores a 50 on the Wiemann scale and Teri scores a 66 on the Monge et al scale. Which student is more competent? As both scores stem from different distributions we cannot compare the scores directly. First, a comparison has to consider how far a score falls below or above the mean. Marcia's score lies 10 points below the mean and Teri's scores 24 points below the mean. Thus, it seems that Teri's score represents a lower score. However, we also have to consider the standard deviation of all scores in a distribution. If, more or less, all people obtain the same average CC score (low standard deviation), then individual deviations from the mean CC score are more important. Dividing deviations of scores from their mean by the scores' standard deviation reflects this idea and is defined as z-transformation which results in z-scores or standardized scores. To compare the two students in our example, we would calculate z-scores as illustrated in Table 2.13.

Table 2.13. Calculating z-scores

$z_i = \frac{y_i - \overline{Y}}{s}$				
Marcia:	$y_i = 50$	$\overline{Y} = 60$	$s = 10;$	$\Rightarrow z = -1$
Teri:	$y_i = 66$	$\overline{Y} = 90;$	$s = 8;$	$\Rightarrow z = -3$
As smaller values stand for lower communication competence, we conclude that Teri is less competent than Marcia.				

Z-scores can also be interpreted as standard deviation units. Given the CC distributions in our example, Marcia's CC scores falls one standard deviation unit below the mean, but Teri's scores falls three standard deviation units below the mean. Hence, Teri's CC score is clearly smaller than Marcia's.

Measures of Position

Percentiles and Quartiles

The pth percentile is a number such that $p\%$ of all values in a dataset fall below it and $100 - p\%$ of all values fall above it. The 25th percentile is called the lower quartile (Q1). The 75th percentile is called the upper quartile (Q3). One quarter of all observations fall below the lower quartile (Q1) and three quarters of all observations fall above Q1. One quarter of all observations fall above the upper quartile (Q3) and three quarters of all observations fall below Q3. The 50th percentile equals the median of values as it is a number such that 50% fall above and 50% fall below it, i.e., the 50th percentile, like the median, splits a distribution into two equal parts. Similarly, the lower quartile (Q1) can be considered as the median for the observations that fall below the 50th percentile (or overall median). Accordingly, the upper quartile (Q3) is also the median for the observations that fall above the overall median. Figure 2.7 demonstrates how these values can be easily found on a stem-and-leaf plot.

Figure 2.7. Stem-and-leaf plot of hours of TV watched

Stem	Leaf
0	0 4 5 6 7 8 9
1	0 2 2 3 3 **9**
2	0 0 0 2 3 4 4 4 5 5 5 5 6 7 7 8 9
3	2 3 3 3 7 8 8 **8**
4	1 1 2 2 3 4 5 6 6 7
5	4 6
n = 50 observations	
The lower quartile (Q1) is the median of the first 50% (=25) ordered observations. The median of the first 25 ordered observations is the observation with index (25 + 1) / 2; i.e., the 13th observation, which is 19. The upper quartile (Q3) is the median of the last 50% (=25) ordered observations. The median of the last 25 ordered observations is the (25+1)/2; i.e., the 13th largest observation, or 38.	

The distance between the lower and upper quartile (Q3–Q1) is defined as the interquartile range (IQR). For the example in Figure 2.7, the interquartile range is 38–19, or 19. An advantage of IQR over range or standard deviation is its insensitivity to outliers. For bell-shaped distributions, the distance between the mean and either quartile is about 2/3 standard deviations, and the IQR is roughly 4/3 standard deviations.

Boxplots

Boxplots are a powerful tool to display measures of position together with measures of variation and central tendency in one plot. Boxplots provide information on the lower and upper quartile, the mean, median, minimum and maximum, range, and outliers. Figure 2.8 illustrates a boxplot of hours of TV watched.

An observation is defined as an outlier if it is more than 1.5 IQR above Q3, or more than 1.5 IQR below Q1. For example, consider Figure 2.9, which shows a boxplot of self-reported text messages received from students in the last hour.

Figure 2.8. Boxplot of hours of TV watched

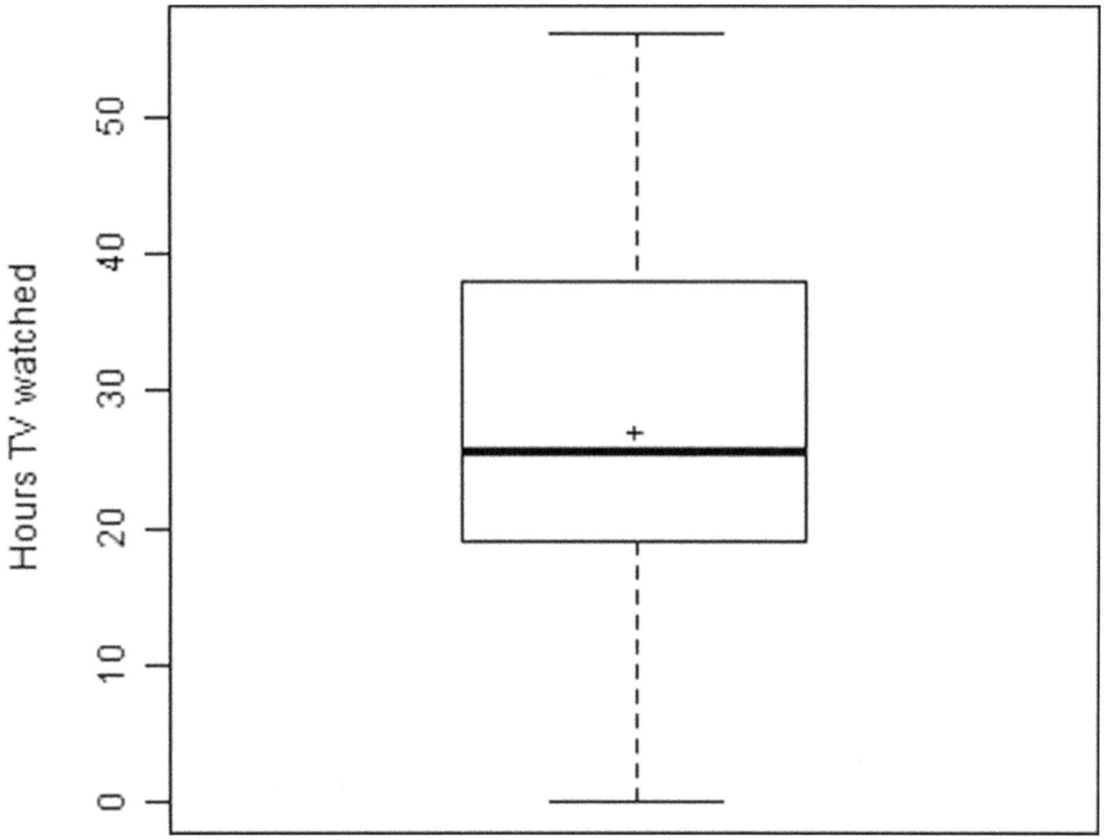

The horizontal line at the bottom (the lower whisker) indicates the minimum value (0). The second horizontal line from the bottom stands for the lower quartile (Q1 = 19). The third line in the middle (in bold) informs about the median (25.5). The cross (+) right above that line indicates the mean (27.22), the fourth horizontal line the upper quartile (Q3 = 38), and the horizontal line at the top of the boxplot stands for the maximum value (56). The range is given by the distance from the bottom to the top line and the interquartile range is given by the distance from the Q1 line to the Q3 line. As the mean is greater than the median and the distance from the median line to the maximum line is longer than the distance from the minimum line to the median line, the distribution of hours TV watched is slightly skewed to the right.

Figure 2.9. Boxplot of text messages students received in the last hour

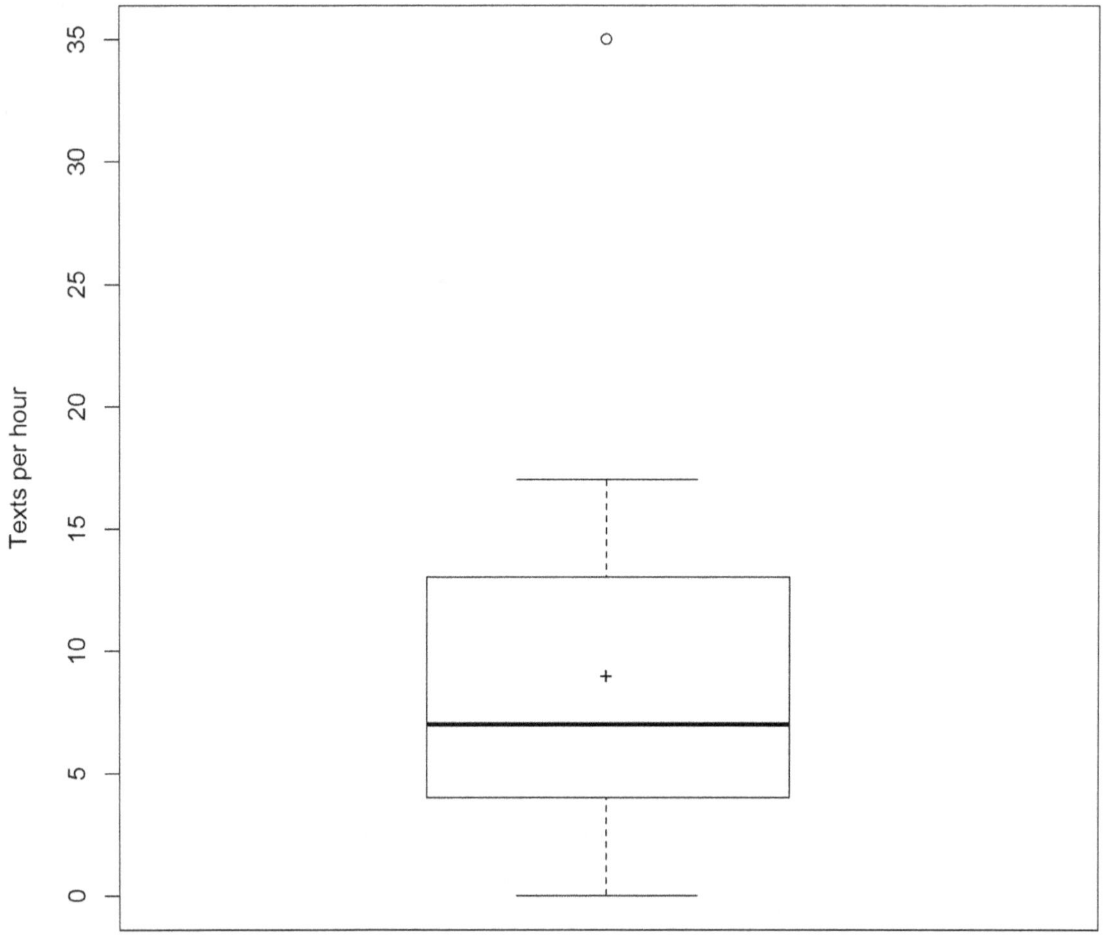

A group of 26 students were asked to report the number of text messages they had received in the last hour. The dot at the top represents an outlier; that is, it is more than 1.5 IQR above the upper quartile.

SUMMARY

The purpose of this chapter was to introduce descriptive statistics with tabular, graphical, and numerical methods. A description of data with respect to distribution characteristics is usually the first step of any data analysis. Of particular importance are descriptive measures of central tendency and measures of variation. Measures of central tendency indicate a typical or representative data value. Measures of variation provide information on how much values deviate from a typical value, or in other words, how well a measure of central tendency indeed represents all values in a dataset. Different measurement levels require different descriptive measures.

TERMS

- **Frequency distribution** shows all possible values for a variable, and includes the number of observations at each value.
- **Relative frequency distribution** lists all the possible values together with the proportions or percentages.
- **Bar chart** visualizes categorical data by drawing a bar over each category; the height of the bar represents the proportion or percentage of that category.
- **Pie chart** visualizes categorical data in the shape of a pie; segments represent the percentages or the proportions of the categories.
- **Histogram** visualizes data at that interval/ratio-level; a bar is drawn over each interval of numbers, with the height of the bar representing the number of observations in that interval; the bars are not separated in order to show that the histogram represents the distribution of an interval/ratio variable.
- **Stem-and-leaf plot** visualizes data at the interval/ratio-level; observations are arranged from smallest to largest; each observation is represented by its leading digit, which defines the stem, and by its final digit, which defines the leaf.
- **Measures of central tendency** are numerical summaries of data, and include the mean, median, and mode.
- **Mean,** or average, is the sum of all measurements divided by the number of measurements (= sample size).
- **Median** splits a distribution into two equal parts, when observations are ordered from lowest to highest. The median is the middle number in a sequence of numbers.
- **Percentile** is a number such that $p\%$ of the scores fall below it and $100 - p\%$ of scores fall above it. The 25th percentile is also called the lower quartile (Q1). The 50th percentile is the same as the median. The 75th percentile is also called the upper quartile (Q3).
- **Mode** is the most frequent value that occurs in a data set.
- **Measures of variation** provide information about the spread of data.
- **Range** is the difference between the maximum and minimum value.
- **Interquartile range** is the difference between the upper quartile (Q3) and the lower quartile (Q1).
- **Variance** is the average of squared deviations of values from the mean.
- **Standard deviation** is the square root of the variance.
- **Empirical rule** states that, assuming a normal distribution, about 68% of observations fall within $\pm 1s$, about 95% fall within $\pm 2s$, and all or nearly all fall within $\pm 3s$.
- **Outlier** is a data point that falls 1.5 IQR above Q3, or 1.5 IQR below Q1.
- ***z*-scores** are standardized scores and can be used to compare scores that stem from different distributions. Z-scores are also known as standard deviation units.

CHAPTER 2 PROBLEM SET

1. A researcher is interested in attitudes about Twitter and obtained the following responses from a random sample of American adults: 1) An important tool (n = 170); A fad that will go away (n = 420); Heard of it but don't know what it really is (n = 375); Never heard of Twitter (n = 70).
   a. Construct a table with answers/values, frequency, proportion, and percentage.
   b. Construct a bar chart with responses.
2. A researcher wants to know whether there is a relationship between students' time spent on the Internet and their performance in the class. She asks her students to record how many hours per week they spend on the Internet and obtains the following 51 data points: 15, 63, 32, 53, 44, 80, 19, 79, 1, 6, 81, 6, 51, 79, 23, 9, 9, 15, 45, 23, 71, 17, 28, 41, 19, 31, 78, 51, 71, 14, 41, 59, 31, 17, 52, 26, 42, 84, 79, 63, 39, 58, 15, 54, 28, 45, 81, 68, 17, 82, 52.
   a. Construct a table with the intervals, frequency, relative frequency, percentage, and cumulative frequency.
   b. Using the intervals from problem 2a, construct a histogram.
3. What is the difference between a bar chart and a histogram?
4. A researcher wants to know whether there is a relationship between Internet use at work and worker productivity. She obtains the following data from a survey (minutes of recreational Internet use per day at work): 39, 82, 34, 53, 30, 27, 36, 16, 11, 33, 37, 75, 31, 24, 90, 34, 98, 30, 31, 84, 27, 45, 39, 69, 24, 49, 22, 61, 80, 68.
   a. Construct a stem-and-leaf plot.
   b. Identify the 1st, 2nd (median), and 3rd quartiles in the stem-and-leaf plot.
5. How are stem-and-leaf plots and histograms similar to and different from each other?
6. A researcher is trying to find out about how age affects TV watching. The ages of the participants in her study are: 9, 10, 2, 22, 45, 65, 33, 55, 87, 2, 23, 5, 9, 7, 34, 52, 9, 4, 37, 55, 13, 22, 25, 31.
   a. Find the mean.
   b. Find the median.
   c. Find the mode.
   d. Based on these values, what can you surmise about the shape of the distribution? Why?
7. If the mean is greater than the median, and the mode is smaller than the median, what is the shape of the distribution?
8. If the median is larger than the mean but smaller than the mode, what is the shape of the distribution?
9. A researcher is studying how the number of hours of television that children watch affects school performance. The following are the number of hours that children reported watching television for this study: 3, 5, 5, 6, 8, 8, 8, 9, 11
   a. Calculate the range.
   b. Calculate the variance.
   c. Calculate the standard deviation.
   d. Calculate the sum of squares.

10. To measure variation:
    a. Why is the standard deviation usually preferred over the range?
    b. The IQR (Inter Quartile Range) is sometimes preferred to the standard deviation when there are some extreme outliers. Why?
11. A researcher is interested in the extent to which individuals are willing to express dissenting opinions in their organizations, using Kassing's Organizational Dissent Scale. In study 1, Group A had a mean of 24 and standard deviation of 3.5. In study 2, Group B had a mean of 27 and a standard deviation of 3. Higher scores indicate higher willingness to express dissent. John is from Group A and has a mean dissent score of 26. Chris is from Group B, and has a mean dissent score of 28. Who is more willing to express dissent in their organization? Calculate the z-scores for each and compare.
12. A marketing manager ran a pilot study of two different email campaigns driving users to the company's home page. She is interested in the visitor loyalty to the site (measured in number of visits per user). She has data on the following:
    Campaign 1: 17, 20, 19, 15, 19, 17, 16, 18, 11, 15, 16, 16, 19, 17, 17, 17, 11, 16, 15, 10, 11, 17, 17, 18, 10, 13, 11, 15, 13
    Campaign 2: 13, 8, 11, 13, 15, 12, 13, 10, 9, 10, 10, 15, 11, 14, 11, 9, 11, 11, 12, 9, 10, 13, 10, 14, 14, 15, 11, 11, 14
    a. Find the mean, median, and mode for each campaign.
    b. How would you describe the distribution of each campaign?
    c. Calculate the range, variance, and standard deviation for each campaign.
    d. Construct boxplots (side-by-side) for each campaign.
    e. Based on your analysis, which campaign would you recommend implementing?

REFERENCES

Wiemann, J. M. (1977). Explication and test of a model of communication competence. *Human Communication Research,* 3, 195–213.

Monge, P. R., Backman, S. G., Dillard, J. P., & Eisenberg, E. M. (1982). Communicator competence in the workplace: Model testing and scale development. *Communication Yearbook,* 3, 505–528.

PROBABILITY AND PROBABILITY DISTRIBUTIONS

CHAPTER 3

As a researcher or communication practitioner, you will often want to know about probabilities—the likelihood that an event will occur. Moreover, given that we are often working with limited resources (time and money), you will want to describe your population of interest as efficiently and accurately as possible. Importantly, good sampling methods can help you accomplish this goal. Probability, you will see, is an important linking pin to sampling and inferential statistics. This chapter provides an overview of probability and probability density distributions.

PROBABILITY

Probability is one of the central concepts of statistical methods. Probability may seem like a challenging concept but it is something that we encounter every day. Few things in life are certain, and we act according to probabilities. Thinking in probabilities seems natural to us. Ask somebody how likely it is to become a victim of a crime (for example, in cultivation theory experiments; Gerbner & Gross, 1976) and you will get a fast answer, even if this answer is biased in people heavily exposed to crime dramas on TV. Interestingly, calculating with probabilities seems far less natural to us and many people perceive probability calculations as difficult and counterintuitive. The first step in learning something new, like a new language, is to establish the vocabulary we need. Likewise, before we learn how to calculate with probabilities, we need to define some terms.

We distinguish between subjective probability and objective probability. Subjective probabilities are figures that indicate an individual's belief that an event occurs. In our example above, an individual's belief about becoming a victim of a crime could be considered as a subjective probability. Although there is a whole class of advanced statistical approaches that incorporate subjective probabilities in calculations (Bayesian Statistics; Dienes, 2011), in this text, we do not deal with subjective probabilities. Rather, we focus on objective probabilities that distinguish empirical probability and theoretical probability. But first, how can we calculate the objective probability of an event? There are two basic approaches.

We calculate objective probabilities through either many independent random experiments with one object or through one random experiment with many objects. For example, imagine we are interested in the probability of rolling a number 3 with a die. We could throw one die many times (a theoretically infinite number of times). Each throw of the die is independent (meaning the resulting roll is not contingent on previous roles). We count the number of times we got the number 3. This frequency divided by the total number of throws defines the relative frequency of the number 3 and we would use this relative frequency as the best estimate of the probability to throw the number 3 with this die. Alternatively, we could also throw many dice (theoretically infinite dice) at once and count how many times we got the number 3. Again, this frequency, this time divided by the total number of dice, defines the relative frequency of the number 3, which we would use as an estimate of the probability to throw the number 3. In both cases, the probability we calculated would be an empirical probability because our calculations are based on an experiment with finite throws. Empirical probabilities approximate theoretical probabilities. The higher the number of rolls (or dice) the more accurate is the approximation. In the case of an infinite number of rolls (or dice), which is not possible in real experiments, the relative frequency would equal the theoretical, true probability of an event. Tables 3.1 and 3.2 provide the mathematical definition of empirical and theoretical probability.

Table 3.1. Empirical probability of an event
We define the empirical probability of event A as $p(A) = \frac{n_A}{n}$.

Table 3.2. Theoretical probability of an event
We define the theoretical probability of event A as $\lim_{n\to\infty} p(A) = \lim_{n\to\infty} \frac{n_A}{n} = \pi(A)$.
This formula states that as the number of trials (or objects) approaches infinity, the empirical probability of event A is equal to the theoretical probability.

In random experiments where we know all possible outcomes (usually games of chance) the theoretical probability of one single event is pre-determined by 1 divided by the number of all possible outcomes. Assuming a die is not biased, the probability of rolling the number 3 (or any other number) is 1/6 = 0.16666 = 16.6%.

Probability is never negative or greater than 1. Stated mathematically, the probability of an event A falls between 0 (an impossible event) and 1 (a certain event); i.e., $0 \leq P(A) \leq 1$. Moreover, the sum of probabilities of all possible elements, which defines a certain event, is equal to 1. Sometimes, probabilities are multiplied by 100 and then expressed as percentages. For example, the certain event is an event with the probability of 1 or 100%.

In the next section, we will provide calculation rules for probabilities, which simplify computations. Before discussing rules of probability, we state the following assumptions definitions:

- Two events are mutually exclusive if they have no elements in common.
- The probability that event A occurs, given that event B has occurred, is called conditional probability. The conditional probability of A, given B, is denoted by the symbol $P(A|B)$.
- The probability that event A will <u>not</u> occur is denoted by $P(A')$.

Calculation Rules for Probabilities

As we mentioned above, the probability of an event ranges from 0 to 1, and the sum of all probabilities is equal to 1. From these statements, the rule of subtraction is a logical deduction and can be defined as follows (see Table 3.3).

Table 3.3. Rule of subtraction

$$P(A) = 1 - P(A')$$

The probability that event A will occur is equal to 1 minus the probability that event A will not occur.

The rule of addition (also called the "either-or" rule) is applied to the following situation. We have two events from the same sample space, and we want to know the probability that *either* event occurs. Table 3.4 formally states the rule of addition.

Table 3.4. Rule of addition

$$P(A \cup B) = P(A) + P(B) - P(A \cap B) \qquad \text{(not mutually exclusive events)}$$

If events A and B come from the same sample space, the probability that either A or B occurs equals the probability that event A occurs plus the probability that event B occurs minus the probability that both events A and B occur; i.e., events A and B are not mutually exclusive.

In the case of mutually exclusive events; i.e., events A and B cannot occur at the same time, the rule of addition simplifies to:

$$P(A \cup B) = P(A) + P(B) \qquad \text{(mutually exclusive events)}$$

As an example, let's say you are playing a game of chance. You win $20 if the card is *either* a Diamond *or* a King. What is the probability that you win? Table 3.5 provides the solution.

Hint: If you are able to re-word a problem into an "either-or" sentence, then you need to apply the rule of addition in order to solve the problem.

Table 3.5. Rule of addition example (not mutually exclusive events)

Solution: Let D = the event that the card is a Diamond; and let K = the event the card is a King

- There are 52 cards in the deck
- There are 13 Diamonds, so $P(D) = 13/52$
- There are 4 Kings, so $P(K) = 4/52$
- There is 1 King that is also a Diamond, so $P(D \cap K) = 1/52$

Therefore, based on the rule of addition:

$$P(D \cup K) = P(D) + P(K) - P(D \cap K)$$
$$P(D \cup K) = 13/52 + 4/52 - 1/52 = 16/52 = 4/13 = 0.308 = 30.8\%$$

An example for exclusive events would be the following: You win \$20 if the card you draw is *either* a King *or* a Queen. What is the probability that you win? Table 3.6 provides the solution.

Table 3.6. Rule of addition example (mutually exclusive events)

Solution: Let K = the event that the card is a King; and let Q = the event the card is a Queen

- There are 52 cards in the deck
- There are 4 Kings, so $P(K) = 4/52$
- There are 4 Queens, so $P(Q) = 4/52$
- A King can NOT be a Queen at the same time; i.e., the event King and the event Queen are mutually exclusive, so $P(K \cap Q) = 0$

Therefore, based on the rule of addition:

$$P(K \cup Q) = P(K) + P(Q)$$
$$P(K \cup Q) = 4/52 + 4/52 = 8/52 = 2/13 = 0.154 = 15.4\%$$

The rule of multiplication (also called the "both-and" rule) is applied to the following situation. We have two events in the same sample space, and we want to know the probability that *both* events occur. Table 3.7 formally states the rule of multiplication.

Table 3.7. Rule of multiplication

$$P(A \cap B) = P(A) * P(B|A) \qquad \text{(Dependent events)}$$

If events A and B come from the same sample space, the probability that both A and B occur equals the probability that event A occurs times the probability that B occurs under condition A has occurred. The phrase "B occurs under condition A has occurred" is also referred to as a *dependent* event as the probability of B depends on whether or not A has occurred.

In the case of independent events; i.e., the probability of B is unaffected by whether A has occurred or $P(B|A) = P(B)$, the rule of addition simplifies to:

$$P(A \cap B) = P(A) * P(B) \qquad \text{(Independent events)}$$

Let's say you have an urn (which you cannot see into) that contains 4 green marbles and 6 black marbles. Two marbles are drawn from the urn *without replacement.* What is the probability that 2 marbles are green? Note: If you do not put back the first marble before you draw the second marble then the probability of the second marble being green depends on whether you obtained a green marble in your first draw—the second draw's probability *depends* on the first draw. The solution is in Table 3.8 below.

Table 3.8. Rule of multiplication example (dependent events)

Solution: Let $G1$ = the event that the first marble is green; and let $G2$ = the event that the second marble is green. We know that there are 10 marbles, 4 are green and 6 are black. Thus,

$$P(G1) = 4/10$$

After selection, there are 9 marbles remaining; 3 are green and 6 are black. Thus,

$$P(G2|G1) = 3/9$$

Based on the rule of multiplication:

$$P(G1 \cap G2) = P(G1) * P(G2|G1)$$
$$P(G1 \cap G2) = 4/10 * (3/9) = 12/90 = 2/15 = 0.133 = 13.3\%$$

Now, we repeat the prior experiment, except this time we select marbles *with replacement;* that is, that the selected marble returns to the urn. What is the probability that both marbles are green? In this case the second draw's probability is *independent* from the first draw. The solution is in Table 3.9 below.

Table 3.9. Rule of multiplication example (independent events)

Solution: Let $G1$ = the event that the first marble is green; and let $G2$ = the event that the second marble is green. We know that there are 10 marbles, 4 are green and 6 are black. Thus,

$$P(G1) = 4/10$$

After selection, we note the color of the marble and return it to the urn. Thus, there are 10 marbles remaining; 4 are green and 6 are black. Thus,

$$P(G2|G1) = P(G2) = 4/10$$

Based on the rule of multiplication:

$$P(G1 \cap G2) = P(G1) * P(G2)$$
$$P(G1 \cap G2) = 4/10 * (4/10) = 16/100 = 4/25 = 0.16 = 16\%$$

To summarize, we have the addition rule for mutually exclusive and non-exclusive events (either-or problems) and the multiplication rule for dependent and independent events (both-and problems). Both the rule of addition and the rule of multiplication are important concepts for the understanding of probability distributions and statistical tests (e.g., the chi-square test of dependence, Chapter 7).

PROBABILITY DISTRIBUTIONS

Probability distributions inform us about the probability of events or combinations of events. It is important to distinguish between probability distributions for discrete variables and continuous variables. Discrete variables are those that cannot take on all values within the limits of the variable. For example, the numbers on a die can only take on the values 1, 2, 3, 4, 5, and 6. There is no 1.5, for instance. Variables that were measured on a nominal or ordinal level are usually discrete variables. A variable such as an individual's weight can take on any value within the limits of this variable and is therefore called a continuous variable. Variables that were measured on an interval or ratio level are usually continuous variables. To be precise, probability distributions for continuous variables are actually called probability *density* distributions. The following paragraphs will explain the difference between probability and probability density distributions.

Discrete Variables

A probability distribution of a discrete (nominal or ordinal) variable lists all possible outcomes of this variable with their probabilities. For example, the probability distribution of rolling a die appears in Table 3.10.

Table 3.10. Probability distribution of rolling a die

Outcome	Probability
1	1/6
2	1/6
3	1/6
4	1/6
5	1/6
6	1/6
Total	1

Figure 3.1 provides a graphical representation of a discrete probability distribution for rolling a die.

How would the probability distribution of the experiment rolling two dice look like? Table 3.11 provides the outcomes and Figure 3.2 provides a graphical representation.

Figure 3.1. Probability distribution of rolling a die

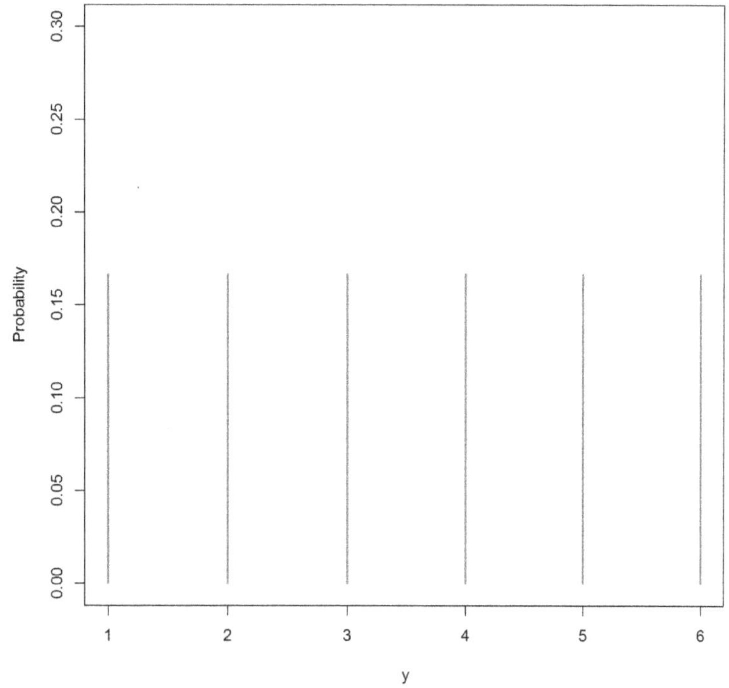

Table 3.11. Probability distribution of rolling two dice

Outcome	Probability
2: (1 ∩ 1)	1/36
3: (1 ∩ 2) ∪ (2 ∩ 1)	2/36, or 1/18
4: (1 ∩ 3) ∪ (2 ∩ 2) ∪ (3 ∩ 1)	3/36, or 1/12
5: (1 ∩ 4) ∪ (4 ∩ 1) ∪ (2 ∩ 3) ∪ (3 ∩ 2)	4/36, or 1/9
6: (1 ∩ 5) ∪ (5 ∩ 1) ∪ (2 ∩ 4) ∪ (4 ∩ 2) ∪ (3 ∩ 3)	5/36
7: (1 ∩ 6) ∪ (6 ∩ 1) ∪ (2 ∩ 5) ∪ (5 ∩ 2) ∪ (3 ∩ 4) ∪ (4 ∩ 3)	6/36, or 1/6
8: (2 ∩ 6) ∪ (6 ∩ 2) ∪ (3 ∩ 5) ∪ (5 ∩ 3) ∪ (4 ∩ 4)	5/36
9: (3 ∩ 6) ∪ (6 ∩ 3) ∪ (4 ∩ 5) ∪ (5 ∩ 4)	4/36, or 1/9
10: (5 ∩ 5) ∪ (6 ∩ 4) ∪ (4 ∩ 6)	3/36, or 1/12
11: (5 ∩ 6) ∪ (6 ∩ 5)	2/36, or 1/18
12: (6 ∩ 6)	1/36
Total	1

Figure 3.2. Probability distribution of experiment of rolling two dice

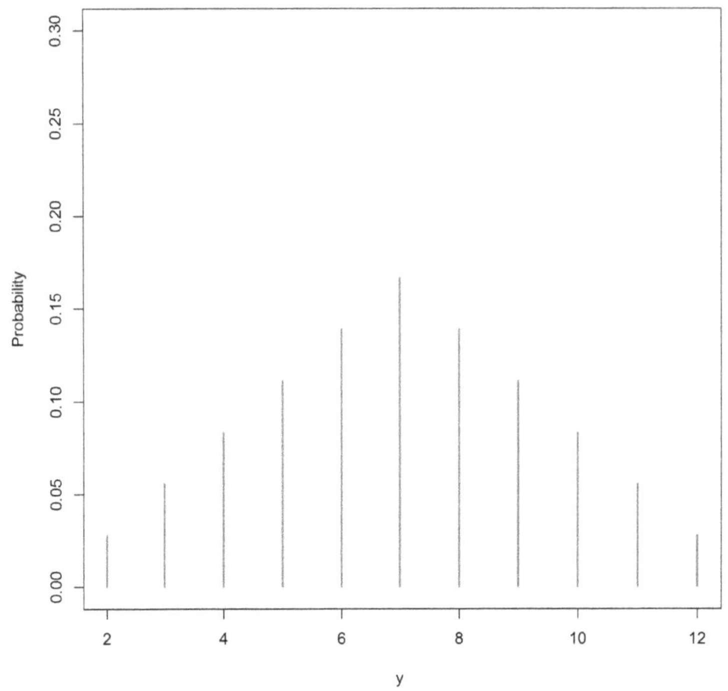

Note: It is interesting that by just adding a second die, the uniform probability distribution in Figure 3.1 changes to a symmetrical, unimodal distribution in Figure 3.2. We will refer to this phenomenon later in this textbook.

The mean, or expected value, of a discrete probability distribution can be calculated and understood as a weighted mean. Table 3.12 provides the mathematical formula for computing the mean of a discrete probability distribution.

Table 3.12. Mean (expected value) of a discrete probability distribution

$$E(Y) = \sum_i y_i P(y_i)$$

The mean or expected value is found by summing the product of the variable's values with their associated probabilities.

To compute the mean of the probability distribution of a rolling die in Figure 3.1 we calculate:

$$E(Y) = 1 * (1/6) + 2(1/6) + 3(1/6) + 4(1/6) + 5(1/6) + 6(1/6) = 21/6 = 3.5$$

Continuous Variables

A continuous variable can take on any value within the limits of the variable and thus is characterized by an *infinite* amount of possible values. As a consequence, the probability of a single event or a single value does not exist or is 0. Therefore, for continuous variables we have to calculate the probability that values of a variable fall within a specific interval. The graph of a continuous variable's probability distribution is a smooth, continuous curve, called probability density distribution or function. Figure 3.3 represents an idealized probability density distribution of communication apprehension (CA) scores (McCroskey, 1982) in the US population, with a mean of 66 and a standard deviation of 15.

Figure 3.3. Idealized probability density distribution of CA Scores in the US Population

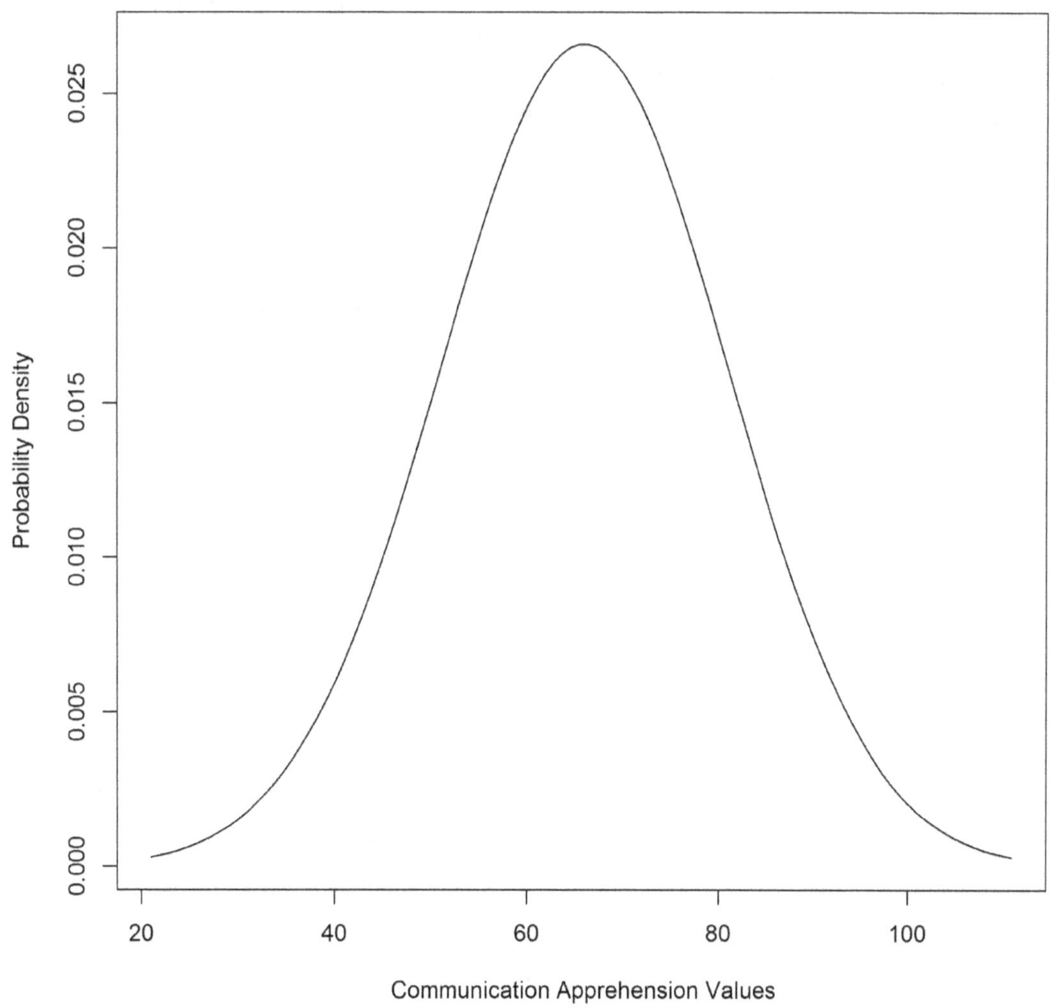

In statistics, there are many different probability density distributions and we need different distributions for various purposes as we will see later in this textbook. One probability density distribution, however, is particularly important—the so-called normal distribution. A good understanding of the normal distribution and the skill to work with normal distributions is essential for the understanding of concepts in inferential statistics (e.g., sampling distributions).

The normal probability density distribution, or just normal distribution, is a symmetric, bell-shaped curve, characterized by its mean μ and standard deviation σ. μ and σ are the two parameters of this distribution. Probabilities in a probability density distribution are defined as areas under the curve (AUC). For example, the probability that a randomly selected US person's communication apprehension score falls between the minimum possible CA value and the maximum possible CA value would be 1 or 100%, which corresponds to the total area under the curve. The probability that a randomly selected US person has a CA score of exactly 85 would be undefined or zero as there is no area under the curve that corresponds to exactly 85 (not 85.000000000 . . . 1 and not 84.999999999 . . . 9).

As Figure 3.4 illustrates, about 68% of all values in a normal distribution fall in an interval of $\mu \pm 1\sigma$, about 95% in an interval of $\mu \pm 2\sigma$, and 99.7% in an interval of $\mu \pm 3\sigma$. In other words, the area under the curve of the interval $\mu \pm 1\sigma$ is about 68%, that of $\mu \pm 2\sigma$ is about 95%, and that of $\mu \pm 3\sigma$ is about 99.7% (see also Figure 2.6).

Figure 3.4. Normal distribution

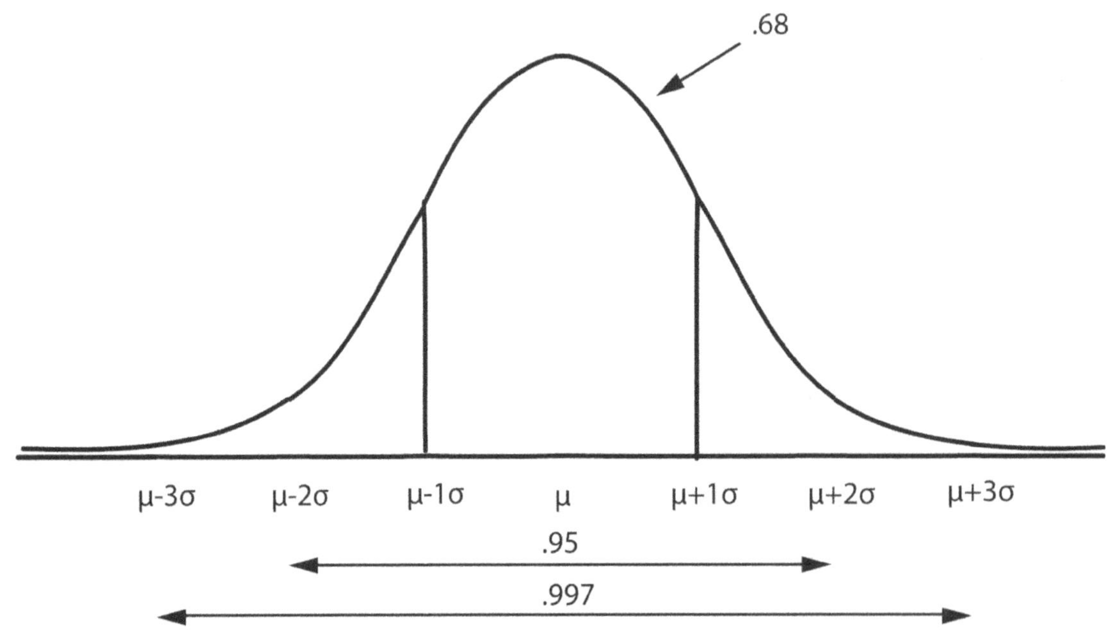

If we wanted to calculate the exact probability that a value y_i of a continuous normally distributed variable falls in an interval from a to b ($a \leq y_i \leq b$), we would have to calculate the interval of a specific normal distribution with the parameters μ and σ between the limits a and b. That is, we would have to calculate the area under the curve between a and b for a normal distribution with the parameters μ and σ. This is difficult to do and most of us would be overwhelmed with this task. For that reason, statisticians have developed a simplification that allows us to easily calculate the area under the curve of any interval in a normal distribution.

Finding Probabilities in a Normal Distribution

Perhaps you remember that values of a variable can be *z*-standardized by subtracting the mean of all values and dividing by the standard deviation of all values (see Table 2.13). To refresh your memory, Table 3.13 provides the *z*-transformation formula again. Instead of sample mean ($\bar{Y}$) and standard deviation (s), however, we use the population mean (μ) and standard deviation (σ).

Table 3.13. *z*-transformation formula

$$z_i = \frac{y_i - \mu}{\sigma}$$

A *z* score is determined by subtracting the mean of all values from an observation and dividing the observation by the standard deviation of all values.

Imagine you applied this transformation to all values of a distribution, what new mean and standard deviation would this *z*-transformed distribution have? This new distribution with z-transformed values would have a mean of zero (you subtracted the mean of all values!) and a standard deviation of 1 (you divided by the standard deviation of all values!). A *z*-transformation of all values would not change the shape of the distribution, but if your values were normally distributed like our CA scores in Figure 3.3 with $\mu = 66$ and $\sigma = 15$ as parameters, then your *z*-transformed CA scores would also be normally distributed with $\mu_z = 0$ and $\sigma_z = 1$. This important new, transformed normal distribution with $\mu = 0$ and $\sigma = 1$ is called *standard normal distribution.* Every normal distribution can be easily converted into a standard normal distribution through a simple *z*-transformation of its values. The big advantage of this procedure is that by doing so, we do not need to integrate normal distributions in order to calculate areas under the curve. We simply standardize (*z*-transform) the limits of the interval and look up the corresponding area under the curve in a prefabricated table that lists all areas under the curve of a standard normal distribution.

For example, let us use our communication apprehension scores in Figure 3.3. Those CA scores are normally distributed with $\mu = 66$ and $\sigma = 15$. We want to know the probability that a randomly selected US person from this population has a communication apprehension score between 72 and 78 (or falls in an interval from 72 to 78). Without *z*-transformation and a tabulated standard normal distribution we would have to integrate the original CA normal distribution with the parameters $\mu = 66$ and $\sigma = 15$ and then calculate the area under the curve between 72 and 78. However, with the help of z-transformation and the standard normal distribution we can do this simply. Table 3.14 demonstrates the necessary calculations:

Table 3.14. Normal distribution and calculating probabilities

Problem: Communication apprehension scores in the US are normally distributed with $\mu = 66$ and $\sigma = 15$. What is the probability that a randomly selected US person has a communication apprehension score between $y_1 = 72$ and $y_2 = 78$; i.e., $72 \leq y \leq 78$?

Step 1: z-transform the limits of the interval

$$z_1 = \frac{y_1 - \mu}{\sigma} = \frac{72 - 66}{15} = 0.4$$

$$z_2 = \frac{y_2 - \mu}{\sigma} = \frac{78 - 66}{15} = 0.8$$

Thus, the probability that a person has a communication apprehension score between $y_1 = 72$ and $y_2 = 78$ in normal distribution with a $\mu = 66$ and $\sigma = 15$ corresponds to the probability that a person has a z-transformed communication apprehension score between $z_1 = 0.4$ and $z_2 = 0.8$ in a *standard* normal distribution with a $\mu = 0$ and $\sigma = 1$.

Step 2: Find the areas under the curve that correspond to the limits of the interval
Table A1 in Appendix A lists areas under the curve (probabilities) of a standard normal distribution. The table provides the areas under the curve *from* a specific z-score to a plus infinity. Thus, the areas under the curve (*AUC*) that correspond to the limits of the above interval are:

$$AUC\ (z \geq 0.8) = 0.2119$$
$$AUC\ (z \geq 0.4) = 0.3446$$

Step 3: Calculate area under the curve of the interval
We are interested in the area under the curve *between* the z-scores 0.4 and 0.8. This area under the curve corresponds to the difference of the areas under the curve in step 2.

$$AUC\ (0.4 \leq z \leq 0.8) = 0.3446 - 0.2119 = 0.1327 = 13.27\%$$

The probability that a randomly selected US person has a communication apprehension score between 72 and 78 is 13.27% given that CA scores in the US are normally distributed with $\mu = 66$ and $\sigma = 15$.

Table 3.14 has shown the calculation of the probability with which an individual's communication apprehension score y falls in an interval above μ (**Case 1:** the interval $72 \leq y \leq 78$ falls above $\mu = 66$ in our example). Figure 3.5 shows the area under the curve for the interval $72 \leq y \leq 78$ in our example.

Figure 3.5. Area under the curve for communication apprehension scores

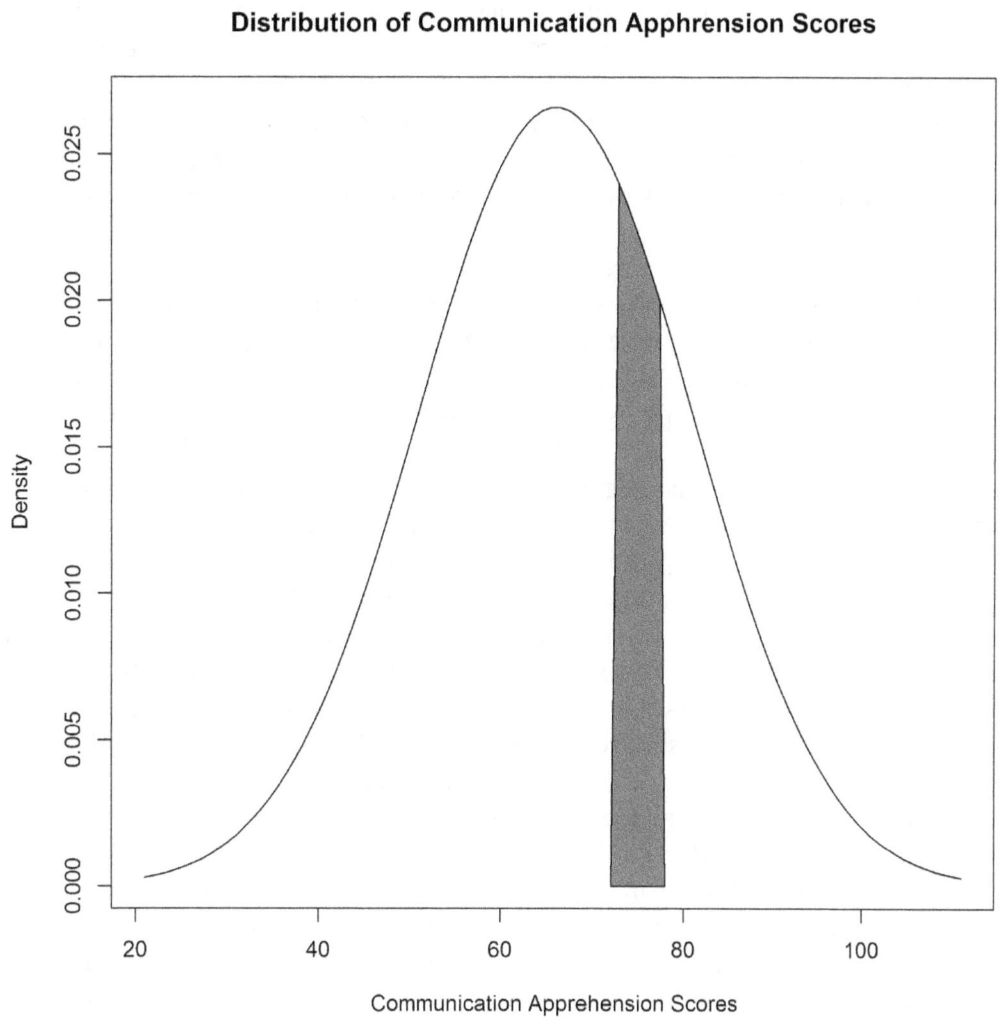

With a similar procedure you are able to calculate the probability that a US person has a CA score . . .

Case 2. of greater than y (for $y > \mu$; positive z-score)
Case 3. of smaller than y (for $y < \mu$; negative z-score)
Case 4. of smaller than y (for $y > \mu$; positive z-score)
Case 5. of greater than y (for $y < \mu$; negative z-score)
Case 6. that falls in an interval below μ (two negative z-scores)
Case 7. that falls in an interval containing μ (one negative, one positive z-score)

These seven cases cover all possible probability calculations you may encounter. After z-transformation, the probability of case (2) can directly be found in the standard normal distribution table as the table lists areas under the curve from a specific, positive z-score to plus infinity. The probability for case (3) cannot directly be read off the table, as the table does not list areas under the curve for negative z-scores. However, the standard normal distribution is a symmetric distribution: areas under the curve for areas under the curve for smaller than $-z$ are identical with areas under the curve for greater than $+z$. The probability for the probability for case (3) is equivalent with the probability of case (2), one just ignores the sign of z, that is, one uses the absolute value of z, which is written $|z|$. Likewise, due to the symmetry of a normal distribution, the probability of case (4) corresponds to 1 (total area under the curve) minus the area under the curve of values greater than y ($y > \mu$) which – after z-transformation – can be found in Table A1. The same applies to case (5), one just ignores the negative sign of the resulting z-score. The z-transformation of values that fall below μ (case 6) will result in two negative z-scores. In this case you can simply ignore the sign of the two negative z-scores and replicate the calculations in table 3.14, the result is identical. Case (7) is the most difficult of all cases, but easy to solve with the skills you have acquired by now. The solution is a combination of case (4) and case (3). If you z-transform one value below the mean and one value above the mean you will obtain one negative and one positive z-score. Let us label the positive z-score with $z1$ and the negative z-score with $z2$. $|z2|$ would stand for the absolute value of $z2$, that is, we simply ignore the negative sign of $z2$. First, you need to find the area under the curve from minus infinity to the positive z-score $z1$ (case 4). This is 1 (total area under the curve) minus the area under the curve of z-scores greater than $z1$ (see above). Then you apply the calculations for case (3) as described above and subtract this result from the case (4) result. That is, the probability of case (7) corresponds to $[1 - AUC(z\text{-scores} \geq z1)] - [AUC(z\text{-scores} \geq |z2|)]$.

Hint: It can be confusing to find the necessary areas under the curve. In our experience it is very helpful to draw a normal distribution on a piece of paper and visualize the area under the curve. Keep in mind that Table A1 in Appendix A only lists areas under the curve *from* a specific, positive z-score to plus infinity and that the normal distribution is a *symmetrical* distribution. With these two simple rules you can find any area under the curve or probability.

In this chapter we only introduced one specific probability density distribution, the normal distribution. While this probability density distribution is of particular importance, in later chapters we will need to work with different probability density distributions, namely with the t-distribution, the chi-square distribution, and the F-distribution. The way you find areas under the curve in these distributions is similar, albeit not identical to the approach we demonstrated above. You can find tabulated versions of these probability density distributions in the appendix. Once we need them, we will explain how to use the tables.

SUMMARY

The purpose of this chapter was to provide an overview of probability theory. Probability is an important concept in statistical methods. In fact, it is one of the foundations of statistical reasoning. We distinguished between probability distributions for discrete variables and probability density distributions for continuous variables. For discrete variables, we can compute the probability with which an event takes on a specific value, whereas for continuous variables we can only compute the probability with which an event falls into an interval of values. Dependent on the type of events, we introduced rules that can be used to calculate with probabilities.

TERMS

- **Normal distribution** is a symmetric, bell-shaped curve, characterized by the mean μ and the standard deviation σ as its parameters.
- **Objective probability** is a figure based on many random experiments or one large experiment with many subjects.
- **Rule of addition** refers to the following rule: If events *A* and *B* come from the same sample space, the probability that either event *A* or event *B* occurs equals the probability that event *A* occurs plus the probability that event *B* occurs minus the probability that both events *A* and *B* occur. The application of this rule depends on whether the events are non-exclusive or exclusive.
- **Rule of multiplication** refers to the following rule: If events *A* and *B* come from the same sample space, the probability that both *A* and *B* occur equals the probability that event *A* occurs times the probability that *B* occurs, given that *A* has occurred. The application of this rule depends on whether the events are dependent or independent.
- **Rule of subtraction** refers to the following rule: The probability that event *A* will occur is equal to 1 minus the probability that event *A* will not occur.
- **Subjective probability** is a figure that indicates an individual's belief that an event occurs.

CHAPTER 3 PROBLEM SET

1. What is the probability of drawing a face card (Jack, Queen, King) or a Spade from a deck of cards?
2. There are 20 large balls and 40 small balls in an urn.
   a. What is the probability of drawing a large ball on 3 draws, *without replacement?*
   b. What is the probability of drawing a large ball on 3 draws, *with replacement?*
3. Construct a probability distribution of 4 coin tosses.
4. What is the difference between probability distributions of discrete variables and continuous variables?
5. We know that individuals' scores on the willingness to censor scale are normally distributed, with a mean $\mu = 1.75$ and standard deviation $\sigma = .45$ (consider the willingness to censor scale as an interval/rating scale in which a 5 indicates the greatest willingness to censor and 1 the lowest).

a. What is the probability that an individual's score falls between 1 and 2?
b. What is the probability that an individual's score falls between 2 and 3?

REFERENCES

Dienes, Z. (2011). Bayesian versus orthodox statistics: Which side are you on? *Perspectives on Psychological Science,* 6(3), 274–290.

Gerbner, G., & Gross, L. (1976). Living with television: The violence profile. *Journal of Communication,* 26, 172–199.

McCroskey, J. C. (1982). *An introduction to rhetorical communication* (4th ed.). Englewood Cliffs, NJ: Prentice Hall.

CHAPTER 4
SAMPLING METHODS

Sampling refers to how researchers select study elements from a population in order to get results that are unbiased. The result of sampling should be a "mini population" that represents those characteristics that are relevant to the study's research interests. The term sample is defined as an unbiased selection of study elements. As such, a sample is—by definition—a representative selection of study elements. The term representative means that the aggregative characteristics of the sample closely approximate those same aggregate characteristics in the population.

If we are able to select all study elements from a population, then we do not need to sample—we would have complete information. However, it is often impossible or impractical to include all members of a population. For example, it would be very time consuming to ask all registered American voters who they think is going to be the next president of the United States. Apart from the fact that it is simply impossible to ask all voters, by the time we would receive the results, the political climate might have already changed. This is why we need sampling procedures, which help us to determine how to optimally select a rather small number of study elements (usually people) and still be able to accurately predict a population parameter, such as the result of an election. There are sampling techniques, which work quite well, allowing for precise estimates of a population's parameter. Other techniques are frequently used, but most likely lead to biased estimates of a population parameter. In the following sections, you will learn what kinds of sampling methods are available and how the selection of a certain technique may influence your results. We distinguish below between two types of sampling methods: nonprobability and probability sampling.

Our overview of sampling methods is far from comprehensive. Others have written entire books on just this issue (e.g., Ardilly & Tille, 2006). Our goal with this chapter is to provide an introduction to this important topic and to highlight some aspects of sampling methods that allow readers to distinguish well from poorly executed sampling procedures.

NONPROBABILITY SAMPLING

Nonprobability sampling provides no information on the probability that a study element or a member of a target population was or will be selected. There are four types of nonprobability sampling procedures overviewed here: convenience, purposive, snowball, and quota sampling.

Convenience Sampling

Convenience sampling leads to a type of nonprobability sample that relies on available participants. An example of this type of sampling is where a researcher asks students in his class to fill out a questionnaire or a journalist on the street asks "randomly" passing people. Results from this type of sample cannot be generalized beyond the classroom or beyond the people who were asked because the researcher or journalist do not know the probability that a member of some larger population (all college students, or American adults) was or will be included in the sample. We recommend that researchers use this technique only when pre-testing a data collection instrument or for a pilot study. Under no circumstances should researchers or communication professionals generalize findings that are based on convenience sampling. It is very likely that those results are biased.

Purposive Sampling

Purposive sampling is a type of nonprobability sampling technique in which the researcher knows information about the target population and exercises her judgment on which study elements to include. For example, it would not make sense if a researcher asked selected study participants for evaluations of a particular soap opera on TV if those participants never have watched any soap operas. In this case a researcher selects individuals for inclusion who have special knowledge about her research topic or who are affected by her findings. Obviously, results that are based on this sampling strategy cannot be generalized beyond expert soap opera viewers.

Snowball Sampling

In snowball sampling, a researcher asks study participants to refer to other similar individuals to participate in the research. Usually, a researcher starts with purposive sampling, and after completion of data collection he asks each participant to identify others who might be available as sample members. This type of sampling is often used when it is difficult to locate members of a special population. For example, a study on entertainment quality perceptions among producers of television entertainment programs requires a sample of show producers in Los Angeles. It is very difficult to obtain the contact information of those producers and to get in touch with them. A researcher who managed to get in contact with one producer could ask this producer for a referral to other producers.

Snowball sampling is also very common in internet-based research. Frequently, study participants are contacted via an email that contains a link to a survey. After completion of the survey (or after refusal to fill-in the survey) participants are encouraged to forward the survey link to their network of friends. Although snowball sampling can easily generate very large samples in a relatively short time, generalizations from a snowball sample are usually quite biased and should be interpreted with extreme caution. Only if the generated sample size comes close to the actual population size or if the distribution of relevant demographic variables in the snowball sample are very similar to their distribution in the population can representativeness be justified.

Quota Sampling

When a researcher draws study participants in proportion to their representation in the known population, she is employing quota sampling. The first step in quota sampling is to describe the population of interest in as much detail as possible, including any and all aspects that will be important to ensuring its representativeness. For example, a researcher is interested in exploring the digital divide; that is, whether individuals have differential access to the informational and communication technologies. We know that access to broadband internet is not likely influenced by hair color, shoe size, or other variables. But we do know that men and women may have differential access to broadband internet. Moreover, we know that access may be influenced by where people live (urban or rural) and personal wealth (rich or poor). Hence, for a good (representative) quota sample, you need to get information on how many men and women are in your population, and how many are poor and rich people, as well as how many live in the countryside or city. In most countries this population information is publicly available and easily accessible (for the US, see http://www.census.gov/). In an oversimplified example, let's assume that you find 50/50% male/female, 60/40% urban/rural, and 70/30% poor/rich in your population and your goal is to sample 1,000 people. Based on this information, you create a quota sample plan according to the rules in Table 4.1.

Table 4.1. Quota Sample Plan

	Male (0.5)		Female (0.5)		
	Rich (0.3)	Poor (0.7)	Rich (0.3)	Poor (0.7)	
Urban (0.6)	90	210	90	210	600
Rural (0.4)	60	140	60	140	400
	150	350	150	350	1,000

To compute the number of people in each cell or combination of sample characteristics, multiply the population proportions for each level of each variable (sex × residence × wealth) by the total sample size. For example, to compute the number of urban, rich, males needed for the sample, multiply: 0.6 (60% urban) × 0.3 (30% rich) × 0.5 (50% male) × 1,000 (sample size) = 90.

Once such a quota sample plan has been created researchers collect data from the number of people having all characteristics of a given cell. For example, with the quota sample plan in Table 4.1, a researcher would collect data from 90 urban, rich, males; from 210 urban, poor, males; etc. The representativeness of a quota sample depends largely on how accurately the actual sample numbers reflect the sample numbers of the quota sample plan. Furthermore, a quota sample can only be labeled representative if all important variables for representativeness have been considered. Sometimes it may be unclear which variables are relevant and which variables are irrelevant for representativeness. It can also be difficult to get accurate information on the relevant variables' distribution in a population, but in most cases it is easier than assumed. You should at least try to construct a quota sample in your projects as it is always better than any of the other previously mentioned nonprobability sampling procedures. Quota samples are frequently used in private

Table 4.2. Nonprobability Sampling—Example Studies

Reference	Sampling type	Rationale for sampling type
Li, X., & Nergazde, N. (2009). Deterrence effect of four legal and extralegal factors on online copyright infringement. *Journal of Computer-Mediated Communication, 14*(2), 307–327.	*Convenience:* Students in student union were asked to fill out a questionnaire on online filesharing	▪ Student union is largest site where students gather ▪ Study is exploratory, and intended to test applicability of deterrence theory
O'Connor, A., & Shumate, M. (2006). An economic industry and institutional level of analysis of corporate social responsibility communication. *Management Communication Quarterly, 24*(4), 529–551.	*Purposive:* 158 corporate web sites from the 2005 online version of US Fortune 500.	▪ The researchers were interested in a specific type of corporation—one that provided a justification for its corporate social responsibility
Johnson, T. J., Kaye, B. K., Bichard, S. L., & Wong, W. J. (2007). Every blog has its day: Politically-interested internet users' perceptions of blog credibility. *Journal of Computer-Mediated Communication, 13*(1), 100–122.	*Snowball:* Politically-motivated Internet users were first purposively selected and asked to fill out a questionnaire; they were then asked to forward the survey to politically-interested friends.	▪ Politically-motivated Internet users are a special group of Internet users; no known list of such users exists
Scharrer, E. (2002). Third-person perception and television violence: The role of out-group stereotyping in perceptions of susceptibility to effects. *Communication Research, 29*(6), 681–704.	*Quota:* The authors gathered data from individuals representing different geographic regions (urban, rural, and rural/suburban), sex (male, female), and age groups (18–35, 36–53, and 54-and older) in order to determine differences in third-person effect, specifically beliefs about other individuals' susceptibility to negative media effects	▪ Sample needed to be representative for the larger community (i.e., beyond college students) ▪ Drawing a random sample was impractical; i.e., too expensive and time-consuming

market surveys and political polls because of their efficiency and low costs. If quota samples are carefully designed and interviewers well trained, experience shows that they are not necessarily inferior to classical probability samples. However, a major criticism of quota samples is the fact that the ultimate selection of the respondent is up to the judgment of the interviewer, which may lead to a biased selection of study participants. A historical discussion of quota sampling (and other sampling procedures) can be found in Frankel and Frankel (1987).

To give you an idea of how nonprobability sampling techniques have been used in communication research, Table 4.2 provides some example studies as well as the rationale for each procedure.

PROBABILITY SAMPLING

Probability sampling provides information on the probability that a member of a target population was or will be selected as a study element. The main problem with nonprobability samples is that they might be biased because a person with subjective judgments and habits selects study elements. A person might have some prejudice toward specific people and prefers to not select and talk with them; or a researcher may prefer to select people to whom he has easy access; or a researcher may prefer to select study elements that support his hypothesis. One way to avoid such biases is to not have individuals select study elements. Rather, a researcher (usually via a computer algorithm) employs a random procedure, which cannot be biased if properly conducted. There are four main types of probability sampling procedures overviewed here: Simple random, systematic random, stratified, and cluster sampling.

A core principle of probability sampling is that a sample will be representative of a population from which it is selected if (1) all members of the population have initially an equal chance of being selected for the sample and (2) the selection of a study element occurs at random. Samples with these characteristics are called equal probability of selection method (EPSEM) samples. All following sampling procedures lead to EPSEM samples if properly executed.

Simple Random Sampling

Simple random sampling is a type of probability (EPSEM) sampling procedure in which the researcher has a complete list of elements in the population of interest, and applies a random procedure such as selecting cases that correspond to a list of randomly generated numbers. A convenient way to generate random numbers is to use the internet. The website http://www.random.org, for example, provides easy to use tools for the generation of random numbers. Let us assume your complete list of elements in the population contains 10 million potential study elements and you would like to draw a sample with sample size of $n = 100$. You could go to http://www.random.org/integers/, enter "100" in the field "generate random integers," and ask for values between 1 and 10 million. Afterwards you select those study elements from your list whose case number correspond to the generated random numbers. Common statistic or database programs (e.g., SPSS©, Statistical Package for the Social Sciences) provide simple procedures and easy-to-use menus for simple random sampling.

Systematic Random Sampling

Systematic random sampling is very similar to simple random sampling. The main difference is that this procedure requires a list of study elements in *random order* and the researcher systematically selects every k^{th} element of the population of interest. If the population size is N and the sample size should be n, then k is N/n. For example, if the result should be a $n = 100$ sample size of a population with $N = 10$ million, a researcher would select every $k = 100{,}000^{th}$ study element in a randomly ordered list of study elements. In practice it is usually easier to generate a list of random numbers as required in simple random sampling than to randomize the order of a long list of study elements.

Stratified Sampling

Stratified sampling is similar to quota sampling. The first step in this type of probability sampling is to describe the population of interest in as much detail as possible, including any and all aspects that will be important to ensuring its representativeness. Thus, a researcher first divides the population into so-called strata that are considered relevant for representativeness. Let us use again our example above (see under quota sampling): In order to study the digital divide the researcher would stratify the population into the strata of males/females, rich/poor, and urban/rural. Unlike in quota sampling, however, the researcher would then apply a random procedure to select study elements within each stratum and would *not* leave the selection up to his or her interviewers. Stratified sampling can be viewed as a combination of quota and simple (or systematic) random sampling. The main difference is that randomness in this type of sampling procedure is a result of a random procedure and not a result of individuals' decisions.

Dividing a population into relevant strata prior to applying simple or systematic random sampling has one essential advantage. If the variables that are used to define the strata are indeed relevant for representativeness (like gender and income in our example), then those variables are important predictors for the dependent variable(s) of interest (like frequency with which people use information technology). If the strata are important predictors for the dependent variable(s), then study participants *within* a stratum are more homogeneous with respect to the dependent variable. For example, if income is an important variable to predict how many hours per week people use computers then we would expect that both within rich people and within poor people the *variance* of hours of computer use per week is smaller than the *variance* of hours of computer use for all people together (poor and rich). Drawing random samples from homogeneous rather than heterogeneous sub-populations, that is from sub-populations with a smaller variance, leads to a smaller sampling error and more accurate estimates for parameters. The concept of sampling errors will be discussed in more detail in Chapter 5.

A researcher may choose to stratify samples proportionally or disproportionately. For proportional stratified sampling, the proportion of sample sizes across the strata reflects the proportions in the population of interest. For example, if we know that 60% of the population lives in urban and 40% in rural areas, then about 60% of the sample should stem from urban and 40% from rural areas. The sample plan of proportional stratified sampling looks much like the quota sample plan in Table 4.1. For disproportional stratified sampling, the sample size proportions across strata differ from the population proportions. This technique is typically used if the resulting sample sizes would be too small for certain strata otherwise. For example, a researcher may know that there are only 1% of Native Americans in the population. Applying proportional stratified sampling would lead to just a handful of Native Americans in a sample and with this small sample size it would be difficult to predict the frequency of computer use among this group accurately. In this example, the researcher would choose to sample a higher proportion of Native Americans and later correct the introduced bias with an adequate weighting procedure (see Ardilly & Tille, 2006 for further details).

All types of probability sampling procedures discussed so far require that the researcher has access to all members of a population for selection purposes. Usually this access is provided by some sort of list or a database in which all members of a population are stored.

Cluster Sampling

Cluster sampling can be used if it is not practical or even possible to compile a list of all study elements comprising the target population. In the US for example, compared to most European countries, it is extremely difficult to draw a simple random sample that is representative of all Americans. For various reasons there is no easily accessible list of all Americans that is available for research. Unlike in European countries, in the US there is no law that demands people have to have a passport or a drivers' license, or report to the authorities their new address when they move. As a result, there simply is no complete and up-to-date list of all Americans and their addresses to which researcher could mail surveys, for instance. In such a case, a researcher could apply a random selection procedure *not* targeting the individual study elements but aggregates or clusters of study elements that are easier to access. For example, imagine a researcher interested in civic participation. Results of the researcher's study should speak for the population of major US cities, like Los Angeles, New York, Chicago, Atlanta, etc. Hence, the researcher needs a sample that is representative for US citizens of major US cities. Unfortunately, there is no complete list of all citizens in major US cities available for this research. However, city maps with a list of all city blocks, streets within blocks, house numbers, even apartment numbers are available and accessible online. Therefore, instead of sampling people as study elements, the researcher decides to draw a simple random sample among clusters of study elements, for example city blocks within major US cities, and then contact all people within the selected city blocks via personal visits.

If the researcher applies simple random sampling among clusters *one time* then we speak of single-stage cluster sampling. Single-stage cluster sampling typically leads to rather large clusters with many study elements that all have to be contacted. In contrast, cluster sampling can be applied in multiple stages, which leads to smaller and more selected clusters. In multistage

Table 4.3. Probability Sample—Example Studies

Reference	Sample type	Rationale for sampling type
Tsfati, Y. (2007). Hostile media perceptions, presumed media influence, and minority alienation: The case of Arabs in Israel. *Human Communication Research, 57*(4), 632–651.	*Simple random sample:* The authors had a complete listing of Arab households in Israel available.	▪ The authors wanted a representative sample of Israeli Arabs, in order to determine their perceptions of hostile media. Which variables are crucial for representativeness was unclear.
Liebler, C. M., Schwartz, J., & Harper, T. (2009). Queer tales of morality: The press, same-sex marriage, and hegemonic framing. *Journal of Communication, 59*(4), 653–675.	*Systematic sample:* The authors used two systematic samples in order to do an analysis of news coverage of same-sex marriage over time. The authors had a complete list of newspaper and wire stories on same-sex/gay marriage published in February 2004 and February 2005, sampling every 5th story in February 2004, and every 3rd story in February 2005.	▪ Systematic sampling was an easy-to-apply procedure for this sampling problem. Stories were listed in random order. Thus, the selection of every 5th and every 3rd story resulted in a random selection of stories.
Schultz, T. (1999). Interactive options in online journalism: A content analysis of 100 US newspapers. *Journal of Computer-Mediated Communication, 5*(1).	*Stratified sample:* The author obtained a list of newspapers (US general circulation dailies) and identified circulation number as an important variable for representativeness. Subsequently, the author randomly selected 100 newspapers within four circulation strata (less than 25,000; 25,001 to 50,000; 50,001 to 100,000; more than 100,000)	▪ The author has prior knowledge of what affects the sample's representativeness. ▪ The use of a stratified sample reduced the sampling error, providing more accurate estimates.
Sohn, D., & Lee, B. (2005). Dimensions of interactivity: Differential effects of social and psychological factors. *Journal of Computer-Mediated Communication, 10*(3).	*Cluster sample:* First, 8 out of 50 US states were randomly selected, then 30 cities were randomly selected from each state (total of 240 cities). Next, all available email addresses of prospective participant living in these cities were found using email databases of Yahoo and Lycos.	▪ Simple random sampling from the population of internet users is difficult because users are geographically dispersed.

cluster sampling, a researcher draws simple random samples from clusters of a population's study elements in more than one stage. For example, our same researcher samples city blocks in stage one, lists all streets in the selected blocks, and then draws a sample of streets in stage two. Finally, the researcher would contact all people that live in the selected streets. Of course, the researcher can add additional stages such as the house numbers within the streets, apartment numbers within houses, and order of age of all people living in an apartment. The interviewer would then receive very detailed instructions about who to contact in a particular city. For instance, the instructions could be: Go to specific address at one of the major US cities, go to apartment C at this address, try to contact the second oldest member of the household that lives in this apartment (if only one person lives in this apartment, select this person).

A major pitfall of cluster sampling is the tendency to end up with *artificially* homogeneous samples, i.e., samples that do not represent the true diversity of study elements in a population. As a consequence, cluster sampling usually leads to lower accuracy when estimating population parameter compared to simple or systematic random samples. This usually requires a researcher to draw more study elements for inclusion and to apply complicated weighting algorithms. Moreover, cluster sampling can be quite complicated and time-consuming. Table 4.3 provides example studies that use probability samples.

Response Rates

A major drawback of all probability sampling procedures is the requirement that once study participants have been selected, the researcher needs to be able to contact all of those participants and convince them to participate in the study. This, however, can almost never be accomplished in practice. The smaller the ratio of contacted people and actual study participants is (response rate), the more biased is the sample. Response rates depend on many factors. Typical response rates in probability sampling without incentives for participation and large, heterogeneous populations range between 5–15% with declining trend. In cases with such low response rates a researcher may only assume representativeness if she can demonstrate that responders do not significantly differ from non-responders. The problem of low response rates in probability sampling, with sample bias as a likely consequence, is often the reason why researchers decide to apply a nonprobability sampling procedure.

SUMMARY

Sampling procedures include both non-probability and probability sampling techniques. Given that researchers who use quantitative research methods generally want to utilize samples—i.e., a representative selection of study elements from a population of interest—it is important to choose techniques that accomplish this. Otherwise a study's findings may be biased and therefore questionable. We introduced the most important non-probability and probability sampling techniques together with examples and a brief discussion of possible drawbacks.

TERMS

- **Cluster sampling** is a type of probability sampling technique that is used when it is not practical to compile a list of study elements comprising the target population.
- **Convenience sampling** is a type of nonprobability sampling technique that relies on available participants.
- **Nonprobability sampling** provides no information on the probability that a member of a target population was or will be selected.
- **Probability sampling** provides information on the probability that a member of a target population was or will be selected.
- **Purposive sampling** is a type of nonprobability sampling technique in which the researcher knows information about the target population and exercises her judgment on which study elements to include.
- **Quota sampling** is a type of nonprobability sampling technique in which the researcher draws study participants in proportion to their representation in a known population.
- **Representative** means that the aggregative characteristics of the sample closely approximate those same aggregate characteristics in the population.
- **Sample** is an unbiased selection of study elements. As such, a sample is—by definition—a representative selection of study elements.
- **Sampling** refers to how researchers select study elements from a population in order to get results that are unbiased.
- **Simple random sampling** is a type of probability sampling technique in which the researcher has a complete list of elements in the population of interest, and applies a random procedure such as selecting cases that correspond to a list of randomly generated numbers.
- **Snowball sampling** is a type of nonprobability sampling technique in which the researcher asks study elements to refer other similar individuals to participate in the research.
- **Systematic random sampling** is a type of probability sampling technique in which the researcher selects every k^{th} study element in a randomly ordered list of all study elements.
- **Stratified sampling** is a type of probability sampling technique in which the researcher randomly draws study elements from homogeneous subsets (strata) of that population.

CHAPTER 4 PROBLEM SET

1. Why do we need sampling methods?
2. What is the difference between non-probability and probability sampling?
3. Why are we concerned with a researcher making decisions about which study elements to include in a sample?
4. What benefit do we get from employing stratified sampling?
5. A researcher is interested in exploring whether there are differences among sex, age, and education for media uses and gratifications. She has the following information about her population of people older than eighteen, and wants to employ quota sampling: sex (male = 49%;

female = 51%); age (18–34 = 41%; 35–49 = 24%; 50-and-up = 35%); highest educational attainment (high school = 32%; some college = 37%; and bachelor's degree or higher = 31%). She wants a total sample size of 1,000. Complete the quota matrix below:

	Male			**Female**			
	18–34	35–49	50-over	18–34	35–49	50-over	
High school							
Some college							
BA or higher							
							1,000

REFERENCES

Ardilly, P., & Tille, Y. (2006). *Sampling methods.* New York: Springer.

Frankel, M. R., & Frankel, L. R. (1987). Fifty years of survey sampling in the United Stated. *Public Opinion Quarterly,* 51, 127–138.

SAMPLING DISTRIBUTIONS AND CENTRAL LIMIT THEOREM

CHAPTER 5

In previous chapters, we focused on topics related to descriptive statistics and on topics that provided the tools for statistical inference. As a researcher or communication practitioner, you will often want to move beyond simply describing samples. For example, a communication researcher might want to find out whether the average level of communication apprehension (cf., McCroskey, 2006) found in a sample of people can be generalized to the population of interest. Or a public relations professional need to know whether the average image score of her company (cf., Gregory & Wiechmann, 1999) in a sample can be generalized to all customers. This information can be crucial for the survival of a company. With the introduction of sampling distributions and the central limit theorem (CLT), we enter the exciting realm of inferential statistics, which aids researchers and professionals in generalizing findings from a sample to a population. Without the concept of sampling distributions and the CLT, researchers and professionals would not be able to make data-driven generalizations and predictions, which are essential for social-scientific reasoning and economic prosperity. One can only imagine what our world would look like if statisticians and mathematicians had never discovered sampling distributions and the CLT.

In inferential statistics, we distinguish between two divisions: estimating and testing. We will cover these topics in more depth in Chapter 6, but mention them very briefly now as an introduction to sampling distributions. In estimating, we are interested in how and how precisely we can estimate or predict parameters (e.g., the mean) of variables in a population by using a sample. In other words, we try to generalize findings in a sample to the population. By contrast, in testing, we are interested in how hypotheses that are related to conditions in a population stand the confrontation with empirical data in a sample. Sampling distributions are important in inferential statistics because they help researchers predict how close statistics fall to the parameters that they estimate or hypothesize. Specifically, sampling distributions portray the variability that occurs when we collect and use sample statistics to estimate or test parameters.

The concept of sampling distributions assumes that we deal with a (true) sample, that is, a representative selection of study elements that is based on some sort of random selection procedure (see Chapter 4). From now on we always assume that researchers operate with samples. The quantification of biases that are introduced by selection errors, that is, by a non-representative selection of study elements, is beyond the scope of this and all following chapters.

SAMPLING DISTRIBUTIONS

Let us begin with a definition: A sampling distribution of a statistic is the probability (density) distribution that specifies probabilities for the possible values the statistic can take. The probabilities are not for individual observations but for possible values of a statistic (e.g., the mean) computed from observations. In the following, it is important for your understanding to distinguish between the distribution of individual values and the distribution of an aggregate of individual values summarized in a statistic. The former refers to distributions outlined in Chapter 3; the latter refers to sampling distributions in this chapter. As you may sense, probability distributions and sampling distributions are related in some way, but how? How can a statistic such as the mean of a number of individual values in a sample have a distribution? To answer this question we will invite you to participate in a thought experiment—a theoretical scenario that will explain the concept of a sampling distribution. We will use the statistic "mean" to guide you through our thought experiment, but the core ideas apply to all sorts of statistics. Later in this chapter we will transfer the core principles to other frequently used statistics such as the median and percentages.

Sample Means

When we analyze data and compute the sample mean as a summary description of the data in a sample, we are uncertain about how close this mean is to the population mean. Why? Because the selection of study elements in a sample is based on a random process (see previous) and, therefore, sample means need to be considered random variables. That is, sample means vary from sample to sample randomly. However, it is possible to specify the distribution of sample means, which, as we will see, makes it possible to determine the probability with which sample means fall into intervals around the population mean.

As an example, imagine that you are interested in the average (mean) amount of "Web slacking" each day, measured by the minutes of Internet for personal use at work. Of course, you cannot collect information from all working Americans on their personal Internet use habits (for review on sampling, see Chapter 4), and so you draw a random sample with a sample size of $n = 500$ working Americans and compute the sample mean as an estimate for the population mean of minutes. In this sample you find a mean of sixty minutes. Would you feel confident to say that the true population mean; that is, the average amount of time working Americans spend with "Web slacking" each day, is sixty minutes? Why? Why not?

Let us assume that you repeat the whole procedure. That is, you draw a new random sample ($n = 500$ working Americans) from the same population and again calculate the mean of that sample. Do you think you will get the exact same mean as in the previous sample? Perhaps, but this is unlikely, because, as explained, the mean is a random variable that depends on chance. Most likely you will get a slightly different mean. This time, for example, you find that the mean number of minutes Americans spend using the Internet for personal use at work is sixty-two.

Now, assume that you do this again and again. Imagine that you do this millions of times. Nobody does it millions of times, just imagine you would, and each time you would record and plot your results in a histogram (see Chapter 2, Figure 2.2). If you look at this histogram you will discover that its shape—amazingly—follows a normal distribution (see Chapter 3) even if the distribution of the original variable—"Minutes of Internet for personal use"—is not normally distributed. This distribution is called the sampling distribution of the mean or just mean sampling distribution. Think about this! It does not matter how the variable "Minutes of Internet for personal use" is distributed. This distribution can be heavily skewed to the right, bimodal, or follow a uniform distribution, it does not matter. If we compute the mean of "Minutes of Internet for personal use" in samples, and look at the distribution of sample means, this distribution is a normal distribution. Figure 5.1 provides an illustration of an empirical sampling distribution of means.

Figure 5.1. Empirical sampling distribution of means

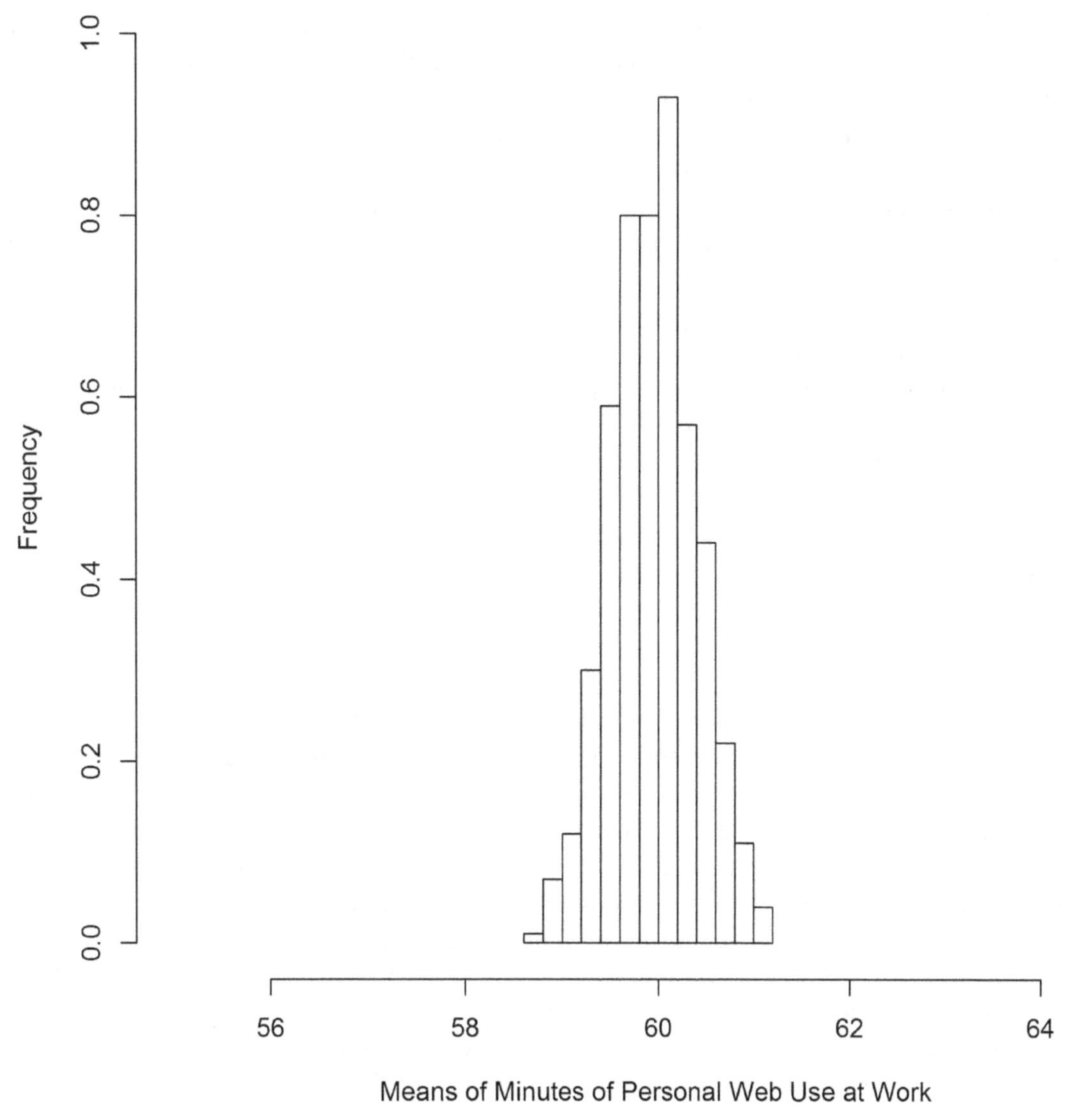

And you learn even more when we look at this normal distribution. What would you answer if someone asked: "What did you learn about average Web slacking in your initial population? What is the mean number of minutes the Internet is used for personal purposes *in the population* of working Americans?" Of course you would intuitively point to the mean of your distribution of means or say that the mean that most frequently occurred in all your samples (i.e., the mean with the highest density in your mean distribution) corresponds to the true mean in your population. It would be irrational to point to any other mean in your distribution of means.

We now know the shape and central tendency/position of this mean distribution. If we also had information on the dispersion of this mean distribution we would have all the ingredients for the calculation of probabilities with which sample means fall in intervals around the true population parameter. This information—as we will see—will be crucial information for the following chapter.

Let's continue with our thought experiment to explore your mean distribution's dispersion or variation. What do you think influences the variation of your means from sample to sample drawn out of the same population?

First, the variation of your variable in the population influences the variation of means. Why? Assume that all working people in the population used the Internet for the exact same amount of time for personal use. This is an unreal assumption, but just imagine. Under this assumption the variable "Minutes of Internet for personal use" would not vary, the variance of "Minutes of Internet for personal use," for example, would be zero. Hence, in each sample that you can draw from this (unreal) population, you would end up with the same number for "Minutes of Internet for personal use," and the mean of each sample would be identical and equivalent to the population mean. In other words, the variation of means from sample to sample, for example the variance of sample means, would be zero. Thus, the smaller/higher the variation of the original variable the smaller/higher the variation of the corresponding mean sampling distribution—the variation of the original variable and the variation of the corresponding mean sampling distribution stand in a proportional relationship.

Second, the sample size influences the variation of means. Why? Assume that in each sample you draw you ask all working Americans in the population. Again, this is an unrealistic assumption, but just imagine. Under this assumption the mean of the variable "Minutes of Internet for personal use" would not vary from sample to sample, because in fact you are collecting data from the entire population in each sample and the population mean is fixed. That is, the population mean that you compute in each sample does not vary from sample to sample. If you operated with a sample size of just $n = 1$ in each sample, then the mean in each sample would be equivalent with the variable's measurement, that is, the mean sampling distribution would have the same variation as the original variable's distribution. Thus, the smaller/higher the sample size the higher/smaller the variation of the corresponding mean sampling distribution—sample size and the variation of the corresponding mean sampling distribution stand in inverse relationship.

In Chapter 2 we introduced various statistics that can be used to capture the variation of a distribution. The most common measures of variation are the variance and standard deviation. In order to distinguish the variation from the original variable (here "Minutes of Internet for personal use") from the variation of means of the original variable from sample to sample, that is, from the variation of the corresponding mean sampling distribution, we use a new term called "standard error." Standard errors refer to the variation of sampling distributions and specify the standard deviation (see Chapter 2) of a statistic's sampling distribution. Hence, the standard error of the mean is defined as (see Table 5.1):

Table 5.1. Standard error of the mean

$$\sigma_{\bar{y}}^2 = \frac{\sigma^2}{n} \Rightarrow \sigma_{\bar{y}} = \sqrt{\frac{\sigma^2}{n}} = \frac{\sigma}{\sqrt{n}}$$

$\sigma_{\bar{y}}^2$:	Variance of means of the original variable in a sampling distribution
$\sigma_{\bar{y}}$:	Standard deviation of means of the original variable (= Standard error of the mean)
σ^2:	Variance of the original variable in the population
σ:	Standard deviation of the original variable in the population
n:	Sample size

Estimated Standard Error of the Mean

As we demonstrated, we can calculate the standard error of the mean if we know the sample size and the variance (or standard deviation) of the variable in the population. However, in most empirical studies the variance of the variable in the population is unknown. For example, if it is your first study to investigate "Web slacking," you usually do not know the variance of "Minutes of Internet for personal use" in the population of working Americans. Thus, there is often no other way than to estimate the variable's population variance with the data of the sample and with the formula in Table 5.2.

Table 5.2. Estimated standard error of the mean

$\hat{\sigma}^2 = \frac{\sum_{i=1}^{n}(y_i - \bar{Y})^2}{n-1}$	Population variance estimated with data in the sample
$\hat{\sigma}_{\bar{y}} = \sqrt{\frac{\hat{\sigma}^2}{n}}$	Estimated standard error of the mean, see Figure 5.1
$\hat{\sigma}_{\bar{y}} = \sqrt{\frac{\sum_{i=1}^{n}(y_i - \bar{Y})^2}{n \cdot (n-1)}}$	Standard error of the mean estimated with data in the sample

Observant readers of this textbook may wonder why we divided by "$n - 1$" rather than by "n" as suggested in Chapter 2 (Table 2.10) when estimating the variable's population variance. While not central to the understanding of sampling distributions in this chapter, the reasons are as follow: Like means in samples form a sampling distribution of means, variances in samples form a sampling distribution of variances. As explained above, the best, that is, unbiased estimation of a population mean is the mean of sample means. It can be shown that this—unfortunately—is not the case for variances. If we estimated the (true) population variance as the mean of sample variances then the estimation would come with a bias of exactly "$(n - 1) / n$." In order to reverse this bias and to estimate the (true) population variance in an unbiased way, we simply multiply the variance in Table 2.10 with the factor "$n / (n - 1)$" resulting in the first formula in Table 5.2. Thus, for purely descriptive, explorative purposes, we would still use the formula in Chapter 2 (Table 2.10). If we calculate a variance to *estimate* a population variance, however, we would use the first formula in Table 5.2. While this might be correct (albeit confusing), we have to admit that this detail is not very important in practice: Dividing by "$n - 1$" rather than by "n" makes only a difference for very small sample sizes that are rarely used in communication research practice. In the case of large samples, the difference between the true and estimated variance and, as a consequence, the difference between the true and estimated standard error is very small.

Central Limit Theorem

The ideas of our thought experiment described above can be summarized in a theorem known as the Central Limit Theorem. This theorem is characterized by the following premises:

- Means from random samples that stem from the same population follow a normal distribution.
- The greater the sample size, the better the approximation of a normal distribution. Sample sizes of thirty and higher provide a reasonable approximation of a normal distribution.
- That means are normally distributed is independent of the original variable's distribution.

In a brief, mathematical form of expression, the Central Limit Theorem states that the random variable mean $\overline{Y}$ is normally distributed for samples with infinite sample size ($n \rightarrow \infty$) with μ (the true mean in the population) and $\sigma_{\bar{y}}$ (standard error of the mean, that is, the standard deviation of means in the sampling distribution of means) as the distribution's parameter:

Table 5.3. Central Limit Theorem
$\overline{Y} \approx N(\mu, \sigma_{\bar{y}}) \qquad \text{for } n \rightarrow \infty$

Figure 5.2 illustrates the Central Limit Theorem. The figure displays sampling distributions of the mean for four different shapes for the population distribution and for different sample sizes.

The Central Limit Theorem is called central because it is of central importance for the understanding of statistical inference. We recommend spending some extra time with simulating the theorem's implications. There are various simulations available in the internet. We found the simulation available at "http://www.intuitor.com/statistics/CentralLim.html" very helpful. The simulation generates differently shaped population distributions and to draw a large number of samples with different sample sizes from these populations. You should begin your simulation

Figure 5.2. Population distributions and their corresponding mean sampling distributions

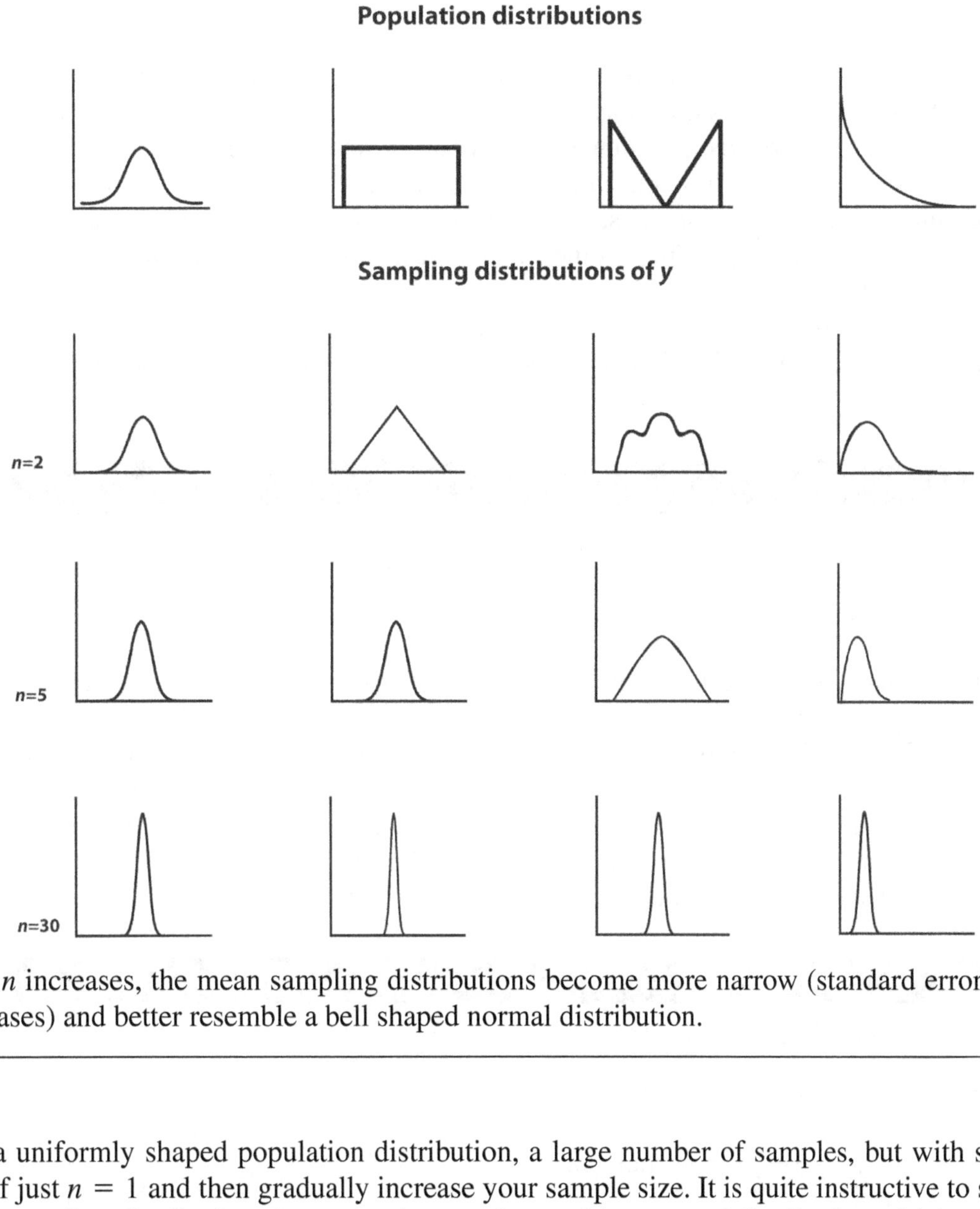

As *n* increases, the mean sampling distributions become more narrow (standard error decreases) and better resemble a bell shaped normal distribution.

with a uniformly shaped population distribution, a large number of samples, but with sample size of just $n = 1$ and then gradually increase your sample size. It is quite instructive to see the mean sampling distribution change its shape and resemble a normal distribution with increasing sample size. Also, look at the (true) population mean in this simulation (upper window) and at the estimated population mean (lower window) from your samples. When using a sample size of thirty and higher you will see that the true mean and the estimate is almost identical.

The Central Limit Theorem provides one of the most important foundations for statistical inference. As many variables follow a unique distribution, it would be difficult to generalize findings from samples to populations, to predict outcomes, or to test hypotheses if the Central Limited Theorem did not exist. There are other foundations that can be used for estimation and testing, but these are beyond the scope of this introductory textbook.

Example

The Central Limit Theorem's premises can be exemplified as follows: The variable "number of friends" of a social network's users has a distribution that is heavily skewed to the right, that is, most users have a few friends, but a few users have an extreme number of friends. The distribution of "number of friends" has a mean of $\mu = 250$ and a standard deviation of $\sigma = 80$. A researcher, unaware of these parameters, draws a sample of $n = 100$ and wants to use the sample mean friends $\bar{y}$ to estimate μ. Where is the sample mean likely to fall relative to the population mean of $\mu = 250$?

Although the distribution is skewed, because of the Central Limit Theorem, we know that sample means are normally distributed. The standard error of the mean distribution is calculated in Table 5.1:

Table 5.4. Example—Standard error of the mean

$$\sigma_{\bar{y}} = \sqrt{\frac{\sigma^2}{n}} = \frac{\sigma}{\sqrt{n}} = \frac{80}{\sqrt{100}} = 8$$

Considering that sample means are normally distributed we know that it is highly likely (see Chapter 3, Figure 3.4) that sample means fall in an interval of $\mu \pm 3$ standard errors, that is, between 226 and 274. The probability that sample means fall between 226 and 274 is precisely 99.7% (see Figure 3.4). For training purposes, let us assume that you are interested in knowing more precisely the probability with which you draw a sample mean of 260 friends or higher in a sample out of this population? With the knowledge of Chapter 3 (and particularly with the knowledge summarized in Table 3.14) this question can be easily answered as Table 5.5 demonstrates:

Table 5.5. Example—finding the probability with which sample means occur

Given:	$\mu = 250 \quad \sigma_{\bar{y}} = 8,$
Sought:	Probability $(\bar{y} > 260)$
z-score:	$z_i = \frac{y_i - \mu}{\sigma}$ (see Table 3.13)
z-score of $\bar{y}$:	$z = (260 - 250)/8 = 1.25$

Using the standard normal distribution Table in Appendix A we find that the corresponding right-tail probability above 1.25 is 0.1056 or 10.56%. Thus, the probability that you draw a sample from this population in which the mean number of friends is 260 or higher is 10.56%.

Other Statistics

So far, we have only focused on the sampling distribution of the mean. Of course, there are other important statistics that communication researchers or communication professionals need to interpret from an inferential statistical perspective and, therefore, need to consider the characteristics of those statistics' sampling distribution. In general and regardless of the specific statistic under focus, there are always three questions to answer:

1. What is the sampling distribution's shape?
2. What is the sampling distribution's expected or most frequent value?
3. What is the sampling distribution's standard deviation (or what is the statistic's standard error)?

The answer to the first question can be a normal distribution or another distribution. As we will see in later chapters, there are statistics that follow a different probability distribution, for instance a t-distribution, an F-distribution, or a χ^2-distribution. There are even statistics for which the sampling distribution is still unknown and, so far, can only be approximated by computer simulations. Specifying the sampling distribution of complex statistics is one goal of statisticians' and mathematicians' research. The answer to the second and third question provides us with information about the central tendency and the dispersion of a statistic. As demonstrated above, we use the answer to all three questions to determine the probability with which a statistic falls into an interval around its population parameter. In practical terms this means that if a statistic follows a different sampling distribution, we apply the new expected value and standard error to the statistic's z-standardization (see Table 3.13), and we simply use another standardized distribution table in the appendix of this textbook in order to find probabilities.

Knowing the answer to the three questions above is essential for the understanding of parameter estimation and statistical tests. Every parameter estimation and statistical test in this book requires exactly the information in the answers to the three questions above, no more and no less. In fact, with an answer to all the three questions above and with the knowledge of the following chapter you are able to conduct any (parametric) statistical test and parameter estimation. So, if you are asked to conduct a test or to understand the parameter estimation for a statistic you are unfamiliar with, just think about (or ask) the three questions above and apply the rules of the next chapter. A good understanding of statistics can be easier than you think. We will do the same now and simply answer the three questions for two frequently used statistics in communication research and communication practice—the median and the percentage.

We assume that with reading of this chapter it is the first time that you have become acquainted with the concept of sampling distributions and with the Central Limit Theorem. Therefore, in this chapter, we will only discuss the characteristic of sampling distributions for very large samples as this simplifies the introduction and supports your understanding of the topic. As you will see in the following chapters, the shape of the sampling distributions changes if you operate with small samples and/or more complex statistics such as the difference of two sample means.

Median

The shape of the median's sampling distribution is—similar to the mean—also a normal distribution for large samples ($n > 30$). Likewise, the most occurring median value of this sampling distribution—that is, the mean of all medians—corresponds to the (true) median in the population. In other words, the expected value of the median's sampling distribution estimates the (true) median in the population. However, the standard error of the median and the estimated standard error of the median differs slightly from the standard error of the mean (see Tables 5.6 and 5.7):

Table 5.6. Standard error of median

$$\sigma_{Md} = 1.25 \cdot \sqrt{\frac{\sigma^2}{n}} = 1.25 \cdot \frac{\sigma}{\sqrt{n}}$$

σ_{Md}:	Standard deviation of medians of the original variable (= Standard error of the median)
σ^2:	Variance of the original variable in the population
σ:	Standard deviation of the original variable in the population
n:	Sample size

Table 5.7. Estimated standard error of median

$$\hat{\sigma}_{Md} = 1.25 \cdot \sqrt{\frac{\hat{\sigma}^2}{n}} = 1.25 \cdot \frac{\hat{\sigma}}{\sqrt{n}}$$

$\hat{\sigma}^2 = \frac{\sum_{i=1}^{n}(y_i - \bar{Y})^2}{n - 1}$	Population variance estimated with data in the sample
$\hat{\sigma}_{Md} = 1.25 \cdot \sqrt{\frac{\sum_{i=1}^{n}(y_i - \bar{Y})^2}{n \cdot (n - 1)}}$	Standard error of the median estimated with data in the sample

As you can see, the standard error for the median is almost the same as the standard error of the mean except that the standard error for the median is 1.25 units higher. This means that medians—compared to means—vary slightly more around their parameter, and as a consequence estimations come with slightly higher uncertainty. Still, the mean of all medians—like the mean of all means in a sampling distribution—estimates the true median without a bias. Statisticians say that median-estimates are unbiased, but less efficient than mean-estimates.

Percentage

The sampling distribution of a percentage, like the sampling distribution for the mean and the median, follows a normal distribution for large samples. The condition that needs to be met in order to assume a large sample, and thus normally distributed sample percentages, is provided below. The most occurring percentage of this sampling distribution—that is, the mean of all percentages—corresponds to the (true) percentage of interest in the population. In other words, the expected value of percentages' sampling distribution estimates the (true) percentage in the population. However, the standard error of the percentage, and thus the estimated standard error of the percentage, is different. Tables 5.8 and 5.9 provide the formula for the standard error and the estimated standard error of percentages.

Table 5.8. Standard error of a percentage

$$\sigma_{\%} = \sqrt{\frac{\pi(100 - \pi)}{n}}$$

π: The (true) population percentage
n: Sample size

Table 5.9. Estimated standard error of a percentage

$$\hat{\sigma}_{\%} = \sqrt{\frac{P(100 - P)}{n}}$$

P: Sample percentage as an unbiased estimator of the population percentage π
n: Sample size

As mentioned above, we need to operate with large samples in order to assume a normal distribution as sampling distribution of percentages (for small samples the sampling distribution of percentages would follow a so-called binomial distribution). We can be confident that our sample size is large enough if the following condition is met: $n \cdot (P/100) \cdot ((100 - P)/100) \geq 9$. Let's assume you collected data with a sample size of $n = 100$ and found that $P = 45\%$ in your sample own a mobile video game platform. Can you assume normally distributed percentages? Let's check: $100 \cdot (45/100) \cdot ((100 - 45)/100) = 100 \cdot 0.45 \cdot 0.55 = 24.75 \geq 9$. Yes, the sample is large enough to approximate the sampling distribution of percentages in this example with a normal distribution.

Example

The 2010 US Census reported that 65% in the population of Americans over the age of eighteen had access to broadband Internet, whereas 35% did not. How likely is it in a sample of

$n = 500$ Americans that the sample percentage of Americans who do not have access to broadband Internet access is below 33%? Table 5.10 provides the solution.

Table 5.10. Example—finding the probability with which sample percentages occur

Given: $\pi = 35\%$ $100 - \pi = 65\%$ $n = 500$

Sought: Probability ($P < 33\%$)

Standard error for percentages: $\sigma_{\%} = \sqrt{\frac{\pi(100 - \pi)}{n}} = \sqrt{\frac{35\% \cdot 65\%}{500}} = 2.13\%$

z-score: $z_i = \frac{P - \pi}{\sigma_{\%}} = \frac{33\% - 35\%}{2.13\%} = -0.94$

Using the standard normal distribution Table in Appendix A we find that the corresponding left-tail probability below −0.94 (or right-tail probability above +0.94) is 0.1736 or 17.36%. Thus, the probability that you draw a sample from this population in which the percentage of Americans who do not have access to broadband is below 33% is 17.36%.

SUMMARY

With the introduction of sampling distributions and the Central Limit Theorem, we enter the exciting realm of inferential statistics. Sampling distributions are important in inferential statistics because they help researchers predict how close statistics fall to their parameters, that is, how accurately statistics estimate their parameter. The Central Limit Theorem explains the conditions under which we observe a specific and centrally important sampling distribution—the sampling distribution of the mean statistic. The development of any (parametric) statistical estimation procedure and statistical test is based on knowing the shape of a statistic's sampling distribution, the sampling distribution's expected or most occurring value, and the sampling distribution's standard deviation (also called the statistic's standard error). Different statistics follow different sampling distributions with different expected values and standard deviations. In this chapter, we discussed the sampling distributions of the statistics mean, median, and percentage.

TERMS

- **Sampling distribution** of a statistic is the probability (density) distribution that specifies probabilities for the possible values the statistic can take. The sampling distribution describes the variability of a statistic that occurs among samples of a certain size.

- **Standard error** is the standard deviation of statistic's sampling distribution.
- **Central Limit Theorem** describes conditions under which we observe the sampling distribution of the mean statistic. It states that the random variable mean is normally distributed for large samples with the true mean in the population and the standard error of the mean as the distribution's parameter.

CHAPTER 5 PROBLEM SET

1. What is the distinction between population distribution, sample distribution, and sampling distribution?
2. What conditions need to be met in order to ensure that the sampling distribution of the mean is approximately normally distributed?
3. How is the standard error of the mean related to a variable's standard deviation?
4. Why do we usually need to estimate a standard error?
5. A researcher knows that the mean number of minutes watching TV each day is 180 and the standard deviation is 60.
   a. Compute the standard error of the mean for sample sizes of 2, 5, and 30. What do you notice?
   b. Now assume that the sample size is $n = 100$. Compute the probability that $\bar{y}$ in a sample falls above 190.
6. A researcher knows that the median number of minutes spent using the Internet for personal reasons at work is 60 and the standard deviation is 10. Compute the standard error of the median for a sample of $n = 100$.
7. A researcher knows that receivers' accuracy judgment for detecting deception is 52 percent. Compute the standard error for the percentage for a sample of $n = 500$.

REFERENCES

McCroskey (1977). Oral communication apprehension: A summary of recent theory and research. *Human Communication Research, 1,* 78–96.

Gregory, J. R., & Wiechmann, J. (1999). *Marketing corporate image.* Chicago, IL: NTC Publishing Group.

CHAPTER 6

ESTIMATING AND TESTING

As a communication professional, for example a campaign manager for a political candidate, you will often want to know whether your candidate truly leads the field when opinion polls indicate that your candidate is 2% ahead of all other candidates, or whether this estimation of the election result is too imprecise to draw this conclusion. If you should decide as campaign manager to spend money on additional ads because your hypothesis is that more ads lead to increased visibility of your candidate, you will want to test whether the money has been well spent and increasing poll numbers are not just due to random or other factors. Likewise, as a communication researcher, you and the people affected by your research (e.g., employers, fellow researchers, policymakers, or members of the population of interest) will want to know answers to two important questions: (1) how precise are your parameter estimates, and (2) how well do your hypotheses stand the confrontation of data in your sample?

In Chapter 5, we introduced statistical inference, specifically sampling distributions and the Central Limit Theorem (CLT). In this chapter, we introduce two major divisions of statistical inference: estimating and testing, which rely on the principles of the CLT. In estimating, we are interested in how and how precisely we can estimate or predict parameters of variables in a population by using a sample. In testing, we are interested in how well hypotheses that are related to conditions in a population face empirical data in a sample.

ESTIMATION OF PARAMETERS

As stated, in estimating, we are interested in how and how precisely we can estimate or predict parameters (e.g., the mean) of variables in a population by using a sample. In other words, we try to generalize findings in a sample to the population.

Point Estimation

In some statistics textbooks, authors make the distinction between point estimation and interval estimation. We address this briefly here although you can confidently skip this section and move on to the next section "interval estimation."

A point estimation is a sample statistic's value (e.g., mean value) that is used to estimate the value of a corresponding parameter (e.g., the population mean). Therefore, sample statistics are also called point estimators. In general, we want point estimators that are unbiased and efficient. What does this mean?

Recall from Chapter 5 the notion of sampling distributions of sample statistics (e.g., means, medians, etc.). We say that a point estimator is unbiased if its sampling distribution centers around the corresponding parameter. For example, we have learned in Chapter 5 that sample means are normally distributed around the population mean. Hence, the point estimator "sample mean" is an unbiased estimator. Biased estimators tend to either under- or overestimate the parameter on average. The sample variance would be an example. If we used the center of a variance's sampling distribution, we could show (not here!) that we would estimate the population's variance with a bias of $(n - 1) / n$ with n as sample size. As we will later see, this bias needs to be corrected by multiplying the sample variance with $n / (n - 1)$ if an unbiased estimation of a population variance is important.

An estimator whose standard error is smaller than those of other potential estimators is said to be efficient. For example, as noted in Chapter 5, the median's sampling distribution has a 25% increased standard error compared to the mean. In a normally distributed population, both the mean and the median would be unbiased estimators of the population mean, but the median would be considered 25% less efficient. Hence, we usually use the sample mean to estimate the population mean and not the sample median.

Interval Estimation

In Chapter 5, we established that precise point estimations from a sample are actually not possible. Estimations can only be calculated with some variation and with some extent of confidence. Up until this point, much of what we have discussed about inferential statistics is based on thought experiments and theoretical considerations. In practice, nobody draws many samples repeatedly out of the same population of interest in order to estimate the parameter of a population. In the following example, we learn how to estimate parameters in the population of interest on the basis of one sample, which corresponds to what researchers or communication professionals actually do. Consider the following example.

Let's assume that μ, the mean parameter of a population, is known, for example $\mu = 100$. Based on the Central Limit Theorem (Chapter 5), we know that means of random samples fall in the interval of $\mu \pm 2 \cdot \sigma_{\bar{y}}$ with a probability of 95.5% (check the probability with Table 1, Appendix A). If we assume for the mean standard error $\sigma_{\bar{y}} = 5$, then this interval is: [90;110]. Let's call this interval mean-values-interval of $\mu = 100$. If our sampling procedure results in a sample mean of $\bar{y} = 95$, then this sample mean falls in the mean-values-interval of $\mu = 100$, and thus, $\mu = 100$ is a plausible population mean parameter that generated this sample mean of $\bar{y} = 95$. However, $\bar{y} = 95$ also could have occurred if the population's mean were $\mu = 90$, because the mean-values-interval of that parameter is [80;100]. This interval also includes $\bar{y}$ of 95.

What if the population mean were $\mu = 70$, could we see a sample mean of $\bar{y} = 95$ in a random sample of that population?

At first no, because the mean-values-interval of $\mu = 70$ is [60;80] and does not include $\bar{y} = 95$. However, this interval was calculated with the assumption 95.5% of the sample means are included (see above). If we extend the mean-values-interval to an interval of $\mu \pm 6 \cdot \sigma_{\bar{y}}$ then 99.99% of all means fall in this interval and in this case we indeed could see a sample mean of $\bar{y} = 95$ even if $\mu = 70$. Because of the normal distribution's asymptotic character practically every μ could be possible if we accept higher probabilities, which means larger mean-values-intervals.

However, and this is the important idea, if we are content with a probability of 95.5%, or in other words, predicting what is plausible as population mean parameter with 95.5% confidence is good enough for us, then certain population mean parameters are eliminated as plausible originators of a sample mean, $\bar{y}$. Apparently, these are just the ones that are more distant from $\bar{y}$ than $2 \cdot \sigma_{\bar{y}}$. If $\bar{y}$ in a sample is 95 and the standard error of the mean is $\sigma_{\bar{y}} = 5$ then $\mu < 85$ and $\mu > 105$ can be eliminated as plausible originators with a probability, or confidence, of 95.5%.

If $\bar{y}$ in a sample is 95 and the standard error of the mean is $\sigma_{\bar{y}} = 5$ then the population parameter μ falls in an interval of [85,105] with a probability, or confidence, of 95.5%. This interval is called a confidence interval. Confidence intervals answer the important question of what we know about a population parameter (here the population mean) if we just draw one sample out of a population, which, of course, is relevant in practice. As we demonstrated, based on one sample mean we are able to narrow down the interval in which the population mean parameter falls—a precise (point) prediction of the population mean parameter, however, is not possible.

In sum, a confidence interval is a range of values that contains an unknown parameter with a certain probability. The probability that the confidence interval contains the parameter is called the confidence coefficient.

Confidence coefficients are usually defined as $(1 - \alpha) = 95\%$ or 99%, which means $\alpha = 5\%$ or 1%. We talk about a 99% or 95% confidence interval. In our example, we have used a 95.5% confidence interval because the standard normal distribution table (Table 1, Appendix A) tells us that the area under the curve that corresponds to the interval "mean $\pm$ (z = 2)" is 95.5%. As the standard normal distribution table reveals, the 95% confidence intervals requires a z-value of 1.96; i.e., the z-value that exactly cuts α / 2% or 2.5% on each side of the standard normal distribution. Correspondingly, the 99% confidence interval requires a z-value of 2.58; i.e., the z-value that cuts α / 2% or 0.5% on each side of the standard normal distribution.

Confidence Intervals for Means

Table 6.1 provides the formula to compute confidence intervals for means (P stands for "Probability," $\hat{\sigma}_{\bar{y}}$ stands for the *estimated* standard error of the mean, all other variables have been mentioned in the above text).

A researcher (c.f., Afifi & Weiner, 2006) is interested in how often college students seek information about their romantic partners' sexual health (on a scale of 1= Don't ask questions to

Table 6.1. Confidence interval for means

$$P(\bar{y} - z_{(\alpha/2)} \cdot \hat{\sigma}_{\bar{y}} \leq \mu \leq \bar{y} + z_{(\alpha/2)} \cdot \hat{\sigma}_{\bar{y}}) = 1 - \alpha$$

$$\text{or simplified: } \bar{y} \pm z_{(\alpha/2)} \cdot \hat{\sigma}_{\bar{y}}$$

$$\hat{\sigma}_{\bar{y}} = \sqrt{\frac{\hat{\sigma}^2}{n}}$$

Estimated population variance $\hat{\sigma}^2$ needs correction $\frac{n}{n-1}$ (see "point estimation")

$$\hat{\sigma}^2 = \frac{\sum_i (y_i - \bar{y})^2}{n} \cdot \frac{n}{n-1} = \frac{\sum_i (y_i - \bar{y})^2}{n-1}$$

thus

$$\hat{\sigma}_{\bar{y}} = \sqrt{\frac{\sum_i (y_i - \bar{y})^2}{n(n-1)}}$$

<u>Confidence Interval:</u>

Based on a sample mean of $\bar{y}$, μ falls in the following interval with a confidence of $(1-\alpha)\%$.

$$\bar{y} \pm z_{(\alpha/2)} \cdot \sqrt{\frac{\hat{\sigma}^2}{n}} = \bar{y} \pm z_{(\alpha/2)} \cdot \sqrt{\frac{\sum_i (\bar{y}_1 - \bar{y})^2}{n \cdot (n-1)}}$$

7 = Asks a lot of questions). The researcher has $n = 200$ students, and gets a mean $\bar{y} = 4.71$ and an estimated standard deviation of $\hat{\sigma} = 1.87$. Calculate the confidence interval for the mean with a confidence coefficient of 95%, and a confidence coefficient of 99%. Table 6.2 shows the step-by-step instructions.

Small Samples

For simplification, our examples have assumed large samples, specifically sample sizes of at least $n = 30$. Sometimes we do not have those large samples in practical research. If we have small sample sizes ($n \leq 30$) we have to assume that standard errors vary from sample to sample. Because of the small sample size we can show mathematically (not here!) that sample means do not follow a normal distribution, but have a so-called student- or t-distribution as their

Table 6.2. Confidence interval for means—example "partner sexual health"

95% Confidence Interval

$$P(\bar{y} \pm z_{(\alpha/2)} \cdot \hat{\sigma}_{\bar{y}}) = 1 - \alpha$$

$$P(4.71 \pm z_{(\alpha/2)} \cdot \hat{\sigma}_{\bar{y}}) = 1 - 0.05 = 0.95 = 95\%$$

$$P\left(4.71 \pm 1.96 \cdot \sqrt{\frac{\hat{\sigma}^2}{n}}\right) = 95\%$$

$$P\left(4.71 \pm 1.96 \cdot \sqrt{\frac{1.87^2}{200}}\right) = 95\%$$

$$P(4.71 \pm 0.26) = 95\%$$

CI 95%: [4.45; 4.97]

Based on this confidence interval we can say that the population mean falls in an interval from 4.45 to 4.97 with 95% confidence.

99% Confidence Interval

$$P(\bar{y} \pm z_{(\alpha/2)} \cdot \hat{\sigma}_{\bar{y}}) = 1 - \alpha$$

$$P(4.71 \pm z_{(\alpha/2)} \cdot \hat{\sigma}_{\bar{y}}) = 1 - 0.01 = 0.99 = 99\%$$

$$P\left(4.71 \pm 2.58 \cdot \sqrt{\frac{\hat{\sigma}^2}{n}}\right) = 99\%$$

$$P\left(4.71 \pm 2.58 \cdot \sqrt{\frac{1.87^2}{200}}\right) = 99\%$$

$$P(4.71 \pm 0.34) = 99\%$$

CI 99%: [4.37; 5.05]

Based on this confidence interval we can say that the population mean falls in an interval from 4.37 to 5.05 with 99% confidence.

sampling distribution. For the calculation of confidence intervals this means that we just have to use another distribution table (see Appendix A for *t*-distribution). Instead of looking for a *z*-value that includes $(\alpha / 2)\%$ we are looking for a *t*-value with $(n - 1)$ degrees of freedom that includes $(\alpha / 2)\%$. Let's consider our example in Table 6.2 with a 95% confidence interval, although this time with a sample size of $n = 20$ students, which is in Table 6.3.

As you can see in Table 6.3, a decrease in sample size is met with an increase in the width of the confidence interval. This makes sense: A sample of $n = 20$ provides considerably less information than a sample of $n = 200$, and as a consequence the prediction of a plausible mean parameter is necessarily less precise.

Table 6.3. Confidence interval for means—example "partner sexual health" with $n = 20$

95% Confidence Interval

$$P(\bar{y} \pm t_{\alpha/2} \cdot \hat{\sigma}_{\bar{y}}) = 1 - \alpha$$

$$P(4.71 \pm t_{\alpha/2} \cdot \hat{\sigma}_{\bar{y}}) = 1 - 0.05 = 0.95 = 95\%$$

Degrees of freedom $df = n - 1 = 20 - 1 = 19$

$$t_{2.5\%} \text{ with } (df = 19) = 2.093$$

$$P\left(4.71 \pm 2.093 \cdot \sqrt{\frac{\hat{\sigma}^2}{n}}\right) = 95\%$$

$$P\left(4.71 \pm 2.093 \cdot \sqrt{\frac{1.87^2}{20}}\right) = 95\%$$

$$P(4.71 \pm 0.88) = 95\%$$

CI 95%: [3.83; 5.59]

Based on this confidence interval we can say that the population mean falls in an interval from 3.83 to 5.59 with 95% confidence.

In practice we will rarely estimate a parameter on the basis of a sample of less than 30 study participants. With such a small sample size parameter estimations are usually too imprecise for their use in practice. In statistics exams, however, you need to check in a problem if you are dealing with a large or a small sample. Depending on these circumstances, you will need to use either the table of the standard normal distribution or the *t*-distribution (Tables 1 and 2, respectively, in Appendix A). By examining the table of the standard normal and *t*-distribution, you will notice that *t*-values are greater than *z*-values for small samples. For large samples, or large degrees of freedom, however, the differences are small. For large samples ($n > 30$) it does not matter whether you use the standard normal or the *t*-distribution table, as the *z*- or *t*-values are identical.

Sample Size Calculations

Before we conduct a study we may only know the confidence coefficient and an acceptable error with which we want to estimate a population parameter. With this information, we are able to compute the sample size that we need to conduct our study within these requirements.

For example, in a study about adolescents' exposure to sexual content, a researcher wants to estimate the population's mean of exposure to sexual content with a maximum error of $E = \pm 0.25$ rating points. The confidence coefficient should be 99%. From previous research (c.f., Chia, 2006) it is known that the standard deviation of ratings for exposure to sexual content is 1.19. What sample size does the researcher need to conduct this study within these requirements? Table 6.4 provides the calculations.

Table 6.4. Computing sample size based on CI width and CI coefficient

A researcher is interested in the amount of sexual content on TV that adolescents are exposed to. Based on prior research (c.f., Chia, 2006), the researcher knows that the standard deviation for late adolescents exposure is 1.19. The researcher would like to be 99% confident that the estimate is accurate within ± 0.25 rating points. What is the minimum sample size required?

Error = 0.25 ⇒ Confidence Interval Width = 2 · Error

$$\text{Error(Mean } CI) = z_{(\alpha/2)} \cdot \hat{\sigma}_{\bar{y}} = z_{(\alpha/2)} \cdot \sqrt{\frac{\hat{\sigma}^2}{n}} \Rightarrow n = \hat{\sigma}^2 \cdot \left(\frac{z_{(\alpha/2)}}{\text{Error}}\right)^2$$

Confidence Coefficient = 99% ⇒ z-value = 2.58

What is the required sample size?

$$n = (1.19)^2 \cdot \left(\frac{2.58}{0.25}\right)^2$$

$$n \approx 151$$

Confidence Intervals for Percentages

What we've learned so far can be transferred to all sample statistics that are approximately normally distributed (or t-distributed) in their sampling distributions, including proportions and medians.

For example, we know that percentages are approximately normally distributed. Remember that the standard error of a percentage is as defined in Table 6.5 (see also Table 5.8 in Chapter 5).

To compute a confidence interval for a percentage, we use the familiar formula for the mean confidence interval analogously. This is provided in Table 6.6.

Table 6.5. Standard error of a percentage

$$\hat{\sigma}_{\%} = \sqrt{\frac{\hat{\pi}(100 - \hat{\pi})}{n}}$$

$\hat{\pi}$: The (estimated) population percentage
n: Sample size

Table 6.6. Confidence interval for percentages

$$P(\hat{\pi} - z_{(\alpha/2)} \cdot \hat{\sigma}_{\%} \leq \pi \leq \hat{\pi} + z_{(\alpha/2)} \cdot \hat{\sigma}_{\%}) = 1 - \alpha$$

$$\text{or simplified:} \quad \hat{\pi} \pm z_{(\alpha/2)} \cdot \hat{\sigma}_{\%}$$

$$\hat{\sigma}_{\%} = \sqrt{\frac{\hat{\pi}(100 - \hat{\pi})}{n}}$$

Confidence Interval:

Based on a sample percentage of $\hat{\pi}$ that serves as an estimation for the population percentage, π falls in the following interval with a confidence of $(1 - \alpha)\%$.

$$\hat{\pi} \pm z_{(\alpha/2)} \cdot \sqrt{\frac{\hat{\pi}(100 - \hat{\pi})}{n}}$$

Consider, for example, that a researcher (c.f., Levine, Kim, & Blair, 2010) is interested in the extent to which individuals correctly identify false confessions as lies. Based on a sample of $n = 456$ the researcher finds that $\hat{\pi} = 11.6\%$ of individuals correctly identify false confessions as lies. With this information we know that the estimated standard error of the percentage is 1.5% (see the formula for $\hat{\sigma}_{\%}$ in Table 6.6). Compute a 95% confidence interval. Table 6.7 provides the results of the computation.

Table 6.7. Confidence interval for percentages—example "false confessions."

95% Confidence Interval

$$P(\hat{\pi} \pm z_{(\alpha/2)} \cdot \hat{\sigma}_{\%}) = 1 - \alpha$$

$$P(11.6\% \pm z_{(\alpha/2)} \cdot \hat{\sigma}_{\%}) = 1 - 0.05 = 0.95 = 95\%$$

$$P(11.6\% \pm 1.96 \cdot 1.5\%) = 95\%$$

$$P(11.6\% \pm 2.94\%) = 95\%$$

CI 95%: [8.66%; 14.54%]

Based on this confidence interval we can say that the population percentage falls in an interval from 8.66% to 14.54% with 95% confidence.

Sample Size Calculations

The same researcher is interested in the extent to which individuals correctly identify false denials. Based on previous research (c.f., Levine, Kim, & Blair, 2010), the researcher knows that individuals accurately identify false denials 52.6% of the time, and wants to predict this with a maximum error of ±4% and a confidence coefficient of 95%. With this information the researcher is able to compute the sample size necessary in order to conduct the study. Table 6.8 provides the formula for computing the required sample size.

Table 6.8. Confidence interval for percentages—computing the required sample size.

$$\text{Error} = z_{(\alpha/2)} \cdot \hat{\sigma}_{\%} = z_{(\alpha/2)} \cdot \sqrt{\frac{\hat{\pi}(100 - \hat{\pi})}{n}}$$

$$n = \hat{\pi} \cdot (100 - \hat{\pi}) \cdot \left(\frac{z_{(\alpha/2)}}{\text{Error}}\right)^2$$

$$n = 52.6\% \cdot (100 - 52.6\%) \cdot \left(\frac{1.96}{4\%}\right)^2 \approx 599$$

As we can see, the researcher needs 599 participants in order to estimate accuracy for false denials within plus or minus 4 percent. The next section shifts now to the second major division of statistical inference: testing.

TESTING OF HYPOTHESES

In testing, we are primarily interested in how hypotheses that are related to conditions in a population stand the confrontation with empirical data in a sample. We already know that results in samples are affected by randomness, so the following question arises: How can we use a result in a sample to decide about the accuracy of a hypothesis? In order to answer this question, we use a pair of hypotheses, the null hypothesis (H_0) and the alternative or research hypothesis (H_a or H_1).

Research Hypotheses

In general, hypotheses are statements about characteristics of variables in the population of interest. Research hypotheses define new ideas and assumptions that extend the current knowledge in a research field. Usually, but not necessarily, they are derived from theory or prior research findings. In the following, we introduce three specific examples that we will consider throughout this section on significance testing.

First, an instructional communication researcher (c.f., McCroskey, 1982) believes that his training method to reduce communication apprehension among university professors is better than prior methods. The researcher defines a research hypothesis about the difference between an old and new training method for addressing communication.

Second, a media researcher (c.f., Saito, 2007) believes that heavy television viewers are more likely to hold traditional gender-role attitudes than light television viewers. The researcher defines a research hypothesis about the difference between heavy and light television viewers.

Third, an education researcher (c.f., Goodboy, 2011) believes that teacher misbehaviors are positively associated with student reports of instructional dissent. The researcher defines a research hypothesis about the dependency between the variables "teacher misbehavior" and "instructional dissent."

These examples describe different versions of research hypotheses. The first two researchers define research hypotheses about a difference, whereas the third defines a hypothesis about a dependency. For hypotheses about differences (Chapter 7), we will be applying tests for differences between means or frequencies (e.g., *t*-tests, tests based on an Analysis of Variance (ANOVA) procedure, and chi-square tests). For hypotheses about dependencies (Chapter 8), we will be applying correlation techniques and regression analysis.

Moreover, we distinguish hypotheses on two dimensions: directionality and specificity. All three of the above mentioned examples are directional. The researcher interested in communication apprehension believes that university professors will have *lower* communication apprehension scores as a result of a new training method. The media researcher believes that heavy television viewers have *more* traditional gender-role attitudes. The education researcher interested in students' expressive action assumes that students will express their concerns *more* as perceived teacher misbehavior increases. Non-directional hypotheses would assume that there is just a difference or dependence with no information on the direction.

None of the hypotheses in our examples provides information about the size of the effect, i.e., the size of a difference or a dependency. If we have enough information from previous research and can define an effect size, we speak of specific research hypotheses. For example, in the communication apprehension example, if the researcher said his method decreased communication apprehension scores on average by 5 points, then this would define a specific (directional) research hypothesis. Specific, directional hypotheses provide the highest amount of information and therefore contain the highest information value in empirical research. The least information requires a non-directional, unspecific research hypothesis: here we posit only a difference or dependency without direction (higher/lower) or effect size. Unfortunately, specific hypotheses are quite rare in communication research, because many theories are not (yet) advanced enough to warrant effect-size information. Specific non-directional hypotheses are theoretically possible, but do not occur in practice. If a theory does not provide enough information to predict the direction of an effect, then usually this theory is also silent on answering the important question of effect size.

As a side note, the inability of many communication theories to specify what effect size the theory actually predicts, that is the dominance of non-specific research hypotheses in our field, is one reason for flawed applications of statistical testing in communication research (and in empirical research in general). As a result, scientific progress is impeded (c.f., Levine, Weber, Hullet, Park, & Massi-Lindsay, 2008a; Levine, Weber, Park, & Hullett, 2008b; Dienes, 2011).

Null Hypotheses

The second hypothesis of the mentioned pair is the null hypothesis. The null hypothesis, or H_0, claims that the statement in the research hypothesis is not true, or that there is no difference or dependency between variables. The null hypothesis provides no additional information; it is just a negation of the research hypothesis.

For the researcher interested in communication apprehension the null hypothesis claims that professors who use the new training method are equally apprehensive or even more apprehensive than professors who used the old method. In the case that the research hypothesis was non-directional, the null hypothesis simply assumes no difference between professors using the new method and professors using the old method.

In the example of the media researcher, the null hypothesis claims that heavy viewers hold equal or even less than traditional gender-role attitudes (that is, they are more progressive and believe in greater equality among the men and women). In the case of a non-directional research hypothesis, the null hypothesis would assume there is no difference.

In the example of the education researcher, the null hypothesis would assume there is no or even a negative relationship between teacher misbehavior and instructional dissent. In the case of a non-directional research hypothesis, the null hypothesis would simply assume that there is no relationship.

Statistical Null and Research Hypotheses

So far we have defined null and research hypotheses in words. Now we will define this important pair of hypotheses with symbols and mathematical operators. Table 6.9 provides the null and research hypotheses for the communication apprehension example.

Table 6.9. Statistical null and research hypotheses for "communication apprehension"

$H_1 : \mu_0 > \mu_1$	$\Leftrightarrow$	$H_0 : \mu_0 \leq \mu_1$ (directional, unspecific)
$H_1 : \mu_0 \neq \mu_1$	$\Leftrightarrow$	$H_0 : \mu_0 = \mu_1$ (non-directional, unspecific)

Table 6.10 provides the null and research hypotheses for the example of television viewing and gender-role attitudes.

Table 6.10. Statistical null and research hypotheses for "gender-role attitudes"

$H_1 : \mu_{lightTV} < \mu_{heavyTV}$	$\Leftrightarrow$	$H_0 : \mu_{lightTV} \geq \mu_{heavyTV}$ (directional, unspecific)
$H_1 : \mu_{lightTV} \neq \mu_{heavyTV}$	$\Leftrightarrow$	$H_0 : \mu_{lightTV} = \mu_{heavyTV}$ (non-directional, unspecific)

Table 6.11 provides the null and research hypotheses for the example of teacher misbehavior and instructional dissent. Note: In Chapter 8 we will introduce a statistic that provides information on the strength of a dependency between two variables—the Pearson correlation coefficient "r" (or ρ when we refer to a Pearson correlation in a population). $\rho > 0$ defines a positive dependency, that is, as teacher misbehavior increases, instructional dissent increases. $\rho < 0$ defines a negative dependency, that is, as teacher misbehavior increases, instructional dissent decreases. $\rho = 0$ indicates "no dependency," that is, teacher misbehavior and instructional dissent are unrelated to one another (see Chapter 8 for more details).

Table 6.11. Statistical null and research hypotheses for "teacher misbehavior and dissent"

$H_1 : \rho > 0 \quad H_0 : \rho \leq 0$ (directional, unspecific)
$H_1 : \rho \neq 0 \quad H_0 : \rho = 0$ (non-directional, unspecific)

At this point, we could ask: Why do we need this complicated and artificial definition of two competing hypotheses? What is the purpose of these complicated definitions? As we will understand later, in significance testing the null hypothesis provides the information whether we reject or do not reject the null hypothesis. As the null hypothesis is the negation of the research hypothesis, rejecting a null hypothesis suggests empirical evidence for the research hypothesis. The logic is as follows: If we cannot explain the data in our sample with the assumption that our null hypothesis is true, then we decide in favor of the research hypothesis, which contradicts the null hypothesis. Significance testing is testing by contradiction! Keep this important notion in mind and you will be much more capable to grasp the concept of any significance test.

Decisions and Types of Errors in Testing Hypotheses

After we have defined the pair of hypotheses we can begin data gathering. As previously mentioned, based on sample data we shall decide which of the two hypotheses is more plausible in the population. But how can we be confident that our decision is valid? Our data is based on a sample after all, and sample results vary from sample to sample as they are drawn randomly from a population (see Chapter 5). This puts researchers in a difficult situation: after data collection, researchers are always confronted with the following decision-making situation, as illustrated in Table 6.12.

As you can see, due to randomness in your data there are two possible wrong decisions: If you reject the null hypothesis based on your sample results when it is true in the population (i.e., you should not have rejected the null hypothesis) and decide for the research hypothesis you commit a Type-I or α error. This situation is like claiming you found a meaningful difference or dependency between variables, but in fact the difference or dependency you found is a random result.

Table 6.12. Type I (α) and type II (β) errors

		Situation in Population	
		H_0 is false	**H_0 is true**
Decision based on Sample Data	**Reject H_0**	Correct	Type-I error (α)
	Do not reject H_0	Type-II error (β)	Correct

If you do not reject the null hypothesis based on your sample results when it is false in the population (i.e., you should have rejected the null hypothesis) and decide for the null hypothesis, you commit a Type-II or β error. This situation is like claiming you found nothing, but in fact there is a difference or dependency in the population and that you missed finding this difference or dependency is a random result.

At this point the question is which error do you want to avoid in research? This question cannot be answered in general—it depends on the consequences of the wrong decision. We will explain this by means of our examples.

Consider our instructional communication researcher who believes that his new training method will lower communication apprehension scores. A Type-I error means we incorrectly assume that the new method for dealing with professors' communication apprehension is better than the old method. What are the potential consequences? This could include retraining of trainers, acquiring new training materials, activities that are actually unjustified and waste money. What would a Type-II error mean in this example? We would assume that the new training method is not better, but in fact the training method leads to lower communication apprehension. As a consequence, trainers would continue to focus on an old method, and would miss an opportunity to help professors to become better communicators and classroom teachers. Which consequence do you think is more relevant? Avoiding the waste of money or avoiding a missed opportunity? Difficult to say as the answer involves an opinion and as such a political statement. As we will later see, it is difficult—and sometimes impossible—to avoid both types of errors at the same time. If a decision is needed, perhaps a majority of people would argue in this example that professors overall were quite effective teachers in the past, even with the old training method, and in times of tight public budgets it is more important not to waste money. If this is your opinion, then you would want to keep the Type-I or α error low in your test decision.

Remember that our media researcher believes that heavy TV viewers hold more traditional gender-role attitudes than light TV viewers. A Type-I error means we wrongly assume that heavy TV viewers have more traditional gender-role attitudes than light TV viewers. What are the consequences of this error? Perhaps media researchers tell parents to limit TV watching, although this has no effect on gender-role attitudes. What does a Type-II error mean? We wrongly assume that heavy TV viewing has no effect on gender-role attitudes. What are the possible consequences? A media researcher tells parents wrongly that heavy TV viewing has nothing to do with gender-role attitudes. As a consequence, parents do not limit the amount of TV watched and traditional gender-role attitudes will continue to be cultivated. Which consequence do you think is more relevant? Again, the answer to this question involves an opinion. Perhaps a majority of people would argue that limiting the amount of TV watched will not lead to any major adverse effects, but increasing the prevalence of traditional gender-role attitudes should be avoided as this may support the discrimination against women in society. If this is your opinion, then you would want to keep the Type-II or β error low in your test decision.

> Note: As we will later see, standard significance tests only provide information on the Type-I or α error of your decision. That is, in situations in which you are actually interested in a low type-II or β error, as in the last example, the standard significance test will not provide you with the information you need.

Recall that our education researcher believes that teacher misbehavior influences instructional dissent (i.e., students speaking to others, to the instructor, or trying to harm the instructor's reputation). A Type-I error translates to an incorrect assumption that student dissent is dependent on teacher misbehavior. What would be the consequences of a Type-I error in this example? Education experts may incorrectly recommend that students' complaints are related to bad behaviors by teachers. This may lead colleges to focus on training teachers on appropriate classroom behaviors, even though this has nothing to do with students' dissenting behavior. A Type-II error means we incorrectly assume that instructional dissent has nothing to do with teacher misbehavior. What are the consequences of this Type-II error? Here, education experts may tell college administrators that instructional dissent has nothing to do with teacher misbehavior, and consequently college administrators may not pay attention to classroom dissent, which may have implications for students' experiences and educational outcomes. Which consequence do you think is more relevant? Again, this depends on your opinion. If you think useless trainings of teachers needs to be avoided, then your focus would be on a low Type-I error. If you think that avoiding negative student experiences and decreasing educational outcomes are more important than avoiding useless trainings, then a low Type-II error is your first priority.

The question that we now have to answer is: "How can we test the extent to which we commit a Type-I or Type-II error if we have information only from a random sample?" The answer to this question is provided by significance statements.

Significance Statements

Let's assume that in a random sample of TV viewers we see more traditional gender-role attitudes. Based on this result, can we say that we found evidence for the rejection of the null hypothesis, that is, evidence for the research hypothesis? No, our sample result could be a random result and we committed a Type-I error if we interpret our findings as evidence for the research hypothesis. Because we do not know the condition in the population we can never be sure whether we make the right or wrong decision (and should therefore avoid statements like "proving" a hypothesis to be true). However, as we will see in the following, we do know how likely it is to commit a Type-I or Type-II error. The more unlikely it is that we commit an error the better is the quality of our decision.

In our gender-role attitudes example, we have to answer the following question: What is the probability that heavy TV viewers have the same or even more progressive gender-role attitudes than individuals who are light TV viewers? In other words, how well can we explain the results in our sample if we assume that the null hypothesis is true in the population? This probability corresponds to the probability of a Type-I error. If we decide for the research hypothesis, we accept a Type-I error with the probability that the result could occur under null hypothesis conditions. Significance statements are statements on the probability of a sample result in a random sample with the assumption that the null hypothesis is true in the population. Thus, significance statements are probability statements. The probability of a sample result under null hypothesis assumption is called the *p*-value of a sample statistic and corresponds to the probability of a Type-I error (we will be talking about Type-II errors later).

Calculation of *p*-Values (*z*-Test)

We focus now on the calculation of *p*-values, and return to our example of communication apprehension with a focus on mean differences from a single sample with a known population parameter. You will now become acquainted with the first statistical test in this book—the *z*-test.

The basis of all statistical tests in this chapter is the Central Limit Theorem (Chapter 5). That is, we assume the sampling distribution of the mean for samples with sample sizes of $n \geq 30$. We know that this sampling distribution is normally distributed with μ as the mean parameter and $\hat{\sigma}_{\bar{y}}$ as the known standard error in Table 6.13.

Table 6.13. Mean sampling distribution—standard error of the mean

$$\hat{\sigma}_{\bar{y}} = \sqrt{\frac{\hat{\sigma}^2}{n}} = \sqrt{\frac{\sum_{i=1}^{n}(y_i - \bar{y})^2}{n \cdot (n-1)}}$$

If we know the population parameter under null hypothesis assumption μ_0 (we assume that we know this parameter because thousands of individuals have taken McCroskey's PRCA-24 test; and we also assume that we know the variance of their scores) we can calculate the probability that our sample result can be explained with H_0 conditions. For these calculations we need to ask two essential questions:

Question 1: What does the Null Hypothesis Posit?

The null hypothesis in our example assumes that professors taught with the old training method achieve apprehension scores equal to or even lower than professors taught with the new training method (see Table 6.9). Logically, this means that the research hypothesis' parameter (μ_1) falls to the negative (left) tail of the null hypothesis sampling distribution.

Question 2: What is the Probability of the Sample Result or More Extreme Results?

In Question 2 we want to know the probability of our sample result or results that are even more distant from the null hypothesis parameter μ_0 in the direction of the research hypothesis. This probability corresponds to the area under the curve (AUC) of the null hypothesis sampling distribution from the sample result to minus infinity. This AUC is defined as p-value or probability of a Type-I error because with this probability we can explain the sample result or more extreme results with null hypothesis conditions. Thus, if we decide for the research hypothesis in light of this result, we commit a Type-I error with the probability of p. Figure 6.1 illustrates the calculation of a p value:

Figure 6.1. Calculation of p-values

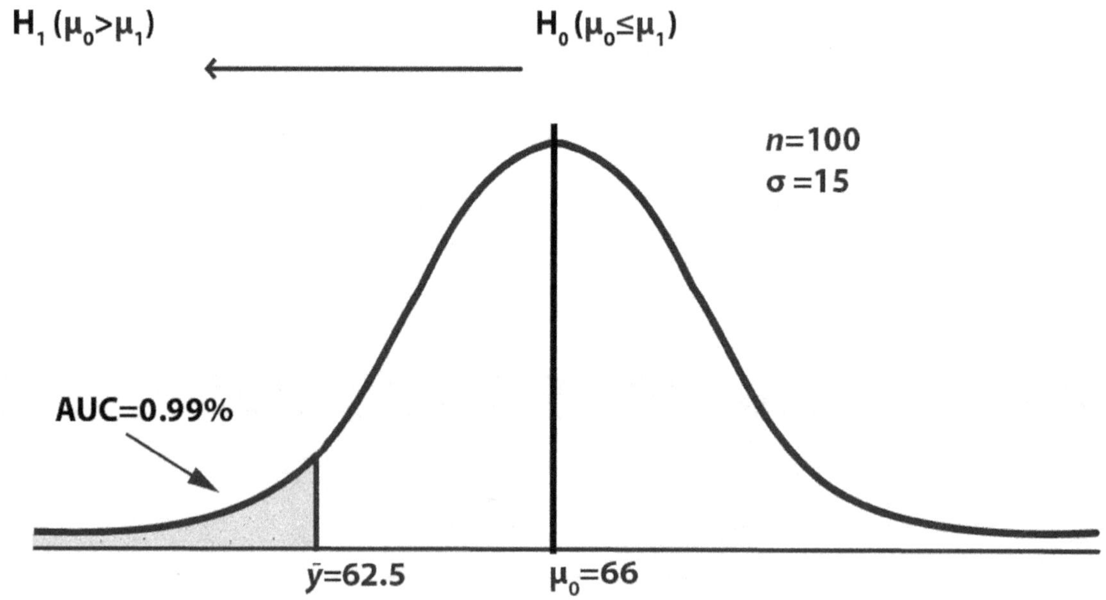

Note: If the research hypothesis stated that scores are expected higher than under null hypothesis ($H_1 : \mu_0 < \mu_1 \Leftrightarrow H_0 : \mu_0 \geq \mu_1$), we would be interested in the AUC of the null hypothesis sampling distribution from the sample result to plus infinity.

Calculations

Let's assume that in our example $\mu_0 = 66$ and $\sigma = 15$. A random sample of $n = 100$ professors was taught with a new training method for dealing with communication apprehension, and achieved a $\bar{y} = 62.5$. What is the p-value of this sample result? The question to ask is: "How likely is this result or results that are even more distant from null hypothesis conditions (in the direction of the research hypothesis), if we assume the null hypothesis is true?" Table 6.14 shows the step-by-step computation of the example.

Table 6.14. Computing the *p*-value (*z*-test)—example "communication apprehension"

Step 1: Calculate the standard error of the mean

$$\sigma_{\bar{y}} = \sqrt{\frac{\sigma^2}{n}} = \frac{15}{\sqrt{100}} = 1.5$$

Step 2: z-transformation of the result

$$z_{\bar{y}} = \frac{\bar{y} - \mu_0}{\sigma_{\bar{y}}} = \frac{62.5 - 66}{1.5} = -2.33$$

Step 3: Find AUC from $z = -2.33$ to $-\infty$ (p-value)

This AUC corresponds to the AUC from $+2.33$ to $+\infty$, see Table 1 in Appendix A
AUC ($z \geq +2.33$) $= 0.0099 \approx 0.99\% \approx 1\%$

From our example, we conclude that it is apparently quite difficult to explain the resulting lower communication apprehension scores in our sample with null hypothesis conditions as we would have expected our result or even lower results under null hypothesis condition with a probability of just 1 percent. As an illustration, if we had drawn 10,000 random samples with a sample size of $n = 100$ and would have trained the professors in our samples with the old method, in only 99 cases would we have seen a communication apprehension score of 62.5 or lower. It can be argued that this is not good evidence for the null hypothesis in the population.

With the calculation of a p-value the statistical test procedure (here z-test) ends. Now the researcher has to evaluate whether the evidence is good or bad to decide in favor or against the null hypothesis. This decision is based on conventions and not on any statistical-mathematical reasoning. Those conventions are called alpha-levels.

Alpha Levels

In order to be able to compare statistical decisions researchers agreed on a convention. If p-values are smaller than $\alpha = 0.05$ (5%) or $\alpha = 0.01$ (1%) we decide in favor of the research hypothesis and we say that our sample result is significant on the 5% (or 1%) level. In general, if the probability of a given result or even more extreme results under the assumption of the null hypothesis (p-value) is smaller than 0.05 or 0.01 then we say a result is significant on the 5% or 1% level. If p-values are greater than $\alpha = 0.05$ or $\alpha = 0.01$ we decide in favor of the null hypothesis (i.e., we do not reject the null hypothesis) and we say that our sample result is non-significant. The numbers 0.05 and 0.01 are called alpha-levels. The decision to use 5% or 1% depends on the context of the study. If the consequences of Type-I error are severe (e.g., in pharmaceutical tests the consequences of a wrong decision would result in dying patients), we select an alpha level of 1% or even lower. The standard choice for alpha-levels in communication research is 0.05 or 5%. It is important that the researcher selects alpha-levels *before* collecting and analyzing data. To change alpha levels and hereby the decision whether a result is significant or not after we have information on the sample results is unscientific and must be avoided. Table 6.15 provides a summary of significance and some misunderstandings of *p*-values.

Table 6.15. Significant results and common misunderstandings

In brief, a result is significant if . . .	The concept of significance is very often misunderstood as:
$p(\text{Result} \mid H_0) \leq 5\%/1\%$	■ The probability of a result: $p(\text{result})$ ■ The probability of H0: $p(H_0)$ ■ The probability of H1: $p(H_1)$ ■ $p(H_0 \mid \text{Result})$.

Evaluation of the Hypothesis Test

As demonstrated in the communication apprehension example, we obtained a significant result on the 1% level (see Table 6.14). Thus, we can say that if we reject the null hypothesis and decide for the research hypothesis, the probability of committing a Type-I error is 0.99%, which is very low. However, we can never say that we are certain that the null hypothesis is not true in the population. We can say, however, that it seems very unlikely or the data do not support the assumption of the null hypothesis.

What do you think about this result and interpretation? Very critical people criticize statistical tests for not providing 100% reliable results and dismiss statistical reasoning in general, especially in cases of important decisions such as medical treatments. It is true that statistical methods cannot provide the truth. The above critique would not be fair, though, because statistical decisions make sure that scientific conclusions that are based on samples (and many scientific conclusions *are* based on samples) are not just random, arbitrary, or speculative. In the absence of statistical methods, researchers would make decisions anyway. Those decisions, however, would be rooted in common sense or individual experience. Many studies have shown that the average human uses alpha-levels of about 20% in daily life decisions (varying by age, gender, cultural background, etc.). In other words, if an average person is confronted with two competing hypotheses or ideas (e.g., a research hypothesis and a null hypothesis), this person would decide for the innovative research hypothesis if the evidence for status-quo is less than 20% likely instead of 5%. Considering this, scientific decisions that are based on samples and statistical tests are much more reliable than decisions based on individual experience. Keep this in mind when you critique statistical tests.

However, traditional hypothesis testing is indeed questionable for various reasons, and in recent years statisticians, methodologists, and empirical researchers in communication and cognate disciplines (e.g., psychology) have argued that traditional hypothesis testing should be replaced by inferential confidence intervals (Tryon, 2001; Tryon & Lewis, 2008), effect- and equivalence tests (Levine, Weber, et al., 2008a, 2008b), or by a Bayesian approach (Dienes, 2011). An in-depth discussion of the problems of traditional hypothesis testing is beyond the scope of this introductory textbook. Chapter 9 provides some information and is a good starting point for interested readers. If you are interested in more details, we recommend reading Levine, Weber, et al. (2008a, 2008b).

Regions of Rejection and Non-Rejection—One-sided v. Two-sided Tests

In order to calculate *p*-values we ask: How likely is the result found in our sample, or even more extreme results, under the assumption that the null hypothesis is true? What does "more extreme results" actually mean?

In the communication apprehension example, we had a directed research hypothesis: The new training method is better than the old method; i.e., it leads to lower communication apprehension scores. In this example, more extreme results means that sample means fall in the interval from the result found in our sample toward $-\infty$ (remember, because lower communication apprehension scores are an improvement).

Thus, we determined the area under the curve (AUC) from $\bar{y}$ to $-\infty$ (see Table 6.14). With a slightly different perspective we could define a region in which $\bar{y}$ sample results need to fall in order to indicate a significant result. This region is called "region of null hypothesis rejection" or simply "rejection region." In the case of our communication apprehension example, the rejection region lies at the negative tail of the null hypothesis sampling distribution and begins at the critical $\bar{y}$ value where the corresponding *p*-value equals alpha. Any $\bar{y}$ value smaller than the critical $\bar{y}$ value which would correspond to a *p*-value smaller than alpha indicates a significant result. Because in our communication apprehension example the null and research hypothesis was defined as a directional hypothesis and, thus, the rejection region with $\alpha = 5\%$ lies only at one side (at the negative tail) of the null hypothesis sampling distribution. This test is called a one-sided test. Figure 6.2 shows a one-tailed test for the communication apprehension example with a one-tailed rejection region and a corresponding one-tailed region of null hypothesis non-rejection.

If the rejection region in which we reject the null hypothesis is just in one direction due to a directional null hypothesis and research hypothesis, then we conduct a one-sided statistical test. However, if we didn't have enough prior knowledge on the subject, we may have formulated our null and research hypothesis in a non-directional way: The new training method for communication apprehension is different, either better or worse. If we use a non-directional research hypothesis, it is insufficient to consider only one side of the sampling distribution as the rejection region. In the case of a non-directional null and research hypothesis we would have to split the alpha-level (i.e., the maximum Type-I error we tolerate) for both tails of the distribution. Figure 6.3 shows a two-tail test for the communication apprehension example with a two-tail rejection region and a corresponding two-tail region of null hypothesis non-rejection.

In order to calculate the *p*-value of our sample result we transformed the sample mean into a *z*-score and found the corresponding *p*-value in the table of the standard normal distribution.

Figure 6.2. Region of null hypothesis rejection and non-rejection—one-tailed test

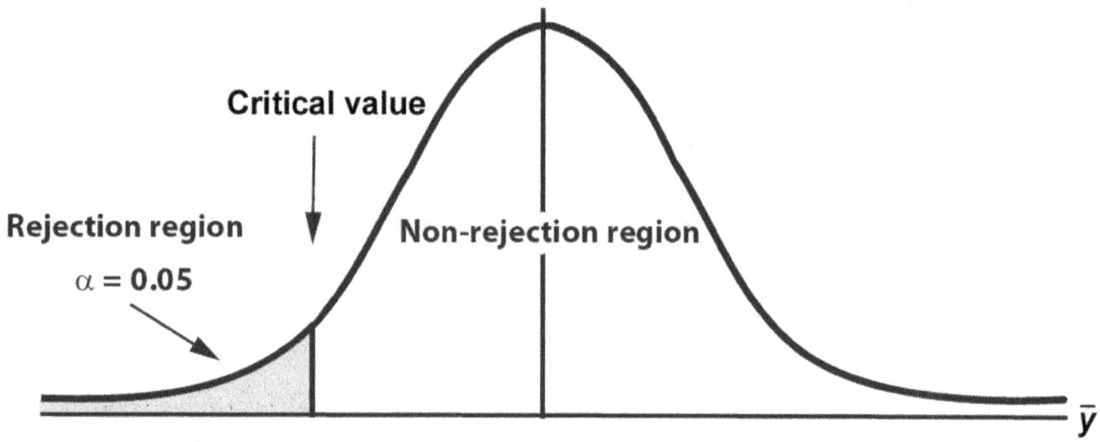

Figure 6.3. Region of null hypothesis rejection and non-rejection—two-tailed test

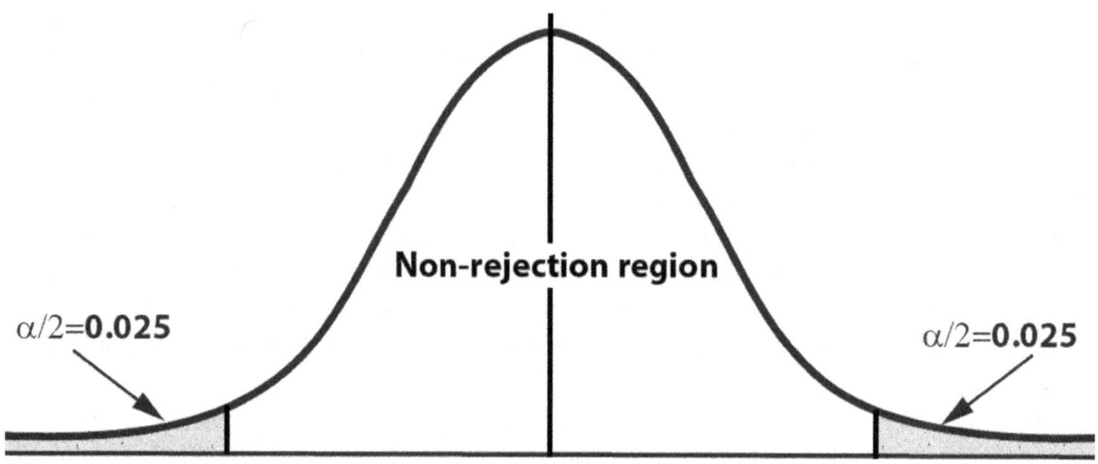

Now we'll try the slightly different approach by calculating a critical $\overline{y}$ value and comparing our sample result with this critical $\overline{y}$ value.

Because *p*-values smaller than $\alpha = 5\%$ (or 1% occasionally) indicate a significant result we find the corresponding critical $\overline{y}$ value by transforming a *z*-score that cuts exactly $\alpha = 5\%$ from the standard normal (sampling) distribution of the mean back into a critical $\overline{y}_{crit}$ test score. This critical mean test score splits the abscissa of the sampling distribution in the rejection and non-rejection region. Table 6.16 shows a worked-out example for our communication apprehension example.

Table 6.16. Critical test value for communication apprehension scores (one-tailed test)

$$z_\alpha = \frac{\bar{y}_{\text{crit}(\alpha)} - \mu_0}{\hat{\sigma}_{\bar{y}}} \quad \Rightarrow \quad \bar{y}_{\text{crit}(\alpha)} = \mu_0 + z_\alpha \cdot \hat{\sigma}_{\bar{y}}$$

$z(5\%)$ at the left tail of the standard normal distribution is -1.65, thus:

$$\bar{y}_{\text{crit}} = \mu_0 + z_\alpha \cdot \hat{\sigma}_{\bar{y}} = 66 - 1.65 \cdot 1.5 = 63.53$$

This means that all sample results that are as low as or lower than the critical test value of 63.53 lead to a rejection of the null hypothesis and all sample results that are larger than 63.53 lead to non-rejection of the null hypothesis. In our sample we got a sample result of 62.5, thus we rejected the null hypothesis with a *p*-value smaller than 5%.

You may ask how the critical test value is affected if we operate with an alpha-level of 1% instead of 5%. The *z*-score for a one-sided test would be 2.33 and the critical value is therefore with $\bar{y}_{crit} = 62.51$ lower; i.e., more distant from μ_0. That is, the "barrier" for the sample result to jump over to get into the region of null hypothesis rejection is higher.

What happens if we are interested in a two-sided test? As we learned, we test a non-directional research hypothesis with a two-sided statistical test. In a two-sided statistical test, both sample statistics that are significantly greater or significantly smaller than the null hypothesis population parameter provide good evidence for the research hypothesis. Consequently, we calculate critical test values on both tails of the sampling distribution, which results in two regions of null hypothesis rejection and one region of null hypothesis non-rejection. The maximum probability for a Type-I error shall still be 5% (alpha-level), which means that we have to calculate critical test values that correspond to $\alpha/2$ instead of α. Table 6.17 provides a worked out example for a two-tail test for communication apprehension.

Table 6.17. Two-tailed critical test for communication apprehension scores

$$\bar{y}_{\text{crit}(\alpha/2)} = \mu_0 + z_{\alpha/2} \cdot \hat{\sigma}_{\bar{y}}$$

$z(2.5\%)$ at the right tail of the standard normal distribution is $+1.96$, thus:

$$\bar{y}_{\text{crit}(\alpha/2)} = \mu_0 + z_{\alpha/2} \cdot \hat{\sigma}_{\bar{y}} = 66 + 1.96 \cdot 1.5 = 68.94$$

$z(2.5\%)$ at the left tail of the standard normal distribution is -1.96, thus:

$$\bar{y}_{\text{crit}(\alpha/2)} = \mu_0 + z_{\alpha/2} \cdot \hat{\sigma}_{\bar{y}} = 66 - 1.96 \cdot 1.5 = 63.06$$

What is the conclusion? Sample means of ≤ 63.06 and ≥ 68.94 result in a rejection of the null hypothesis. The sample mean in our sample was 62.5. This result is therefore significant in a one-tail test as well as a two-tail test.

Now we'd like to compare one-tail and two-tail tests. What would happen if we chose an alpha-level of 1% instead of 5%? The *z*-score, which cuts 0.5% at both tails of the sampling distribution would have been ± 2.58 (See Table 1, Appendix A). The critical test scores can be found in Table 6.18.

Table 6.18. Critical values for a two-tailed test at alpha-level 1%

$$\bar{y}_{\text{crit}(\alpha/2)} = \mu_0 + z_{\alpha/2} \cdot \hat{\sigma}_{\bar{y}}$$

$z(0.5\%)$ at the right tail of the standard normal distribution is +2.58, thus:

$$\bar{y}_{\text{crit}(\alpha/2)} = \mu_0 + z_{\alpha/2} \cdot \hat{\sigma}_{\bar{y}} = 66 + 2.58 \cdot 1.5 = 69.87$$

$z(0.5\%)$ at the left tail of the standard normal distribution is −2.58, thus:

$$\bar{y}_{\text{crit}(\alpha/2)} = \mu_0 + z_{\alpha/2} \cdot \hat{\sigma}_{\bar{y}} = 66 - 2.58 \cdot 1.5 = 62.13$$

Our sample result of 62.5 would now fall in the region of null hypothesis non-rejection. Thus, the result would have been non-significant. How do we deal with this result? Only if we had formulated a theoretically derived, directional null and research hypothesis in advance (i.e., prior to data collection) and had set the appropriate alpha-level to $\alpha = 5\%$ in advance could we reject the null hypothesis. It is important to note that hypotheses and alpha-levels must be formulated before any data collection and have to remain untouched after data have been collected and results been computed. A modification of hypotheses and alpha-levels in consideration of the data is unscientific. Such behavior would lead to random knowledge in the long-term and contribute to contradicting research results in social sciences and in communication research in particular. If we have suggestions about how to modify our initial hypothesis based on our results in a study we may report these suggestions in the discussion section of a research paper. However, modified hypotheses have to be tested with a new sample and a new study in which the modified hypotheses are then used as a priori hypotheses.

Non-significant Results

To conclude with this section, let's discuss what it actually means if a sample result turned out to be non-significant. Does a non-significant result mean that we have confirmed the null hypothesis? No, unfortunately not. It would be correct to say that our study did not provide sufficient evidence to reject the null hypothesis but we cannot draw a conclusion on whether either the research hypothesis or null hypothesis is true in the population. This conclusion sounds strange, but is correct if we consider the following example.

Let's assume that the *p*-value of our sample result in the communication apprehension example would have been 6%, which indicates a non-significant result and lies just above the common alpha-level of 5%. In this case we cannot reject the null hypothesis, but can we assume its confirmation? No, because this *p*-value indicates nothing more than the probability of our sample result (mean) under the assumption that the null hypothesis is true, or even more extreme, results in the direction of the research hypothesis is just 6%. This result is certainly not a ringing endorsement for the null hypothesis. If we are interested in finding good evidence for a null hypothesis we have to find a sample result that suggests rejecting the research hypothesis. This demand leads over to the Type-II error problem, which we will return to shortly.

Sample Size and Practical Significance of Test Results

In our communication apprehension example, a difference of 3.5 score points between the null hypothesis population parameter $\mu_0 = 66$ and the sample result $\overline{y} = 62.5$ was significant in a one-sided test ($n = 100$, p-value $\leq 0.99\%$). What do you think? Would this result have been significant in any other study? What are the factors that determine whether an empirical result (a mean in our example) is significant or not? One important factor is the sample size.

Assume that we had taught $n = 36$ professors with the new method instead of $n = 100$. Table 6.19 provides a worked out example with this new sample size.

Table 6.19. Communication apprehension example—test with a smaller sample

$n = 36$ instead of $n = 100$ (compare with Table 6.14).

Step 1: Calculate the standard error of the mean

$$\sigma_{\overline{y}} = \sqrt{\frac{\sigma^2}{n}} = \frac{15}{\sqrt{36}} = 2.5$$

Step 2: z-transformation of the result

$$z_{\overline{y}} = = \frac{\overline{y} - \mu_0}{\sigma_{\overline{y}}} = \frac{62.5 - 66}{2.5} = -1.4$$

Step 3: Find AUC from $z = -1.4$ to $-\infty$ (p-value)

This AUC corresponds to the AUC from $+1.4$ to $+\infty$, see Table 1 in Appendix A

$$\text{AUC } (z \geq +1.4) = 0.0808 \approx 8.08\%$$

Critical $\overline{y}_{\text{crit}}$ with $n = 36$ and $\alpha = 1\%$ and critical difference d

$$\overline{y}_{crit} = \mu_0 + z_\alpha \cdot \hat{\sigma}_{\overline{y}} = 66 - 2.33 \cdot 2.5 = 60.175$$

$$d = \mu_0 - \overline{y}_{\text{crit}} = 66 - 60.175 = 5.825$$

A p-value of 8.08% is considerably more than 1% or 5%. Thus, the empirical result is no longer significant when we had done a study with $n = 36$ instead of $n = 100$ students. As Table 6.19 demonstrates, in a study with $n = 36$ professors and an alpha-level of 1%, a mean difference of at least 5.825 would have been significant at the 1% level. In the same way you can calculate that with a sample of $n = 1{,}000$ a difference of $d = 1.105$ score points, and with a sample of $n = 10{,}000$ a difference of $d = 0.3495$ communication apprehension score points, would have been significant.

Now think about this: With a sample size of $n = 10{,}000$ even a difference between the new and the old training method of just $d = 0.3495$ for communication apprehension scores would provide a significant result at the $\alpha = 1\%$ level. The calculations demonstrate that if the sample size is large enough, *any* difference in the correct direction of the research hypothesis (if the test is a one-sided test) will become significant even if this difference may be of no practical relevance. A test score difference of $d \approx 0.35$ after a troublesome introduction of a new training method for improving communication apprehension among college professors may not be worth the effort, albeit the improvement in test scores is significant.

Therefore, please carefully pay attention to the following relationship: On the one hand the statement "a result is statistically significant" may be of no practical significance if this statement is based on large samples. On the other hand, results that clearly indicate a practically significant result may be worthless as long as one has not tested whether the result occurred at random. With both of these statements in mind, it is important to see both sides of a statistical result. To deepen your understanding of this important aspect of statistical tests we will try to link the concept of statistical significance with the concept of effect sizes.

Effect Sizes

An effect size is defined as the least difference between two population parameters that can be considered practically relevant. In other words, with an effect size we define how far away we expect the research hypothesis (H_1) parameter from the null hypothesis (H_0) parameter that we can say this distance or difference is relevant in practice. We usually divide the least practically relevant difference of the two parameters by the variable's standard deviation in order to make it possible to compare effect sizes across studies. See Table 6.20 for the definition of effect sizes.

Back to our example: Experienced educators and practitioners could have the opinion that a new training method should be introduced only if the improvement of communication apprehension scores is at least 5 points because only a difference of at least 5 points is considered substantial and relevant in practice. This results in an effect size as provided in Table 6.20.

Table 6.20. Effect size definition—example "communication apprehension"

$$\varepsilon = \frac{|\mu_1 - \mu_0|}{\sigma}$$

$$\varepsilon = \frac{|61 - 66|}{15} = 0.333$$

Please be aware that the determination of an effect size is not a statistical problem—it is a content problem. Statisticians do not determine effect sizes; it is the researcher or practitioner who knows about his or her research field and is able to define an effect size. Effect sizes usually result from theory or from studies that have been conducted before. In order to define an effect size, we have to do a thorough literature review and have to thoroughly think about our research field and about what effect sizes make sense in practice.

These efforts, however, will pay off because as you will see later, if we are able to define a well-founded effect size, then we can calculate an optimal sample size for a concrete study. This may be very important for high quality scientific research. The criteria for ensuring high quality research are: the sample size should not be so large that irrelevant results are significant, but also not so small that relevant effect sizes have no chance to become significant. Before we can determine optimal sample sizes according to these criteria, we have to first understand how to calculate the probability of Type-II errors.

Calculating Type-II Errors

As you will recall, we commit a Type-II error if we wrongly decide for the null hypothesis, that is, when the research hypothesis is true in the population. If we transfer the calculation of a Type-I error to the calculation of a Type-II error we run into the following problem: Like we needed the sampling distribution under null hypothesis conditions in order to calculate the probability of a Type-I error, we now need the sampling distribution under research hypothesis conditions (i.e., μ_1 and standard error) in order to calculate the probability of a Type-II error.

Let's consider the communication apprehension example again. We do know the mean CA score in the population of professors taught with the old training method ($\mu_0 = 66$). Because the directional, non-specific research hypothesis posits that the new training method is supposed to be better than the old method and does not provide any information on how much better we expect the method to be, the assumed population's parameter under the research hypothesis is unknown. Hence, in the case of a non-specific research hypothesis it is impossible to calculate the probability of a Type-II error. In order to calculate the probability of Type-II errors we have to have a specific research hypothesis, that is, more information about the research hypothesis (H_1) parameter.

As mentioned above, experienced teachers and researchers in instructional communication have defined that a difference of at least 5 points is relevant in practice. With this idea of an effect size, the μ_1 parameter in this example would be identified as $\mu_1 = \mu_0 - 5 = 66 - 5 = 61$. If we further assume that the standard deviation of professors who were taught with the new method does not change; i.e., it is still equal to 15, and that the result we obtained in our random sample of professors taught with the new method is still $\bar{y} = 62.5$, then we have all information to calculate the Type-II error of this result. Table 6.21 shows the computation to determine the probability of a Type-II error in our example.

The corresponding area under the curve for a z-value of 1.0 in our example is 15.87%. Specifically, the z-value of 1.0 cuts 15.87% of the research hypothesis sampling distribution's right tail. Note: It helps with your calculations if you draw the sampling distributions for both the research hypothesis and the null hypothesis keeping in mind the type and locations of the problem's hypothesis (i.e., whether the research hypothesis parameter is assumed to be either smaller or greater (one-sided test) or is assumed to be both smaller and greater (two-sided test) than the null hypothesis parameter). If based on the sample result of $\bar{y} = 62.5$ we decide for the null hypothesis in our example, we would commit a Type-II error with 15.87% probability. If we used the same convention as for Type-I errors, i.e., $\beta = 5\%$, the probability of a Type-II error in our example would have been too high to reject the research hypothesis and decide for null hypothesis.

Table 6.21. Calculating Type-II error probabilities—example "communication apprehension"

Step 1: Calculate the standard error of the mean

$$\sigma_{\bar{y}} = \sqrt{\frac{\sigma^2}{n}} = \frac{15}{\sqrt{100}} = 1.5$$

Step 2: z-transformation of the result in the context of the H_1 sampling distribution

$$z_{\bar{y}} = \frac{\bar{y} - \mu_1}{\sigma_{\bar{y}}} = \frac{62.5 - 61}{1.5} = 1.0$$

Step 3: Find area under the curve from $z = 1.0$ to $+\infty$ (Type-II error probability)

AUC $(z \geq +1.0) = 0.1587 \approx 15.87\%$

Selection of β-levels

In contrast to α-levels there is no convention when it comes to β-levels. The choice of a maximal β-level that we accept depends solely upon the context and the consequences of a study. It is important to discuss thoroughly and to decide if the desired result of a study would be the research hypothesis or the null hypothesis. If a researcher is primarily interested in rejecting a null hypothesis (traditional hypothesis testing procedure) then it is a good idea to use for the α-level a value of 5%/1% and for the β-level a value of 20%. In research practice, a result with a Type-I error probability of 5% at maximum and a Type-II error probability of 20% at minimum are considered good evidence for a specific research hypothesis. However, if the null hypothesis is actually the desired result of the study, the α- and β-levels should be reversed; i.e., $\beta \leq 5\%/1\%$ and $\alpha \geq 20\%$. That the null hypothesis is the desired result in a study is relatively rare, but it does happen. Consider for example a study by Ferguson (2007) in which the researcher wanted to establish that exposure to violent video games does <u>not</u> increase violent behaviors (null hypothesis!). Or consider a study by Linz, Paul, and Yao (2006), in which the researchers tried to show that the incidence of crime in the areas surrounding establishments fostering sexual communication (peep shows) was <u>not</u> greater than in areas further away from them.

Relationship between Type-I and Type-II Errors

The relationship between these error types can be stated as follows: The more you want to avoid one error, the more likely you are to commit another. For example, the more you want to avoid a

Type-I error, that is, a wrong decision in favor of the research hypothesis, the more likely you are to commit a Type-II error, that is, a wrong decision in favor of the null hypothesis. This is a dilemma in any statistical decision based on samples and cannot be resolved. Figure 6.4 illustrates this dilemma for the communication apprehension example.

Figure 6.4. Relationship between Type I and Type II errors

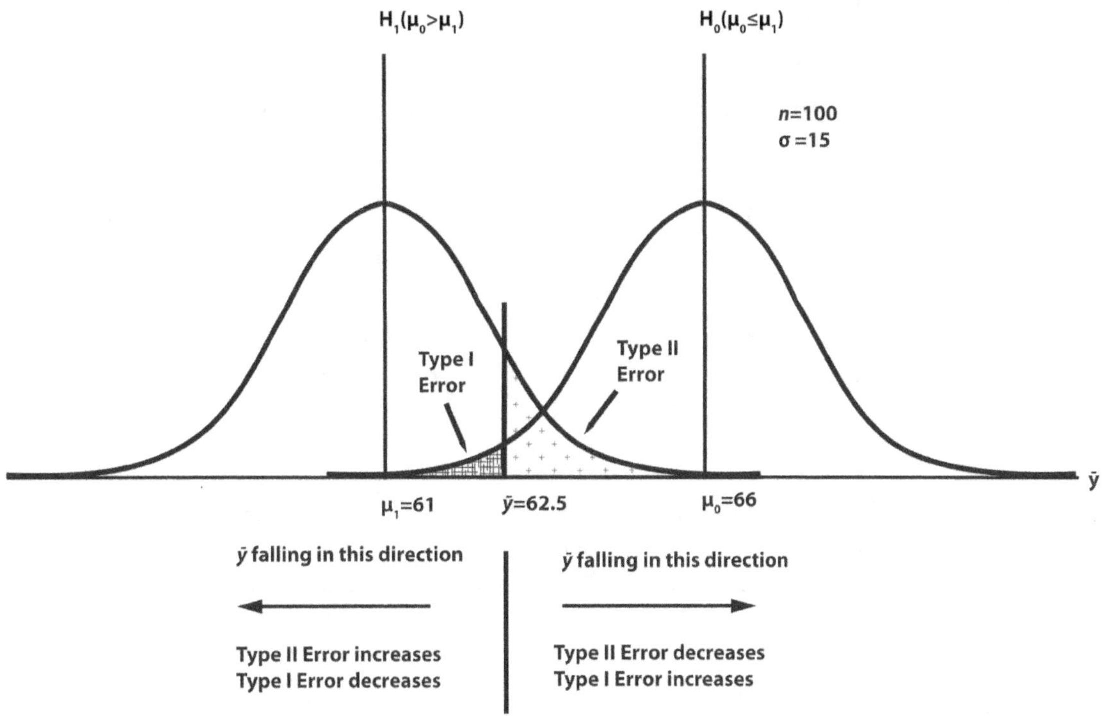

Let's focus now on the last and probably most difficult concept in hypothesis testing that is related to Type-I and Type-II errors—the power of a statistical test. Once you understand the concept of statistical test power, you will be able to scientifically discuss statistically significant results from various aspects. It will help you to be confident when making statistical decisions and arguments.

Power of a Statistical Test

As we have established, the probability of a Type-I error (p-value) describes the probability of rejecting the null hypothesis when in fact the null hypothesis is true in the population. The probability of a Type-II error describes the probability of rejecting the research hypothesis when in fact the research hypothesis is true in the population. Thus, 1 minus the Type-II error probability describes the probability of not rejecting the research hypothesis when in fact the research hypothesis is true in the population. In other words, (1 − Type-II error) provides the probability of deciding for the research hypothesis when the research hypothesis is in fact true. (1 − Type-II error), with a Type-II error we commit when our sample result would just reach significance (i.e., $\bar{y} = \bar{y}_{crit}$), is defined as the power of a statistical test. The power of a statistical test describes the probability of a statistical test to detect significant results when in fact the research hypothesis is true in the population. Figure 6.5 illustrates the relationship between errors and power of the statistical test.

Figure 6.5. Type I and Type II errors and power

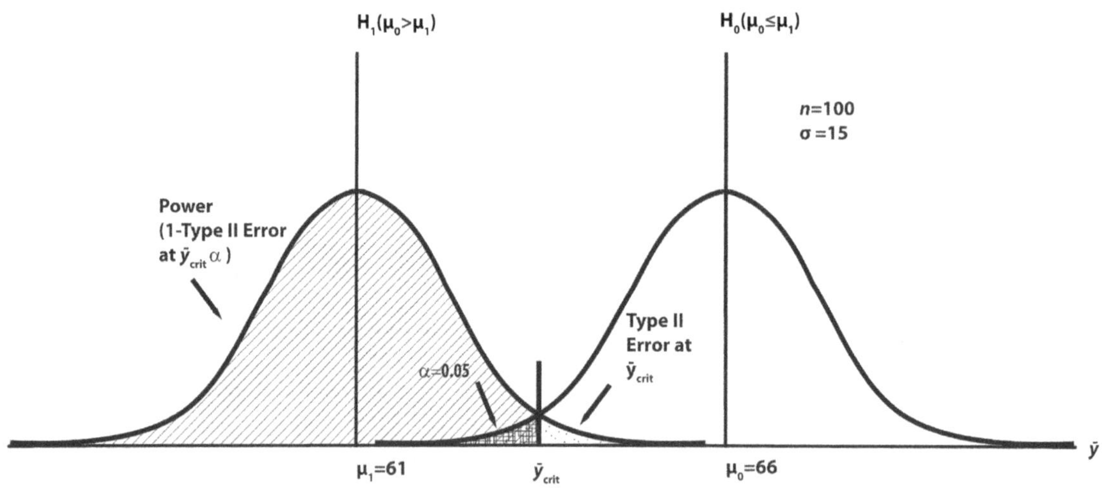

Note that the power of statistical test is (1 − Type-II error) at $\bar{y}_{crit}$ and NOT (1 − Type-II error) at $\bar{y}$. Confusing the critical value (which leads to power) with the empirical result (which would lead to empirical or observed power, a concept that has been found useless in the literature; see O'Keefe, 2007) is frequently the source of misunderstanding the power of a statistical test.

What would we expect at least as the power of any statistical test and why? We would expect a power of at least 50% or higher. If the power of a test were equal to or less than 50%, then the simple flip of a coin would lead to equal or better decisions about competing null and research hypotheses. Research that is based on statistical test power of equal to or less than 50% is nonsense research.

What type of research hypothesis do we need in order to calculate the power of a statistical test? As the calculation of test power requires the calculation of the Type-II error probability, we would need a specific research hypothesis, that is, we need effect size information (see above), which specifies the H_1 parameter. Furthermore, we need information on α, that is, the maximum Type-I error we are prepared to tolerate (usually 5%), the sample size, and the statistic's standard error. All this information is independent of an empirical result in a sample. Thus, the power of a statistical test procedure does not vary between samples of equal size; test power is *not* subject to random variations.

Table 6.22 provides the computation of test power for our communication apprehension example. It is important for your calculations to keep in mind your pair of hypotheses, that is, whether you deal with a directional (one-sided) or non-directional (two-sided) research hypothesis and whether the H1 parameter falls toward the negative or positive tail of the null hypothesis sampling distribution. With H1: $\mu 0 > \mu 1$, and H0: $\mu 0 \leq \mu 1$ (see Table 6.9) in our example, we deal with a directional hypothesis and the H1 parameter falls toward the negative tail of the null hypothesis sampling distribution.

Table 6.22. Calculating test power—example "communication apprehension"

Step 1: Calculate critical test value ($\alpha = 5\%$; left tail of H0; i.e., $z_\alpha = -1.65$)

$$z_\alpha = \frac{\bar{y}_{\text{crit}(\alpha)} - \mu_0}{\hat{\sigma}_{\bar{y}}} \quad \Rightarrow \quad \bar{y}_{\text{crit}(\alpha)} = \mu_0 + z_\alpha \cdot \hat{\sigma}_{\bar{y}} = 66 - 1.65 \cdot 1.5 = 63.525$$

Step 2: Calculate Type-II error at $\bar{y}_{\text{crit}(\alpha)}$

$$z_\beta = \frac{\bar{y}_{\text{crit}(\alpha)} - \mu_1}{\hat{\sigma}_{\bar{y}}} = \frac{63.525 - 61}{1.5} = 1.68$$

AUC ($z \geq +1.68$) = 0.0465 = 4.65%

Step 3: Calculate power

Power = 1 − Type-II error at $\bar{y}_{\text{crit}(\alpha)} = 1 - 0.0465 = 0.9535 = 95.35\%$

Assuming an α-level of 5%, a sample size of 100, a given standard deviation of the variable of 15 (i.e., a standard error of the mean of 1.5), and an effect size of $\varepsilon = 0.333$ or $\mu_1 = 61$, the power of the statistical test we apply would be 95.35%. Hence, if we obtained a sample result of $\bar{y} = 62.5$ in our study on improving communication apprehension among professors, this result would be significant at the 1% level (see Table 6.14) with an empirical Type-II error of 15.87% and a test power of 95.35%. This result is pretty good evidence for the research hypothesis and the involved researcher(s) are able to recommend the new training method with confidence.

Note: In case of a non-directional research hypothesis we need to apply a two-sided test and the power of a two-sided test is (1 − Type-II error), with a Type-II error we commit at $\bar{y}_{\text{crit}(\alpha/2)}$ instead of $\bar{y}_{\text{crit}(\alpha)}$ (if $\alpha = 5\%$, we use $\alpha / 2 = 2.5\%$ to calculate $\bar{y}_{\text{crit}}$ and the corresponding Type-II error.)

Factors that Affect the Power of Statistical Tests

If you consider the information needed to calculate the power of a statistical test, then you know what factors have an impact on test power. Those factors are (1) effect size, (2) sample size, (3) the test variable's standard deviation in the population, (4) alpha-level, and (5) whether you test a directional (one-sided) or a non-directional (two-sided) hypothesis.

With increasing effect size, the research hypothesis's sampling distribution moves toward the lower or upper end of the null hypothesis' sampling distribution resulting in lower Type-II errors at $\bar{y}_{crit}$, which will lead to increased power (1 − Type-II at $\bar{y}_{crit}$). If either the sample size increases or the test variable's standard deviation decreases, the standard error of the mean decreases, leading to more precise parameter estimations and to reduced Type-II errors at $\bar{y}_{crit}$ which, again, will lead to increased power (1 − Type-II at $\bar{y}_{crit}$). Increasing the alpha-level (for example from 1% to 5%), will move $\bar{y}_{crit}$ toward the center of the null hypothesis distribution, which results in a lower Type-II error at $\bar{y}_{crit}$ and increased test power. Similarly, one-sided tests (with $\alpha = 5\%$ falling on one side of the null hypothesis sampling distribution) lead to higher test power than two-sided tests (with $\alpha = 5\%$ falling on both sides of the null hypothesis sampling distribution; i.e., $\alpha = \alpha / 2 = 2.5\%$ on each side).

Calculation of Optimal Sample Sizes

To this point we have learned that a sample result can be statistically significant but not relevant in practice. This may happen because the researcher ran the study with a large sample size. A sample result can also turn out to be non-significant but at the same time indicate an effect size that is substantial and would be relevant in practice. The reason for this could be that the researcher ran the study with too small a sample or in fact the sample result has occurred due to chance.

Given our consideration above regarding Type-I and Type-II errors it may also happen that a researcher rejects the null hypothesis (because of a *p*-value of smaller than the a priori defined α-level) but is not able to confidently decide for the research hypothesis because of the important consequences of a Type-II error and a Type-II error probability of far lower than 20%.

All these results would be unfortunate in a research study. So, what can researchers do to avoid these scenarios? A good solution would be to adjust all factors that affect the decision in favor of H_0 or H_1 such that none of the scenarios can happen. Here is how we would do this:

Based on what we have discussed so far, the following factors affect our test decision: (1) α-level (maximum Type-I error) we tolerate, (2) β-level (maximum Type-II error we tolerate), (3) effect size ε (i.e., expectations for H_0 and H_1 parameter), and (4) the sample size *n*. α-, β-level, and effect size ε are determined by content criteria (consequences of a wrong decision, prior knowledge, etc.) and cannot be arbitrarily defined. The only factor that can be changed by the researcher in light of his research project is the sample size *n*.

As a general rule: The sample size should not be so large that practically irrelevant results become significant, but also not so small that practically relevant effect sizes cannot become significant. This leads to the definition of an optimal sample size: The optimal sample size is defined as a sample size that leads to an unambiguous decision about H_0 and H_1 and to significance for practically relevant results. Table 6.23 provides the calculations for the optimal sample size in our communication apprehension example.

Thus, a sample size of $n=56$ would have been optimal for our communication apprehension researcher. With this sample size, the researcher would have found the sample result significant at the 5% level with a Type-II error probability of at least 20% or a test power of 80% if indeed the research hypothesis is true in the population.

Table 6.23. Calculating optimal sample size—example "communication apprehension"

Step 1: Determine appropriate α- and β-levels for the test decision

α-level: 5%, one-sided, $z_\alpha \approx -1.65$
β-level: 20% (test power = 80%), one-sided, $z_\beta \approx +0.84$

Step 2: Determine effect size (see Table 6.20)

$$\varepsilon = \frac{|\mu_1 - \mu_0|}{\sigma}$$

$$\varepsilon = \frac{|61 - 66|}{15} = 0.333$$

Step 3: Calculate optimal sample size

$$n_{\text{opt}} = \frac{(z_{(\alpha)} - z_\beta)^2}{\varepsilon^2} = \frac{(-1.65 - 0.84)^2}{0.333^2} = 55.91 \approx 56$$

Step 4 (optional): Check test power

$$\sigma_{\bar{y}} = \sqrt{\frac{\sigma^2}{n}} = \frac{15}{\sqrt{56}} = 2.0$$

$$\bar{y}_{\text{crit}(\alpha)} = \mu_0 + z_\alpha \cdot \hat{\sigma}_{\bar{y}} = 66 - 1.65 \cdot 2.0 = 62.7$$

$$z_\beta = \frac{\bar{y}_{\text{crit}(\alpha)} - \mu_1}{\hat{\sigma}_{\bar{y}}} = \frac{62.7 - 61}{2.0} = 0.85$$

AUC ($z \geq +0.84$) $\approx 0.20 = 80\%$

SUMMARY

This chapter provided an overview of the two major divisions of inferential statistics: estimation and significance testing. In estimation, we are interested in how precisely we can predict parameters of variables in the population based on samples. In testing, we are interested in how theoretical predictions about parameters of variables stand up against empirical data in a sample. Furthermore, important concepts for testing hypotheses such as effect size, power of a statistical test, and optimal sample sizes have been discussed. This chapter is the foundation for specific significance tests for differences between variables (Chapter 7) and specific significance tests for relationships among variables (Chapter 8).

TERMS

- **Estimation** is the division of statistics concerned with how and how precisely we can estimate or predict parameters (e.g., the mean) of variables in a population by using a sample.
- **Confidence interval** contains an unknown parameter with a certain probability. The probability that the confidence interval contains the parameter is called the confidence coefficient.
- **Significance testing** is the division of statistics that is concerned with the formulation of hypotheses about variables' characteristics in a population and then determine how hypotheses hold up with sample data.
- **Research hypothesis** is the statement about characteristics of variables in the population of interest. Research hypotheses can be about differences or dependencies. The type of hypothesis determines the statistical test used.
- **Null hypothesis** is the counterpart to the research hypothesis, and claims that the statement in the research hypothesis is not true, that there is no difference or dependency between variables.
- **Type-I error** refers to accepting the research hypothesis when the null hypothesis is true.
- **Type-II error** refers to accepting the null hypothesis when the research hypothesis is true.
- **Effect size** is defined as the least difference between two population parameters that can be considered practically relevant.
- **Power** of a statistical test describes the probability of a statistical test to detect significant results when in fact the research hypothesis is true in the population.
- **Optimal sample size** is defined as a sample size which leads to an unambiguous decision about H_0 and H_1 and to significance for practically relevant results.

CHAPTER 6 PROBLEM SET

1. You administer the PRCA-24 to 100 students in a public speaking course and find a mean of 66 and a standard deviation of 15.
   a. Compute a 95% confidence interval for the mean.
   b. Compute a 99% confidence interval for the mean.
2. You conduct a study on veracity judgments and find in your sample that accuracy for true denials was 56.1%. Your sample size is 130.
   a. Compute the standard error of the percentage.
   b. Compute a 95% confidence interval for the percentage.
   c. Compute a 99% confidence interval for the percentage.
3. What do directionality and specificity mean with regard to hypotheses?
4. Based on a sample, a researcher rejects the null hypothesis and decides for the research hypothesis, but the null hypothesis is true in the population. What type of error has the researcher committed?
5. A researcher does not reject the null hypothesis and decides for no effect, but the research hypothesis is true in the population. What type of error has the researcher committed?

6. What is an effect size? How do researchers determine effect sizes?
7. A media researcher believes that the amount of television watching leads to greater cynicism (a mean-world perspective). However, the researcher finds the opposite, that television watching leads to a more positive world outlook. Rather than reporting these findings, the researcher modifies his hypothesis to the following: The more television people view, the more optimistically they view the world. You overhear this researcher talking about this, what would you say to him and why? What would he need to do in order to test his modified hypothesis?
8. Consider the example of the media researcher in problem 7.
   a. When the media researcher states that "the more television people view, the more optimistically they view the world," what type of research hypothesis does the media researcher posit? Use words and also statistical terms with mathematical operators to characterize the research hypothesis.
   b. What type of null hypothesis does the media researcher posit? Use words and also statistical terms with mathematical operators to characterize the null hypothesis.
9. The following question asks about one- and two-tailed statistical tests.
   a. What is a one-tailed test? Under what circumstance would you use a one-tailed test?
   b. What is a two-tailed test? Under what circumstance would you use a two-tailed test?
10. What is the relationship between a Type-I error and a Type-II error?
11. What type of research hypothesis is needed in order to assess the power of a statistical test?
12. What determines whether you choose to use a z-table or a t-table to determine critical values for a z-test?
13. A researcher is interested in whether a new test method is better than an older method. Assume a known population parameter of $\mu_0 = 40$, an α-level of 5%, a sample size of 100, a given standard deviation of 8, and a meaningful effect size of $\varepsilon = 0.375$, answer the following questions:
    a. What is the test's power?
    b. What is the optimal sample size?
14. A researcher is interested in organizational identification (cf. Cheney, 1983). Assume that $\mu_0 = 4.7$, and $\Sigma = 1$. The sample mean of 100 randomly selected organizational members was 5.0. Calculate the p-value of the sample mean.

REFERENCES

Afifi, W. A., & Weiner, J. L. (2006). Seeking information about sexual health: Applying the theory of motivated information management. *Human Communication Research, 32,* 32–57.

Cheney, G. (1983). On the various and changing meanings of organizational membership: Field study of organizational identification. *Communication Monographs,* 50, 342–362.

Chia, S. C. (2006). How peers mediate media influence on adolescents sexual attitudes and sexual behavior. *Journal of Communication, 56,* 585–606.

Dienes, Z. (2011). Bayesian versus orthodox statistics: Which side are you on? *Perspectives on Psychological Science,* 6(3), 274–290.

Ferguson, C. J. (2007). Evidence for publication bias in video game violence effects literature: A meta-analytic review. *Aggression and Violent Behavior,* 12, 470–482

Goodboy, A. K. (2011). The development and validation of the instructional dissent scale. *Communication Education, 60,* 422–440.

Linz, D., Paul, B., & Yao, M. Z. (2006). Peep show establishments, police activity, public place, and time: A study of secondary effects in San Diego, California. *The Journal of Sex Research, 43,* 182–193.

Levine, T. R., Kim, R. K., & Blair, J. P. (2010). (In)accuracy in detecting true and false confessions and denials: An initial test of a projected motive of veracity judgments. *Human Communication Research, 36,* 82–102.

Levine, T. R., Weber, R., Park, H., & Hullett, C. (2008a). A communication researcher's guide to null hypothesis significance testing and alternatives. *Human Communication Research,* 34, 188–209.

Levine, T. R., Weber, R., Hullett, C., Park, H., & Massi-Lindsey, L. (2008b). A critical assessment of null hypothesis significance testing in quantitative communication research. *Human Communication Research,* 34, 171–187.

McCroskey, J. C. (1982). *An introduction to rhetorical communication* (4th ed.). Englewood Cliffs, NJ: Prentice Hall.

O'Keefe, D. J. (2007). Post hoc power, observed power, a priori power, retrospective power, prospective power, achieved power: Sorting out appropriate uses of statistical power analyses. *Communication Methods and Measures,* 1, 291–299.

Saito, S. (2007). Television and the cultivation of gender-role attitudes in Japan: Does television contribute to the maintenance of the status quo? *Journal of Communication, 57,* 511–531.

Tryon, W. W. (2001). Evaluating statistical difference, equivalence, and indeterminacy using inferential confidence intervals: An integrated alternative method of conducting null hypothesis statistical tests. *Psychological Methods,* 6, 371–386.

Tryon, W. W., & Lewis, C. (2008). An inferential confidence interval method of establishing statistical equivalence that corrects Tryon's (2001) reduction factor. *Psychological Methods,* 13, 272–278.

CHAPTER 7

SPECIFIC SIGNIFICANCE TESTS FOR DIFFERENCES

In Chapter 6, we provided a general introduction to estimating parameters and testing hypotheses. Everything you learned in Chapter 6 is important for your deeper understanding of inferential statistics and will now be applied for statistical tests that are designed specifically to address data-analysis problems communication researchers and professionals are confronted with in their daily work. In the previous chapter we mentioned two types of statistical problems or hypotheses and provided examples for each: (1) hypotheses of differences between variables (or categories of variables; e.g., performance in a communication apprehension test of students trained with a new and an old training method); and (2) hypotheses of relationships among variables (e.g., a relationship among the variables "teacher misbehavior" and "instructional dissent"). The type of problem or hypotheses can be further categorized into different measurement levels of the involved variables. For each combination of problem type (difference/relationship) and variables' measurement level (nominal, ordinal, and interval/ratio level) one can find statistical tests specifically designed to analyze the data.[1] In fact, identifying the type of problem and the measurement level of the involved variables is often the first step in the identification of an appropriate statistical test. If tests are applied that are designed for a problem type and measurement level other than the problem and measurement level at hand, the results and the conclusions of a statistical test procedure can be meaningless.

In this chapter, we focus on significance tests specifically designed to test for differences between variables that are measured on an interval/ratio level (significance tests for mean differences) or that are based on nominal-level data (significance tests for frequency differences; also called chi-square tests). Significance tests for relationships among variables will be discussed in Chapter 8. Significance tests that test hypotheses about variables that are measured on an ordinal level (rank data) are not discussed in this book. Should you be confronted with testing hypotheses that involve ordinal-level data we recommend that you either treat the data as nominal level data (which is possible, but that comes with a loss of information and decreases the power of your statistical test), or you consult a textbook that deals specifically with so-called non-parametric statistical tests and the analysis of ordinal data (e.g., Sprent & Smeeton, 2007).

[1]We use this categorization here for didactical reasons. Advanced statisticians know that all parametric tests discussed here are a special case of the General Linear Model and as such the categorization by type of hypothesis and measurement level is actually not necessary.

TESTS FOR MEAN DIFFERENCES

In Chapter 6 we introduced a very basic statistical test, the z-test. The z-test assumed that the null hypothesis parameter μ_0 in the population is known from prior research. μ_0 was important information that we used to calculate the p-value of an empirical result. As discussed in detail in Chapter 6, if we found the probability of the empirical result (or even more extreme results) under the condition that the null hypothesis is true (i.e., assuming μ_0), to be smaller than 5%, then we said the result is statistically significant.

Now, we modify our assumptions. In practical research we usually know little about the conditions in populations. It is much more common to decide about conditions in populations based on samples. In this chapter, we will learn and understand the background of statistical methods that we use to decide about differences of parameters or relations between parameters if available information is based on samples only. As discussed above, the categorization of those statistical procedures is based on the type of hypothesis they test, the measurement level they require, and also on the type of sample we use.

Independent Samples *t*-Test

The independent samples t-test requires that the outcome variable (dependent variable) is measured at the interval level (e.g., communication apprehension), so that means and variances make sense. Further, with this test, we are interested in testing a hypothesis about a difference between two groups (e.g., professors taught with a new or an old training method). Because the outcome variable has an interval measurement level, we are interested in testing a mean difference between two groups (e.g., mean communication apprehension score in the new versus the old training method groups). Finally, the two groups are assumed to be independent. This means that the two groups we are interested in are sampled such that the selection of study elements (here, research participants) in one group does not influence the selection of study elements in the other group. For example, a random sample of individual professors that are randomly assigned to either the group taught with the new or to the group taught with the old training method would qualify as an independent sample. There is no relationship between the selected professors and, therefore, we say that measurements of individual professors are independent from each other (later we will provide an example of a dependent sample or dependent measurements).

Background

Recall from Chapter 6 that a statistical test requires the formulation of a pair of hypotheses. In the case of the independent samples t-test, our null and research hypotheses (for a directional hypothesis) are:

- H_1: $\mu_1 - \mu_2 < 0$ or $\mu_1 < \mu_2$
- H_0: $\mu_1 - \mu_2 \geq 0$ or $\mu_1 \geq \mu_2$

H_1 states that the means of the two groups are different from each other such that $\mu_1 < \mu_2$ (the hypothesis could also state the opposite direction, that $\mu_1 > \mu_2$). H_0 states that the means of the two groups in the population are equal; that is, both group samples are samples of the same population with the same population parameter $\mu_1 = \mu_2 = \mu$, or the difference is even reversed; i.e., $\mu_1 > \mu_2$. Of course, the hypotheses could also be non-directional, that is H_1: $\mu_1 - \mu_2 \neq 0$ ($\mu_1 \neq \mu_2$) and H_0: $\mu_1 - \mu_2 = 0$ ($\mu_1 = \mu_2$).

In this section, we will show that this test is nothing new and can be easily understood with the background you mastered when we introduced you to the basics of statistical tests in Chapter 6. Essentially, you need four pieces of information or you need to answer four questions in order to conduct a (parametric) statistical test:

Question 1: What is the Statistic We are Interested in?

The statistic we are interested in is given by the null and research hypotheses and ultimately by the indication of an independent samples *t*-test. As discussed above, we use an independent samples *t*-test when we are interested in testing the difference of mean parameters in two groups. Thus, the statistic we are interested in is the difference of sample means in two groups.

Question 2: What is the Statistic's Sampling Distribution's Expected Value (or Mean Value) if the Null Hypothesis is True in the Population?

If the null hypothesis is true in the population we would expect that the mean parameters in the two groups are the same, thus, for the independent samples *t*-test we expect: The parameter of the H_0 sampling distribution is $E(\mu_1 - \mu_2) = 0$ ("E" stands for expected value or simply for "expectation").

Question 3: What is the Statistic's Standard Error (or Standard Deviation of the Statistic's Sampling Distribution)?

The answer to question 3 is a bit more complicated, but simply, here is how we derive a mean difference's standard error (don't panic when you scan the formulas, we provide step-by-step explanations). We already know:

$$\sigma^2{}_{\bar{y}1} = \frac{\sigma_1^2}{n_1} \quad \text{and} \quad \sigma^2{}_{\bar{y}2} = \frac{\sigma_2^2}{n_2}$$

If the two groups/samples are independent, as a rule in stochastic theory, we can just add the two variances. The square root of a variance leads to the standard deviation. Therefore, we get:

$$\sigma_{\bar{y}_1 - \bar{y}_2} = \sqrt{\frac{\sigma_1^2}{n_1} + \frac{\sigma_2^2}{n_2}}$$

If the variances in both groups are not significantly different (i.e., equal in the population) the standard error can be simplified to:

$$\sigma_{\bar{y}_1-\bar{y}_2} = \sqrt{\sigma^2 \cdot \left(\frac{1}{n_1} + \frac{1}{n_2}\right)}$$

Usually, the joint population variance σ is unknown, so we just estimate the population variance with all the data we found in our two samples by simply computing the sample variance (recall that the sample variance is a biased estimator; in order to correct for the bias, we divide by $n - 1$ instead of n; see Chapter 5):

$$\hat{\sigma}^2 = \frac{\sum_{m=1}^{n_1} (y_{m1} - \bar{y}_1)^2 + \sum_{m=1}^{n_2} (y_{m2} - \bar{y}_2)^2}{(n_1 - 1) + (n_2 - 1)}$$

With this, the estimated standard error of mean differences is:

$$\hat{\sigma}_{\bar{y}_1-\bar{y}_2} = \sqrt{\frac{\sum_{m=1}^{n_1} (y_{m1} - \bar{y}_1)^2 + \sum_{m=1}^{n_2} (y_{m2} - \bar{y}_2)^2}{(n_1 - 1) + (n_2 - 1)} \cdot \left(\frac{1}{n_1} + \frac{1}{n_2}\right)}$$

The formula looks complicated, but if you take a closer look, all you need to do is to compute the squared deviations of measurements across the two samples from their sample mean, divide by and multiply with the terms containing the sample sizes, and draw the square root. It requires a little practice if you do the calculations by hand or with a calculator, but it is not difficult.

Finally, if you even knew the shape of the statistic's sampling distribution you would have all information you need to compute the p-value of an empirical result (here: a difference of two sample or group means). Thus, our final question is:

Question 4: What is the Shape of the Statistic's Sampling Distribution?

Because according to the Central Limit Theorem (Chapter 5) means are normally distributed for large samples ($n \geq 30$), differences of means are also normally distributed for large samples. Samples in this context are considered large if $(n_1 + n_2) > 50$. For $(n_1 + n_2 < 50)$, the distribution of means follows a t-distribution with $(n_1 + n_2 - 2)$ degrees of freedom if the dependent variable is normally distributed in the population.

Calculation of p-Values in t-Tests for Two Independent Samples

P-values are calculated as in any other statistical test. Specifically, we ask how likely is the empirical result (here, a mean difference) or even more extreme results (in direction to the research hypothesis) given that the null hypothesis is true? Thus, the calculation of p-values proceeds in three simple steps:

Step 1: We Standardize the Empirical Result (Here: The Difference of Two Group Means)

For this, we need to know the expected or mean value of the statistic's sampling distribution under condition of the null hypothesis (i.e., $\mu_1 - \mu_2 = 0$) as provided in the answer to question 2. We also need to know the standard deviation of the statistic's sampling distribution (standard error), which is provided in the answer to question 3. The standardized empirical mean difference leads to a z-value for large samples (because we use the standard normal distribution in Table 1 (Appendix A), or to a t-value for small samples; i.e., for sample where $n_1 + n_2 < 50$ (because in this case we use the t-distribution in Table 2 of Appendix A). Thus:

$$z_{\text{emp}} \text{ or } t_{\text{emp}} = \frac{(\bar{y}_1 - \bar{y}_2) - (\mu_1 - \mu_2)}{\hat{\sigma}_{\bar{y}_1 - \bar{y}_2}} = \frac{(\bar{y}_1 - \bar{y}_2)}{\hat{\sigma}_{\bar{y}_1 - \bar{y}_2}}$$

Step 2: We Look Up the Corresponding Area Under the Curve (AUC)

We need to find the AUC from the empirical value in direction of the research hypothesis in the correct standardized distribution table (here: either the standard normal distribution or the t-distribution) in order to find the p-value. Figure 7.1 demonstrates the p-value for our

Figure 7.1. Calculating p-value for (one-sided) independent samples t-test

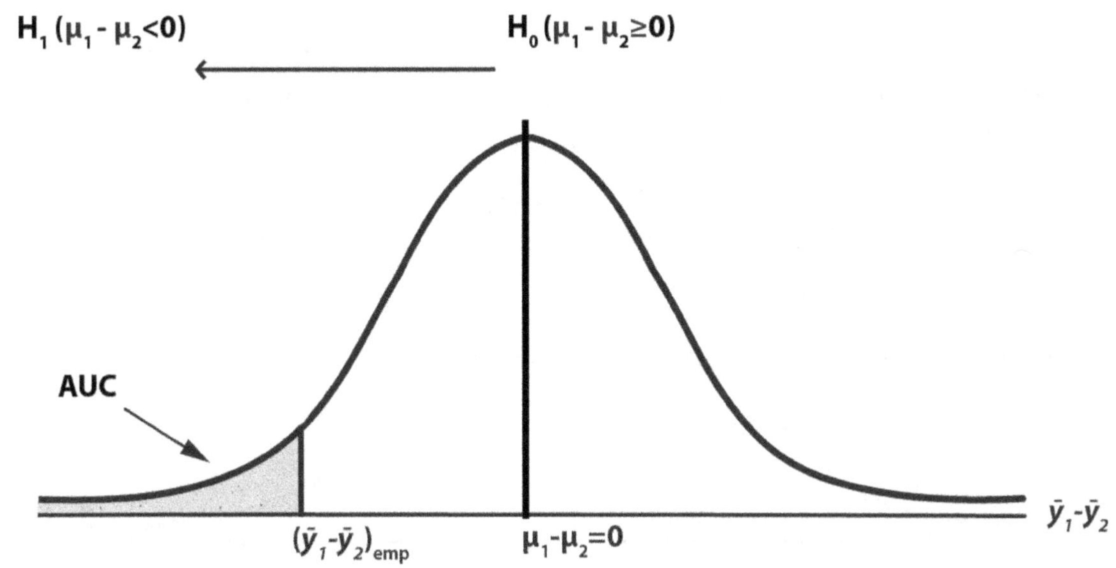

hypotheses here (note: AUC's from a negative value to $-\infty$ are equivalent to AUC's from a positive value to $+\infty$ due to the z- or t-distribution's symmetry.)

Step 3: We Formulate a Test-Decision

If the p-value is smaller than a pre-defined alpha level of $\alpha = 5\%$ or $\alpha = 1\%$ then we identify the result as significant.

As an alternative (which is easier especially if you want to use the t-distribution), you define regions of null hypothesis rejection and null hypothesis non-rejection as introduced in Chapter 6. This simply means that you first look up critical standardized test values (either a critical z-value in Table 1 in case of large samples or a critical t-value in Table 2 in case of small samples) that exactly cut $\alpha = 5\%$ or $\alpha = 1\%$ from the appropriate distribution in case of a one-sided hypothesis or $\alpha/2 = 2.5\%$ or $\alpha/2 = 0.5\%$ in case of a two-sided test (see Chapter 6). Step 2 and step 3 above would be modified as follows:

Step 2a: We Look Up the Critical Value that Corresponds to the α-Level

Dependent on the type of hypothesis (one-sided vs. two-sided; here: one-sided) and the chosen α-level (here: 0.05 or 5%) we look up a critical value using the correct standardized distribution table (here: either the standard normal distribution or the t-distribution).

Step 3a: We Formulate a Test-Decision

If the absolute value (i.e., the value without the sign) of the empirical standardized test value is greater than the critical test value, then we identify the result as significant. We know this because in this case the empirical standardized test value cuts less than $\alpha\%$ from one side or less than $\alpha/2\%$ from both sides of the standardized sampling distribution and the empirical results falls in the area of null hypothesis rejection.

Figure 7.2 visualizes the area of null hypothesis rejection and non-rejection for our hypotheses.

Figure 7.2. Independent samples t-test using rejection and non-rejection regions

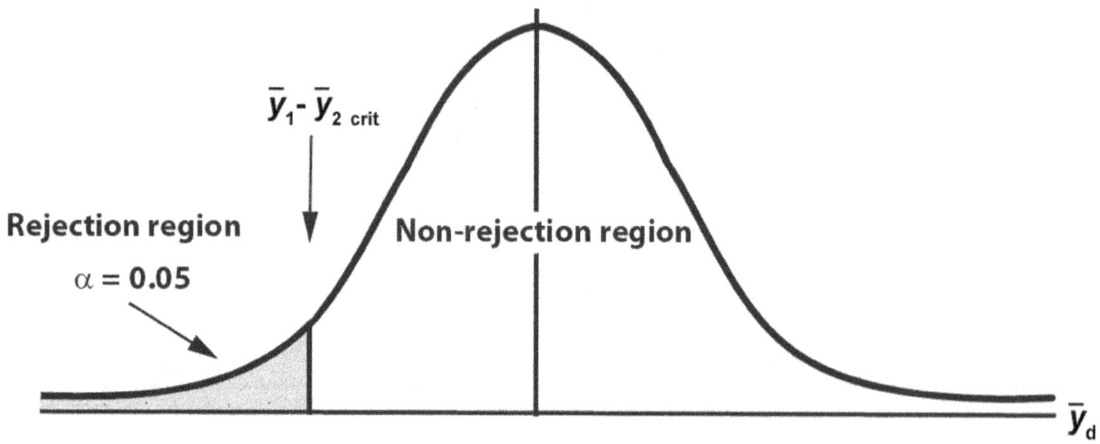

Independent Samples t-Test Example: Discrepancies in Photos on Dating Websites

A researcher (c.f., Hancock & Toma, 2009) is interested in whether there are more discrepancies for women than for men between dating web site profile pictures and photos taken in a lab. The researcher randomly selects fifteen men and fifteen women from a dating web site (the grouping variable is gender with men and women defining the two groups). Independent, trained judges rate discrepancies between a photo from the dating web site and a photo taken in the lab. Ratings are done by means of a rating scale that provides interval-level measurements on various dimensions (weight, age, skin, hair style, hair length, teeth, cropping, retouching, and professional photographs) that are collapsed to a single value. Based on previous research demonstrating that men value physical attractiveness in their mates more than women, the authors hypothesize that women's photos will contain more discrepancies (i.e., are characterized by a higher discrepancy rating) than men's photos. Putting this into words and mathematical operators:

- H_1: Women daters will have more discrepancies in their profile photographs than male daters.
- H_0: Women daters will have equal or even less discrepancies in their profile photographs than male daters.

OR

- H_1: $\mu_{\text{women daters}} - \mu_{\text{male daters}} > 0$
- H_0: $\mu_{\text{women daters}} - \mu_{\text{male daters}} \leq 0$

Table 7.1 contains the data for the problem.

Table 7.2 provides the computation for the independent samples *t*-test. We are applying the three steps explained above in the version of critical test values.

The results in Table 7.2 can be summarized as follows: An independent samples *t*-test was conducted to determine whether female and male online daters differed significantly on their profile picture discrepancies. The results indicated that profile photos of female daters had on average 3.46 discrepancies ($\hat{\sigma}_1^2 = 2.695$), while photos of male daters had 1.4 discrepancies ($\hat{\sigma}_2^2 = 0.4$). The study found a statistically significant difference between male and female daters on their profile photo discrepancies, $t_{\text{emp}}(\text{df} = 28) = 4.547, p \leq 0.05$.

Dependent (Paired) Samples *t*-Test

So far we have assumed and discussed independent samples. The questions now are what changes if we work with dependent samples in a research project, and what constitutes dependent samples (also called paired samples)?

Independent samples means that the two groups that we would like to compare with a *t*-test are sampled such that the selection of study elements (usually research participants) in one group does

Table 7.1. Ratings of profile photo discrepancies for women and men online daters

Rating Women Daters	$(y_{m1} - \bar{y}_1)^2$	Rating Male Daters	$(y_{m2} - \bar{y}_2)^2$
2	2.1316	1	0.16
6	6.4516	2	0.36
2	2.1316	1	0.16
6	6.4516	2	0.36
5	2.3716	2	0.36
6	6.4516	1	0.16
3	0.2116	2	0.36
2	2.1316	2	0.36
2	2.1316	1	0.16
3	0.2116	1	0.16
2	2.1316	2	0.36
3	0.2116	1	0.16
3	0.2116	2	0.36
2	2.1316	0	1.96
5	2.3716	1	0.16
$\bar{y}_1 = 3.46$	$\sum_{m1=1}^{15} (y_{m1} - \bar{y}_1)^2 = 37.73$	$\bar{y}_2 = 1.4$	$\sum_{m2=1}^{20} (y_{m2} - \bar{y}_2)^2 = 5.6$
	$\hat{\sigma}^2_{y1} = \frac{\sum_{m1=1}^{15} (y_{m1} - \bar{y}_1)^2}{n_1 - 1} = \frac{37.73}{14} = 2.695$		$\hat{\sigma}^2_{y2} = \frac{\sum_{m2=1}^{15} (y_{m2} - \bar{y}_2)^2}{n_2 - 1} = \frac{5.6}{14} = 0.4$

not influence the selection of study elements in the other group in any way. Consider our example and data in Table 7.1. Which individual female online dater has been selected for the sample had no implication for the selection of male online daters. Both female and male online daters were selected independently. Now, in dependent samples the selection of study elements in one group indeed influences the selection of study elements in another group. Consider the following examples:

Among married couples, you want to compare husbands and wives in terms of TV watching per day. Or think of a repeated measurement, which is very common in practical research. For instance, you are interested in the change of attitudes in the same people before and after an advertising campaign. Both examples lead to dependent samples. The main statistical difference in these examples is that the variance of the outcome variable (TV watching, attitudes) in one group will affect the variance in the other group. For example, attitudes among smokers after an

Table 7.2. Independent samples *t*-test—Example photo discrepancy among online daters

Step 1: Standardize empirical result:

Compute estimated standard error:

$$\hat{\sigma}_{\bar{y}_1-\bar{y}_2} = \sqrt{\frac{\sum_{m=1}^{n_1}(y_{m1}-\bar{y}_1)^2 + \sum_{m=1}^{n_2}(y_{m2}-\bar{y}_2)^2}{(n_1-1)+(n_2-1)} \cdot \left(\frac{1}{n_1}+\frac{1}{n_2}\right)}$$

$$\hat{\sigma}_{\bar{y}_1-\bar{y}_2} = \sqrt{\frac{37.73+5.6}{(15-1)+(15-1)} \cdot \left(\frac{1}{15}+\frac{1}{15}\right)} = \sqrt{.2053} = .453$$

Standardize mean difference:

$$t_{\text{emp}} = \frac{(\bar{y}_1-\bar{y}_2)-(\mu_1-\mu_2)}{\hat{\sigma}_{\bar{y}_1-\bar{y}_2}} = \frac{(\bar{y}_1-\bar{y}_2)}{\hat{\sigma}_{\bar{y}_1-\bar{y}_2}} = \frac{3.46-1.4}{0.45} = 4.547$$

Step 2a: Determine the critical test value

With a one-sided *t*-test (see the hypotheses), $\alpha = 5\%$ (standard value for alpha), and small samples ($n1 + n2 = 30 < 50$), we find the critical value that cuts exactly 5% from the right tail of a *t*-distribution with $n1 + n2 - 2 = 28$ degrees of freedom in Table 2 of Appendix A. The critical *t*-value is 1.701.

Step 3a: Test decision

As the empirical test value is greater than the critical test value; i.e., $t_{\text{emp}} > t_{\text{crit}}$, the result is significant at the $\alpha = 5\%$ level.

anti-smoking campaign might change but will not be completely independent from the same smokers' attitudes prior to the campaign. TV watching time of husbands will likely affect TV watching time of wives assuming that husbands and wives share similar interests and have developed similar media-use habits as a couple that lives in the same household.

If the measurements of the two samples are dependent or the variances in both samples/groups are not independent, we cannot just add the two variances in both groups as we did in the case of independent samples. The standard error of our mean difference has to be calculated differently. Or, more precisely, we need another statistic to cover dependent samples.

Calculation of p-Values in t-Tests for Two Dependent Samples

The procedure for calculating *p*-values is almost identical to the procedure we applied for the two independent samples *t*-test. In fact, the procedure will not change and is similar for all parametric statistical tests. We ask: how likely is the empirical result or even more extreme results (in direction to the research hypothesis) given that the null hypothesis is true? As mentioned above, in the case of two dependent samples we need to think about the fact that the measurements in one group are not independent from the measurements in the other group. The *t*-test for two dependent samples deals with this problem by comparing differences (or changes) on the level of a measurement pair (e.g., the difference of TV watching between a wife and her husband) or on an individual level (e.g., the change of attitudes toward smoking before and after an anti-smoking campaign). Rather than comparing the difference of two (group) means, as in the case of the two independent samples *t*-test, the *t*-test for two dependent samples compares the mean of (pairwise or individual) differences, symbolized with $\bar{y}_d$. That is, the test statistic contains information about the dependency of the measurements. Correspondingly, the directional hypotheses of the *t*-test for two dependent samples are:

- $\mathrm{H}_1 : \mu_\mathrm{d} > 0$
- $\mathrm{H}_0 : \mu_\mathrm{d} \leq 0$

H_1 states that the mean of pairwise differences across all pairs is greater than zero, that is, the measurements increased from the first to the second measurement. The research hypothesis could also state the opposite direction, that $\mu_\mathrm{d} < 0$. H_0 states that the mean of pairwise differences is assumed to be zero in the population, that is, the first and the second measurement are equivalent in the population. Of course, the hypotheses could also be non-directional; that is, H_1: $\mu_\mathrm{d} \neq 0$ and H_0: $\mu_\mathrm{d} = 0$. Beyond this modification, the calculation of *p*-values for two dependent samples *t*-tests requires answers to the same questions asked when we discussed the two independent samples *t*-test (see previous) and follows the same steps. To simplify, we now integrate the necessary answers with the calculation steps:

Step 1: We Standardize the Empirical Result (Here: The Mean of Pairwise Differences)

A pairwise or individual difference is given by:

$$d_m = y_{m1} - y_{m2}$$

Thus, the mean of all pairwise differences can be calculated with (*n* stands for the number of pairwise measurements):

$$\bar{y}_d = \frac{\sum_{m=1}^{n} d_m}{n_{\text{pairs}}}$$

The mean value of the statistic's sampling distribution under condition of the null hypothesis is $\mu_\mathrm{d} = 0$. The standard deviation of the statistic's sampling distribution (standard error) is given by:

$$\sigma_{\bar{y}_d} = \frac{\hat{\sigma}_d}{\sqrt{n_{\text{pairs}}}}$$

The standard deviation of pairwise differences can be estimated with sample data using:

$$\hat{\sigma}_d = \sqrt{\frac{\sum_{m=1}^{n}(d_m - \bar{y}_d)^2}{n_{\text{pairs}} - 1}}$$

Thus, the standardized empirical result for large samples is given by:

$$z_{\text{emp}} = \frac{\bar{y}_d - \mu_d}{\hat{\sigma}_{\bar{y}_d}} = \frac{\bar{y}_d}{\hat{\sigma}_{\bar{y}_d}}$$

Large samples are defined by $n_{\text{pairs}} \geq 30$. For small samples ($n_{\text{pairs}} < 30$), the standardization leads to an empirical t-value, because in this case we will use the t-distribution to find the corresponding AUC or the critical values (see next step).

Step 2: We Look Up the Corresponding Area Under the Curve (AUC)

We need to find the AUC from the empirical value in direction to the research hypothesis in the correct standardized distribution table in order to find the p-value. Means of pairwise differences are, like all other means (see the Central Limit Theorem), normally distributed for large samples, and t-distributed with ($n_{\text{pairs}} - 1$) degrees of freedom for small samples. Figure 7.3 demonstrates how to find the p-value for our hypotheses above.

Figure 7.3. Calculating the p-value for a dependent samples t-test

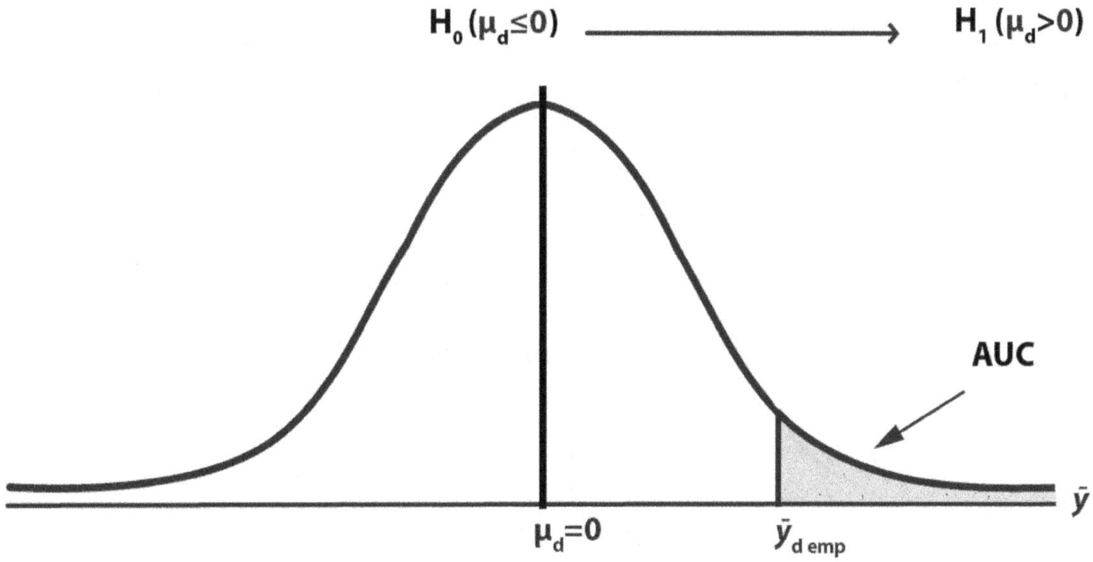

Step 3: We Formulate a Test-Decision

If the *p*-value is smaller than a pre-defined alpha level of $\alpha = 5\%$ or $\alpha = 1\%$ then we identify the result as significant.

As an alternative (which is easier especially if you want to use the *t*-distribution), you define regions of null hypothesis rejection and null hypothesis non-rejection as introduced in Chapter 6. This simply means that you first look up critical standardized test values (either a critical *z*-value in Table 1 in the case of large samples, or a critical *t*-value in Table 2 in the case of small samples) that exactly cut $\alpha = 5\%$ or $\alpha = 1\%$ from the appropriate distribution in the case of a one-sided hypothesis or $\alpha / 2 = 2.5\%$ or $\alpha / 2 = 0.5\%$ in the case of a two-sided test (see Chapter 6). Step 2 and step 3 above would be modified to:

Step 2a: We Look Up the Critical Value that Corresponds to the α-Level

Dependent on the type of hypothesis (one-sided vs. two-sided; here: one-sided) and the chosen α-level (here: 0.05 or 5%) we look up a critical value using the correct standardized distribution table (here: either the standard normal distribution or the *t*-distribution with $df = n_{\text{pairs}} - 1$).

Step 3a: We Formulate a Test-Decision

If the absolute value (i.e., the value without the sign) of the empirical standardized test value is greater than the critical test value, then we identify the result as significant. We know this because in this case the empirical standardized test value cuts less than $\alpha\%$ from one side or less than $\alpha/2\%$ from both sides of the standardized sampling distribution and the empirical results falls in the area of null hypothesis rejection.

Figure 7.4 visualizes the area of null hypothesis rejection and non-rejection for our hypotheses above.

Figure 7.4. Null hypothesis rejection regions for one-sided dependent samples *t*-test

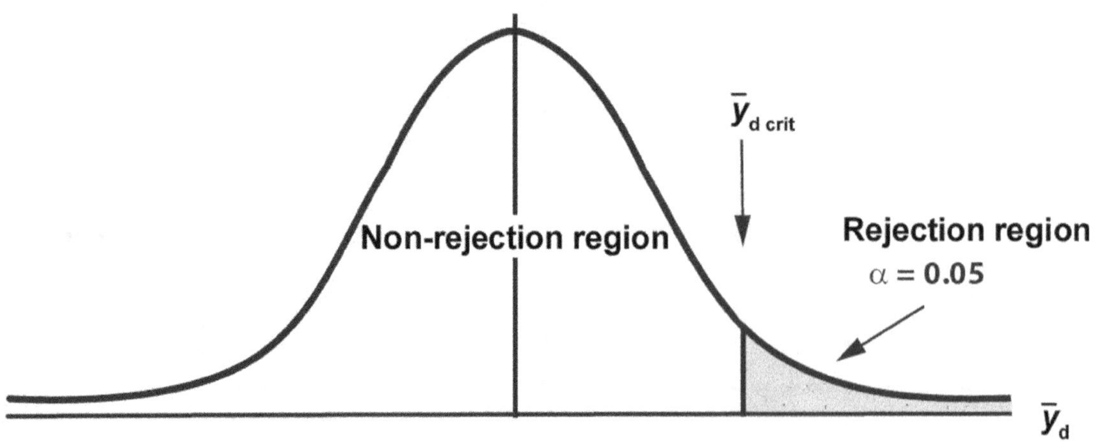

Dependent Samples t-Test Example: Attitudes Before and After an Anti-Smoking Campaign

A researcher is interested in whether an anti-smoking campaign was successful. Success is operationalized as attitude change in smokers such that attitudes toward quitting smoking are more positive after exposure to campaign ads than before. The researcher randomly selects twenty smokers to participate in her pilot study. Before and after exposure to campaign ads smokers are asked to rate how likely it is that they quit smoking (1: very unlikely; 4: very likely). Based on previous research, the researcher hypothesizes that smokers who are exposed to a well-designed anti-smoking campaign will develop more positive attitudes toward non-smoking and quitting smoking, thus:

- $H_1 : \mu_d > 0$ (d = attitude after − attitude before)
- $H_0 : \mu_d \leq 0$ (d = attitude after − attitude before)

Table 7.3 provides the data the researcher collected.

Table 7.4 provides the computation for the dependent samples *t*-test and our anti-smoking example. We are applying the three steps explained above in the version of critical test values.

The result in Table 7.4 can be summarized as follows: A dependent samples *t*-test was computed to determine whether an anti-smoking campaign was successful. Twenty participants were exposed to ads of an anti-smoking campaign and their attitudes toward quitting smoking were measured before and after the campaign. The researcher hypothesized that exposure will lead to more positive attitudes toward quitting smoking. The mean of individual (paired) attitude changes (after − before) was $\overline{y}_d = 1.3$. The result of the study was significant with t_{emp} $(df = 19) = 8.84$, $p \leq 0.05$.

Assumptions of Statistical Tests

So far, we've not discussed assumptions of statistical tests. These are important because only if certain conditions are met will the sampling distributions follow the distribution we use to calculate *p*-values. If assumptions are not met, we calculate incorrect *p*-values or biased *p*-values (and also β- or Type-II error probabilities) and ultimately decide wrongly for one of the hypotheses.

However, it is not the same if we over- or underestimate *p*-values if the assumptions of a statistical test are not met. If *p*-values are overestimated because assumptions are not met then it becomes more difficult to show that an empirical result is significant. A test with this characteristic is called a test that decides *conservatively* if assumptions are violated or simply a conservative test. If a conservative test indicates a significant result, we would have gotten an even "more significant" result if the test had met the assumptions. Conservative test decisions are usually considered less problematic in practical research (in case we'd like to decide for our research hypothesis).

However, if *p*-values are underestimated because assumptions are not met then it becomes easier to show that an empirical result is significant. A test with this characteristic is a liberal test. If a liberal test indicates a significant result it could be that we would have gotten a non-significant result if the test had met the assumptions. Liberal test decisions are usually

Table 7.3. Ratings of 20 smokers: Attitude toward quitting smoking

Participant ID	y_{after}	y_{before}	$d_i = y_{after} - y_{before}$	$(d_i - \bar{y}_d)^2$
1	3	1	2	0.49
2	1	1	0	1.69
3	4	2	2	0.49
4	4	2	2	0.49
5	2	1	1	0.09
6	4	2	1	0.49
7	2	1	0	0.09
8	3	1	2	0.49
9	2	1	1	0.09
10	2	1	1	0.09
11	3	1	2	0.49
12	2	1	1	0.09
13	3	2	1	0.09
14	3	1	2	0.49
15	2	1	1	0.09
16	1	1	0	1.69
17	2	1	1	0.09
18	3	1	2	0.49
19	2	1	1	0.09
20	2	1	0	0.09
			$\bar{y}_d = 1.3$	$\Sigma = 8.2$

considered a problem in practical research (in case we'd like to decide for our research hypothesis).

Finally, if the calculation of p-values is more or less unaffected by whether assumptions are met or not, then we call such a test a robust test. Most (but not all) standard tests provided in this textbook decide robustly, even if assumptions are violated. Hence, we will just mention, but not discuss in detail, how to deal with situations in which assumptions have been violated.

Table 7.4. Dependent samples *t*-test—Example anti-smoking attitudes

Step 1. Standardize empirical result

Compute estimated standard error

$$\hat{\sigma}_d = \sqrt{\frac{\sum_{m=1}^{n}(d_m - \bar{y}_d)^2}{n_{\text{pairs}} - 1}} \qquad \hat{\sigma}_{\bar{y}_d} = \frac{\hat{\sigma}_d}{\sqrt{n_{\text{pairs}}}}$$

$$\hat{\sigma}_d = \sqrt{\frac{8.2}{20 - 1}} = 0.657 \qquad \hat{\sigma}_{\bar{y}_d} = \frac{0.657}{\sqrt{20}} = 0.147$$

Standardize the mean of pairwise differences (small sample)

$$t_{\text{emp}} = \frac{\bar{y}_d - \mu_d}{\hat{\sigma}_{\bar{y}_d}} = \frac{\bar{y}_d}{\hat{\sigma}_{\bar{y}_d}}$$

$$t_{\text{emp}} = \frac{1.3}{0.147} = 8.84$$

Step 2a: Determine the critical test value

With a one-sided *t*-test (see the hypotheses), $\alpha = 5\%$ (standard value for alpha), and small samples ($n_{\text{pairs}} < 30$), we find the critical value that cuts exactly 5% from the right tail of a *t*-distribution with $n_{\text{pairs}} - 1 = 19$ degrees of freedom in Table 2 of Appendix A. The critical *t*-value is 1.73.

Step 3a: Test decision

As the empirical test value is clearly greater than the critical test value; i.e., $t_{\text{emp}} > t_{\text{crit}}$, the result is significant at the $\alpha = 5\%$ level.

Assumptions of the Independent Samples t-Test

The independent samples *t*-test requires that three assumptions are met:

1. We operate with independent samples or measurements. This requirement is usually met when using a sample of individual study elements/research participants that are unrelated, and not a sample of matched study elements or multiple/repeated measurements on the same study elements.
2. The two groups in the population are characterized by homogeneous (i.e., equal) variances. We assume this when calculating the standard error of the mean differences.
3. The test variable (or dependent variable) is normally distributed in the population in case we operate with small samples, which are defined as samples of $n_1 + n_2 < 50$.

What type of test is the independent samples t-test if assumptions of the test are violated? Numerous simulation studies have shown that the t-test for independent samples is an amazingly robust test. As long as your sample size in both groups is equal and large, you needn't worry. If the sample sizes n_1 and n_2 are very different and the variances are homogeneous, the test is still robust. However, if neither the sample sizes nor the variances are homogenous, then you have to expect biased or incorrect test decisions. In this case the t-test will decide liberally if the variance of the smaller sample is significantly greater than the variance in the larger sample, thus $\sigma^2_{\text{smaller sample}} > \sigma^2_{\text{larger sample}}$ (we will learn below how to test variance differences). In other words, the t-test for independent samples favors H_1. Avoid this! The t-test will decide conservatively if $\sigma^2_{\text{smaller sample}} < \sigma^2_{\text{larger sample}}$. In other words, the t-test favors H_0. This is usually considered as less problematic if H_1 is the desired hypothesis in a research project.

In cases in which the assumptions of the t-tests for independent samples are strongly violated one should consider a corresponding nonparametric test (e.g., the Mann-Whitney Test; see Sprent & Smeeton, 2007).

Assumptions of the Dependent (Paired) Samples t-Tests

The dependent samples t-test requires that two essential assumptions are met:

1. We operate with dependent samples or measurements, that is, with measurements that can be paired on an individual level.
2. We assume normally distributed differences of paired variable measurements in the population if we operate with small samples $n_{\text{pairs}} < 30$. Due to the Central Limit Theorem, the distribution of differences in the population is irrelevant if we operate with large samples.

The t-test for dependent samples is an extremely robust test. However, one should test whether high values in one group correspond to small values in the other group or vice versa (negative correlation). In this case, t-tests for dependent samples lose power and one should consider a nonparametric test (e.g., the Wilcoxon Test; see Sprent & Smeeton, 2007).

ANALYSIS OF VARIANCE (ANOVA)

Analysis of variance (ANOVA) can be considered an extension of the t-test. For example, whereas the t-test for independent samples focuses on two groups and one dependent variable, ANOVA analyzes the effect of one or many independent variables on one or many dependent variables. Often times our research questions ask us to compare more than one group, and we will show later why running several t-tests is problematic.

ANOVA was invented in 1919 by Ronald Fisher, a British statistician and geneticist. He developed and applied ANOVAs for the first time to analyze data collected from agricultural experiments. Far later, after the Second World War, ANOVA was discovered by social science researchers and is today one of the most-used statistical procedures for testing hypotheses in the social sciences.

If we analyze one independent variable or factor (measured on a nominal scale with p groups) and one dependent variable (measured on an interval or ratio scale), we speak of a one-way/factorial, one-dimensional ANOVA. If we have more than one independent variable but one dependent variable, we speak of a two-way/factorial, three-way/factorial, etc. one-dimensional ANOVA. If we have more than one factor and more than one dependent variable, we speak of a two-way/factorial, three-way/factorial, etc., two-dimensional, three-dimensional, etc. ANOVA. ANOVAs with one or many factors and one dependent variable are called univariate ANOVAs. ANOVAs with one or many factors and more than one dependent variable are called multivariate ANOVAs or MANOVAs. An ANOVA can be based on independent samples or on dependent samples. ANOVAs based on independent samples are also called independent-groups ANOVAs. ANOVAs that are based on dependent samples are usually called repeated-measures ANOVAs. Because ANOVAs allow the analysis of more than one independent variable or factor, one of those factors can be an independent-groups factor and another can be a repeated-measures factor. ANOVAs with different types of factors are called "mixed-measures ANOVAs" or "mixed-model ANOVAs."

In this introductory textbook, we concentrate on the basic *understanding* of the one-way, one-dimensional, independent-groups ANOVA (in Appendix D you will also find an applied example of a two-way ANOVA). Because all ANOVAs, regardless of their type, follow the same logic, a basic understanding of the ANOVA principle prepares you well for studying other types of ANOVAs. For an advanced, but still applied, introduction to ANOVA that includes all different types of ANOVAs we recommend Turner and Thayer (2001) or Rutherford (2001).

Example: Teacher Confirmation and Student Satisfaction

This section will take you through a one-way, one-dimensional ANOVA. We will do this by way of example: A researcher (c.f., Goodboy & Myers, 2008) is interested in the effect of teacher confirmation (not confirming, somewhat confirming, and confirming) on student satisfaction. The researcher develops a classroom experiment, where she manipulates teacher confirmation in small teaching sections led by a teaching assistant, and then asks students to rate their satisfaction. The factor (independent variable) is teacher confirmation with $p = 3$ groups, and the dependent variable is student satisfaction (the higher the rating, the more satisfied were the students). Table 7.5 has the data for the three groups of the independent variable.

The not-confirming style produces the worst level of student satisfaction and the confirming strategy the best result. Based on these results, would we conclude that teacher confirmation is important for student satisfaction? Perhaps, but we have not tested yet whether the mean differences across the three groups occurred at random or whether the mean differences can indeed be explained by the different student treatment. The t-test for independent samples will not help us here because we need to compare the means of three (and not two) groups. We need an ANOVA that can test hypotheses about mean differences in more than two groups.

Hypotheses

What are the null and research hypotheses in our ANOVA example? The null hypothesis in our ANOVA states that in the population, students in the three conditions of teacher confirmation do

Table 7.5. Example: Effects of teacher confirmation on student satisfaction

Not Confirming	Somewhat Confirming	Confirming
13	15	19
15	12	17
10	14	17
11	19	21
14	16	16
12	15	19
13	13	20
15	18	20
8	15	17
5	16	24
$\overline{A}_1 = 11.6$	$\overline{A}_2 = 15.3$	$\overline{A}_3 = 19$

not differ in their satisfaction scores. In other words, the group means are all identical in the population. That is, there are no group differences in the population. Stated in mathematical operators and symbols:

$$H_0: \mu_1 = \mu_2 = \mu_3 \, (= \ldots\ldots = \mu_p)$$

The research hypothesis states that at least two of the p (here: 3) group mean parameters are different in the population. Stated in mathematical operators and symbols:

$$H_1: \mu_i \neq \mu_i{}'$$

Be careful to note that the research hypothesis does not state that all group means are different, only that at least two groups are different.

The following sub-sections will demonstrate, step-by-step, the development of reasonable test-statistics in an ANOVA as well as the rationale of ANOVAs' statistical inference. In essence, we are answering the same four questions that we introduced when we discussed the background of the independent samples *t*-test.

Group Means and Grand Mean

What are reasonable test-statistics for an ANOVA? We begin by providing necessary terminology to understand ANOVA. Paying attention to the terminology used in ANOVAs is very helpful for understanding more advanced types of ANOVAs, as they all build on the same principle of variance decomposition and use a similar rationale for statistical inference. In our example in Table 7.5 we have a factor "*A*," which has "*p*" groups (3 in our example); the index of factor *A* with p groups is "*i*." y_{mi} indicates the value of the dependent variable (student satisfaction) for a student "*m*" in group "*i*." What does y_{22} mean for our example? y_{22} stands for the satisfaction score of the second student that was in the second group (somewhat confirming), or $y_{22} = 12$.

The first step in an ANOVA is to compute the group means $\overline{A}_i$ and the overall mean, which is also called the grand mean $\overline{G}$. As we will see later, group means and the grand mean are the building blocks of ANOVA's test-statistics. Table 7.6 provides a worked example to compute the mean for each group.

Table 7.6. ANOVA: Group mean and grand mean

Sum of all measurements under factor group "*i*"

$$A_i = \sum_{m=1}^{n} y_{mi} = \sum_{m} y_{mi}$$

$$A_1 = 116; A_2 = 153; A_3 = 190$$

Mean of all measurements under factor group "*i*"

$$\overline{A}_i = \frac{A_i}{n}$$

$$\overline{A}_1 = \frac{116}{10} = 11.6; \overline{A}_2 = \frac{153}{10} = 15.3; \overline{A}_3 = \frac{190}{10} = 19$$

Sum of all measurements (total sum)

$$G = \sum_{m=1}^{n} \sum_{i=1}^{p} y_{mi} = 116 + 153 + 190 = 459$$

Mean of all measurements (grand mean)

$$\overline{G} = \frac{G}{p \cdot n} = \frac{\sum_{m} \sum_{i} y_{mi}}{p \cdot n} = \frac{459}{30} = 15.3$$

Sum of Squares and Variance Decomposition

We mentioned that group means are the building blocks of ANOVA's test-statistics, an idea which we will now develop. The creation of reasonable test-statistics that capture the information we are interested in testing is one of the most important parts of understanding ANOVA. We recommend paying especially close attention to this section.

As we can see in our example in Table 7.5, there is variability among students' satisfaction scores (test scores). In an initial step we quantify this variability by means of a statistic that we already know: the variance. After this initial step, we simply try to answer the following general question: To what extent can we explain the variability of all scores by the treatment groups (that is, the three teacher confirmation conditions) and to what extent by random (or error effects)?

Total Sum of Squares/Total Variance

We already know how to calculate a simple variance as an estimator for a population's variance:

$$\hat{\sigma}^2 = \frac{\sum_i (y_i - \bar{y})^2}{n - 1}$$

In the terminology of ANOVAs (see Table 7.7), this formula is equivalent to:

$$\hat{\sigma}^2_{\text{tot}} = \frac{\sum_i \sum_m (y_{mi} - \overline{G})^2}{n \cdot p - 1} = \frac{\text{Total Sum of Squares (TSS)}}{\text{Total Degrees of Freedom } (df_{\text{tot}})}$$

The numerator of this formula contains the total sum of squares ("sum of squares" is the short version of "sum of squared deviations of values from their mean"). There are two sum signs because we have n ($m = 1 \ldots n$) students in each of the p ($i = 1 \ldots p$) groups. Correspondingly, the sample size is n (per group) times p groups (instead of n). As we explained earlier, we obtain an unbiased estimation of a population variance if we subtract 1 from the sample size; i.e., we divide by the sum of squares' degrees of freedom. As we can see, a variance is nothing else than the sum of squares divided by their degrees of freedom. For this reason, variances are also called "mean squares." Table 7.7 shows the computation of the total sum of squares (TSS). The group means and grand mean are given in Table 7.6.

We can now compute the total sum of squares and the total variance or mean square total. Table 7.8 shows the computation of the total sum of squares, degrees of freedom, and mean square.

Next, we will explain how this variability can be explained by our treatment groups, the three teacher confirmation conditions.

Between Sum of Squares/Treatment Variance

The basic principle of an ANOVA is based on the following essential question: What would the data matrix look like if only factor A (the three teaching conditions) was responsible for the variability of the student satisfaction scores? If this was the case, test scores within a group of factor A should not vary. Thus, we simply assume that scores within the groups do not vary and replace all scores within a group with a constant. But which constant would we use? The most reasonable constant (and the best estimation for the effect of teaching confirmation) would be the means of each group. Therefore, we replace the test scores y_{mi} with the group means $\overline{A}_i$, and again calculate the sum of squares of our test scores in this modified data table (not shown here). Because we compare groups, this sum of squares is called between (participants) sum of squares or treatment variance. As we replaced individual scores y_{mi} with the group means, the formulas for the treatment variance are slightly different (see Table 7.9).

Table 7.7. Computing the total sum of squares

$\overline{G} = 15.3$					
Not Confirming	$(y_{1i} - \overline{G})^2$	**Somewhat**	$(y_{2i} - \overline{G})^2$	**Confirming**	$(y_{3i} - \overline{G})^2$
13	5.29	15	0.09	19	13.69
15	0.09	12	10.89	17	2.89
10	28.09	14	1.69	17	2.89
11	18.49	19	13.69	21	32.49
14	1.69	16	0.49	16	0.49
12	10.89	15	0.09	19	13.69
13	5.29	13	5.29	20	22.09
15	0.09	18	7.29	20	22.09
8	53.29	15	0.09	17	2.89
5	106.09	16	0.49	24	75.69
$\overline{A}_1 = 11.6$	$\text{TSS}_1 = 229.3$	$\overline{A}_{21} = 15.3$	$\text{TSS}_2 = 40.1$	$\overline{A}_3 = 19$	$\text{TSS}_3 = 188.9$

Table 7.8. Total sum of squares, degrees of freedom, and variance

Within a group we get:

$$TSS_i = \sum_m (y_{mi} - \overline{G})^2$$

For all p groups we get:

$$TSS = \sum_i \sum_m (y_{mi} - \overline{G})^2 = 229.3 + 40.1 + 188.9 = 458.3$$

$$df_{\text{tot}} = n \cdot p - 1 = 10 \cdot 3 - 1 = 29$$

$$\hat{\sigma}^2_{\text{tot}} = \frac{\sum_i \sum_m (y_{mi} - \overline{G})^2}{n \cdot p - 1} = \frac{458.3}{29} = 15.08$$

Why does the between sum of squares have $p-1$ degrees of freedom? The explanation is simple. Within a treatment group all scores have been replaced by the group mean. Thus, random has no opportunity to vary scores within a group. Between the groups, random can vary $p - 1$ group means, but not p, because the p group means have to add up to the grand mean. Hence, the between sum of squares has $p - 1$ degrees of freedom.

Table 7.9. Total sum of squares, degrees of freedom, and mean square

Within a group we get:

$$BSS_i = \sum_{m=1}^{n} (\overline{A}_i - \overline{G})^2 = n \cdot (\overline{A}_i - \overline{G})^2 \text{ because } (\overline{A}_i - \overline{G})^2 \text{ is a constant within a group}$$

For all p groups we get:

$$BSS = \sum_i n \cdot (\overline{A}_i - \overline{G})^2 = n \cdot \sum_i (\overline{A}_i - \overline{G})^2$$

$$= 10 \cdot ((11.6 - 15.3)^2 + (15.3 - 15.3)^2 + (19 - 15.3)^2) = 273.8$$

$$df_{\text{treat}} = p - 1 = 3 - 1 = 2$$

$$\hat{\sigma}^2_{\text{treat}} = \frac{BSS}{p-1} = \frac{n \cdot \sum_i (\overline{A}_i - \overline{G})^2}{p-1} = \frac{273.8}{3-1} = 136.9$$

As usual, we obtain the treatment variance (or mean square treatment) by dividing the within sum of squares by the degrees of freedom. With the treatment variance we captured variability in the data that can be explained by the treatment groups. Now let's see what kind of variability remains in the data after we accounted for the treatment effects.

Within Sum of Squares/Error Variance

As we can see in our example (Table 7.5), the scores within a group are not all equal. "Equal values within a group" was just a theoretical assumption that is met IF only the treatment generates variance in our data. In reality, this assumption is never met; values within a group are usually all different. Why? Because there may be other independent variables that have an impact on satisfaction test scores—variables that we did not cover with our independent variable/ factor teacher confirmation. This could be: teacher competence, student ability, class topics, student personality differences, etc. Moreover, satisfaction scores may just randomly vary due to low reliability of the satisfaction scale used in the study. In ANOVA, any variability or variance that cannot be explained by the treatment groups or factor(s) included in the ANOVA is considered error and accounted for via the calculation of within sum of squares or error variance.

Again we ask the essential question: What would the data matrix look like if no treatment group effect and only error was responsible for the variability of the student satisfaction scores? We would see variability within the groups even after we have subtracted the treatment group effect (i.e., the treatment group means), from the data. Thus, we first eliminate the treatment group effect from the data by subtracting the group means. Table 7.10 illustrates this.

Table 7.11 illustrates the computation of within group sum of squares, degrees of freedom, and the error variance or mean square error.

Table 7.10. Computing within group sum of squares

Not Confirming	$(y_{m1} - \bar{A}_1)^2$	Somewhat	$(y_{m2} - \bar{A}_2)^2$	Confirming	$(y_{m3} - \bar{A}_3)^2$
13	1.96	15	0.09	19	0
15	11.56	12	10.89	17	4
10	2.56	14	1.69	17	4
11	0.36	19	13.69	21	4
14	5.76	16	0.49	16	9
12	0.16	15	0.09	19	0
13	1.96	13	5.29	20	1
15	11.56	18	7.29	20	1
8	12.96	15	0.09	17	4
5	43.56	16	0.49	24	25
$\bar{A}_1 = 11.6$	$WSS_1 = 92.4$	$\bar{A}_1 = 15.3$	$WSS_2 = 40.1$	$\bar{A}_1 = 19$	$WSS_3 = 52$

Table 7.11. Within sum of squares, degrees of freedom, and mean square error

Within a group we get:

$$WSS_i = \sum_m (y_{mi} - \bar{A}_i)^2$$

For all p groups we get:

$$WSS = \sum_i \sum_m (y_{mi} - \bar{A}_i)^2 = 92.4 + 40.1 + 52 = 184.5$$

$$df_{\text{error}} = \sum_i df_{\text{error}(i)} = \sum_{i=1}^{p} (n-1) = p \cdot (n-1)$$

$$\hat{\sigma}^2_{\text{error}} = \frac{WSS}{df_{\text{error}}} = \frac{\sum_i \sum_m (x_{mi} - \bar{A}_i)^2}{p \cdot (n-1)} = \frac{92.4 + 40.1 + 52}{3 \cdot (10-1)} = \frac{184.5}{27} = 6.83$$

Why does the within sum of squares have $p \cdot (n - 1)$ degrees of freedom? We use the same rationale as with the between sum of squares. All scores within a treatment group have to add up to the group mean. Thus, within a group, random has $n - 1$ opportunities to vary scores. Overall, we have p treatment groups. Hence, the error degrees of freedom is $p \cdot (n - 1)$.

In order to calculate one error variance (to sum the group specific error variances) we have to assume that all groups are equally affected by error effects. In practice, there are various tests available to test this assumption of equal error variances (homoscedasticity). We will get back to this point later. So far, we have determined:

- TSS = 458.3, BSS = 273.8, and WSS = 184.5
- $df_{\text{tot}} = 29$, $df_{\text{treat}} = 2$, and $df_{\text{error}} = 27$
- $\hat{\sigma}^2_{\text{tot}} = 15.08$, $\hat{\sigma}^2_{\text{treat}} = 136.9$, and $\hat{\sigma}^2_{\text{error}} = 6.83$

Notice that sum of squares and the degrees of freedom are additive, that is: TSS = BSS + WSS and $df_{\text{tot}} = df_{\text{treat}} + df_{\text{error}}$. Variances, however, are not additive. Also note that WSS could have been calculated by TSS − BSS.

Significance Test

So far, we have only quantified the extent of variability that can be explained by our treatment and error and by doing so we have developed empirical statistics that capture information we are interested in. We found that clearly more variance of students' satisfaction scores can be explained by the treatment groups than by other factors, including error. But is this result significant? Let's now consider the inferential side of ANOVA.

As with any other parametric test, the ANOVA test provides the probability of the observed treatment group mean differences (or even more extreme mean differences) under the assumption that H_0 is true in the population (that is, the teacher confirmation conditions do not differ). Recall that our hypotheses were:

$$H_0: \mu_1 = \mu_2 = \mu_3 \; (= \ldots\ldots = \mu_p)$$

$$H_1: \mu_i \neq \mu_{i'}$$

If H_0 is true in the population then μ-parameter are assumed to be identical. However, the group means that estimate the μ-parameters still vary randomly around the grand mean because we operate with random samples. The variance of group means around the grand mean is given by:

$$\hat{\sigma}^2_{\bar{A}} = \frac{\sum_i (\bar{A}_i - \bar{G})^2}{p - 1}$$

If H_0 is true then all p samples/groups belong to the same population with one population mean μ as parameter. In this case we can apply the Central Limit Theorem (see Chapter 5), that is, means of samples drawn from the same population form a sampling distribution of means with the standard error of means as its standard deviation. Thus:

$$\hat{\sigma}_{\bar{A}}^2 = \frac{\sum_i (\bar{A}_i - \bar{G})^2}{p - 1} \quad \Leftrightarrow \quad \sigma_{\bar{y}}^2 = \frac{\hat{\sigma}^2}{n}$$

$$\frac{\sum_i (\bar{A}_i - \bar{G})^2}{p - 1} = \frac{\hat{\sigma}^2}{n}$$

If we multiply with n on both sides we yield:

$$\frac{n \cdot \sum_i (\bar{A}_i - \bar{G})^2}{p - 1} = \hat{\sigma}_{\text{treat}}^2 = \hat{\sigma}^2$$

This means that if the null hypothesis is true in the population, we would expect the treatment variance to equal the variance of the dependent variable under observation (in our example student satisfaction scores). Now, if we assume that the group-specific error variances are homogeneous in the population (this is an assumption of ANOVA procedures), then the error variance serves as an unbiased estimation for the dependent variable's variance in the population (without treatment effects; i.e., under H_0). In other words: If H_0 is true in the population, then we would expect in an ANOVA:

$$\hat{\sigma}_{\text{treat}}^2 = \hat{\sigma}_{\text{error}}^2$$

Therefore, our initial hypotheses that claimed all μ-parameters to be identical can be translated to the following two hypotheses:

$$H_0 : \sigma_{\text{treat}}^2 = \sigma_{\text{error}}^2$$

$$H_1 : \sigma_{\text{treat}}^2 \neq \sigma_{\text{error}}^2 \quad \text{or} \quad \sigma_{\text{treat}}^2 > \sigma_{\text{error}}^2$$

Therefore, we call the procedure analysis of variance although we test mean differences. The null hypothesis of two variances are equivalent can be tested with the F-test. Variance ratios follow an F-distributed sampling distribution with separate degrees of freedom for the numerator and denominator of the ratio. We compute the empirical test statistic as an "F ratio" of our two sample variances:

$$F_{\text{emp}} = \frac{\hat{\sigma}_{\text{treat}}^2}{\hat{\sigma}_{\text{error}}^2} = \frac{136.9}{6.8} = 20.04$$

We compare this empirical F-Value with a critical F-value that has p-1 degrees of freedom for the numerator (this is df_{treat}), $p(n - 1)$ degrees of freedom for the denominator (this is df_{error}), and cuts $\alpha = 1\%$ or $\alpha = 5\%$ from the F-distribution. We find critical values in Appendix A, Table 3. For our example, this value is for $\alpha = 1\%$: $F_{crit}(df_{numerator} = 2, df_{denominator} = 27, \alpha = 1\%) = 5.49$. Since our empirical F value is greater than the critical F value we identify this result as significant at the $\alpha = 1\%$ level. That is, at least two of the three means are significantly different from each other. Stated differently, a treatment variance that is about twenty times greater than the error variance is very bad evidence for the null hypothesis and therefore we feel confident to reject the null hypothesis.

Explained Variance: η^2 (Eta Square)

Subsequent to a significant statistical test result, communication researchers and professionals are usually interested in quantifying the portion of the total variance that can be explained by the treatment. The more variance the treatment explains (i.e., the higher the treatment's effect size), the more important is the treatment in interpretations of the result and in decisions that are based on the test results. The most-used measure for the quantification of "explained variance" is eta square, which is defined as follows:

$$\eta^2 = \frac{BSS}{TSS} \cdot 100\% = \frac{273.8}{458.3} \cdot 100\% = 59.74\%$$

This means in our example that 59.74% of student satisfaction scores' total variability can be explained by the three teacher confirmation conditions. 100% − 59.74% = 40.26% of satisfaction scores' variability is unaccounted for and must be explained by other factors not considered in the ANOVA or by random error.

Comparison of Means (Mean Contrasts)

If we obtain a significant test result in an ANOVA, we know at least two of the p population means are different from each other. However, frequently we are interested in which of the means are significantly different. Sometimes we are even interested in whether the mean of the first two (or more) means differ from the mean in the last (e.g., if the last group was a control group) or if other combinations of means differ from each other. All of these questions can be answered by so-called multiple mean comparison procedures (often simply called mean contrasts).

At this point, you might ask: "Why can't we simply analyze all possible pairwise comparisons of group means by way of simple t-tests? Why would we need an ANOVA with more complicated calculations?" With three groups, we can create three pairwise combinations of two groups, that is, we would conduct three t-tests. The problem with this solution is that the probability that two means are different at random (Type-I error probability), which is the overall hypothesis of an ANOVA, is inflated the more sub-tests we conduct in order to test the overall hypothesis. Why is this the case and how do we deal with this problem in the context of mean contrasts?

α-Error Inflation and α-Error Correction

If we reject H_0 in a one-way ANOVA the maximum probability that we wrongly reject the H_0 is with $\alpha = 0.05$ exactly 5% and the probability that we correctly do not reject H_0 is $(1 - \alpha) = 95\%$.

If we try to test the overall hypothesis of an ANOVA with three groups (as in the case of our teacher confirmation example) with sub-tests of three pairwise mean contrasts and test each contrast with $\alpha = 5\%$, the probability that we correctly do not reject the null hypothesis would be $0.95 \cdot 0.95 \cdot 0.95 = 0.95^3$ or $(1 - \alpha)^3 \approx 0.86$. Correspondingly, the probability that we wrongly reject the null hypothesis in at least one of the three tests would be $\alpha' = 1 - (1 - \alpha)^3 = 1 - 0.86 = 0.14$ or 14%. Thus, our real maximum tolerable Type-I error (alpha level) became inflated from 5% to 14%. In general: If we try to test the overall hypothesis of an ANOVA by means of $m = p \cdot (p - 1)/2$ pairwise mean contrasts our significance level α becomes inflated to $\alpha' = 1 - (1 - \alpha)^{\mathrm{m}}$ (m = number of tests). This is not acceptable and we need to correct for this inflation if we are interested in mean contrasts.

A simple correction strategy is to modify the α-level for each single test so that the inflated α-level does not exceed the desired α level for the overall hypothesis. This can be done by using a modified α-level of $\alpha^* = 1 - (1 - \alpha)^{1/\mathrm{m}}$. For example, for our three pairwise mean contrasts we calculate $\alpha^* = 1 - (1 - 0.05)^{1/3} = 0.017$. If we use this modified α level for our three tests the inflated α level will not be higher than 5%. Test: $\alpha' = 1 - (1 - 0.017)^3 = 0.05 = 5\%$. This correction is known as the Sidak Correction or Sidak Equation.

A simplified way to calculate the modified α level is the so-called Bonferroni Correction. The Bonferroni Correction simply approximates the Sidak Correction for calculating modified α levels. We simply divide the α level for the global hypothesis by the number of tests: $\alpha^* = \alpha / m$. In our example: $0.05/3 = 0.0166$ which is ≈ 0.017 and provides a slightly more conservative α level. If you want to learn more about this important topic (including the controversy around this topic) we recommend Weber (2007).

Calculation of Contrasts

With the problem of alpha error inflation in mind, the calculation of contrasts follows a simple procedure. Compute the modified alpha-error (for example with the Bonferroni Correction) and reject mean contrast null hypotheses (tested with an independent samples *t*-test, see previous) that are significant on the modified alpha level. Table 7.12 provides calculations of contrasts with Bonferroni correction.

Bonferroni and similar α level correction methods are called explicit α level correction methods, because you explicitly modify the level of your test decision. Explicit correction procedures have major drawbacks: One problem is that they invite a subjective judgment about how many sub-tests a researcher wants to conduct with the consequence that alpha-levels depend on decisions in individual studies rather than on a global convention. Another drawback is that explicit correction procedures like Bonferroni lead to overly conservative test decisions. Other correction methods that are less conservative and are not dependent on researchers' subjective judgment use another method that is called an implicit α-level correction.

An implicit α level correction is based on a modification of the test statistic's degrees of freedom rather than on a modification of the α level. In other words, we don't have to modify

Table 7.12. Calculation of contrasts with Bonferroni correction

From our earlier calculations, we know (see Table 7.10):

$$\overline{A}_1 = 11.6; \overline{A}_2 = 15.3; \overline{A}_3 = 19$$

$$\sum_m (y_{m1} - \overline{A}_1)^2 = 92.4; \sum_m (y_{m2} - \overline{A}_2)^2 = 40.1; \sum_m (y_{m3} - \overline{A}_3)^2 = 52$$

Contrast 1: $\overline{A}_1 - \overline{A}_2$

Step 1: Standardize empirical result:
Compute estimated standard error of the mean difference (cf. equations on page 112):

$$\hat{\sigma}_{\overline{A}_1-\overline{A}_2} = \sqrt{\frac{92.4 + 40.1}{(10-1)+(10-1)} \cdot \left(\frac{1}{10} + \frac{1}{10}\right)} = \sqrt{1.472} = 1.213$$

Standardize mean difference:

$$t_{\text{emp}} = \frac{(\overline{A}_1 - \overline{A}_2) - (\mu_1 - \mu_2)}{\hat{\sigma}_{\overline{A}_1-\overline{A}_2}} = \frac{(\overline{A}_1 - \overline{A}_2)}{\hat{\sigma}_{\overline{A}_1-\overline{A}_2}} = \frac{11.6 - 15.3}{1.213} = -3.05$$

For contrast 2 ($\overline{A}_1 - \overline{A}_3$) and contrast 3 ($\overline{A}_2 - \overline{A}_3$) we obtain:

$$\hat{\sigma}_{\overline{A}_1-\overline{A}_2} = \sqrt{\frac{92.4 + 52}{(10-1)+(10-1)} \cdot \left(\frac{1}{10} + \frac{1}{10}\right)} = \sqrt{1.604} = 1.266$$

$$t_{\text{emp}} = \frac{11.6 - 19}{1.266} = -5.84$$

$$\hat{\sigma}_{\overline{A}_2-\overline{A}_3} = \sqrt{\frac{40.1 + 52}{(10-1)+(10-1)} \cdot \left(\frac{1}{10} + \frac{1}{10}\right)} = \sqrt{1.023} = 1.012$$

$$t_{\text{emp}} = \frac{15.3 - 19}{1.012} = -3.65$$

Step 2: Determine the critical test value
A Bonferroni correction with three mean contrasts suggests a modified α-level of $\alpha = 5\%/3 = 1.7\%$ for each test. With a one-sided t-test (see the hypotheses), $\alpha = 1.7\%$, and small samples ($n1 + n2 = 20 < 50$), we find the critical value that cuts 1.7% from the right tail of a t-distribution with $n1 - n2 - 2 = 18$ degrees of freedom. The critical t-value is 2.28.

Step 3: Test decision and interpretation
A mean contrast has to be significant at the modified $\alpha = 1.7\%$ level. As the absolute empirical test value is in each of the three contrasts greater than the critical test value; i.e., $|t_{\text{emp}}| > t_{\text{crit}}$, all three mean contrasts are significant. We can conclude that all three strategies lead to significantly different results with the confirming strategy yielding the highest overall student satisfaction.

the α level, we just use an appropriate (modified) sampling distribution to find our critical values. Popular mean contrast tests that use an implicit α-level correction procedure are the Scheffé or the Tukey Test for mean contrasts. All statistics software packages, including "R," which we use in this textbook (see Appendix D), offer a number of slightly different tests for mean contrasts. A detailed discussion of the various mean contrast tests is beyond the scope of this introductory textbook. In Appendix D (problem 11) we demonstrate how to apply the Tukey Test for mean contrasts. For more background on mean contrast tests we refer to Keppel & Wickens (2004).

As a final note, the mean contrasts we discussed here are also known as post-hoc mean contrasts. This means that researchers had no hypotheses about which of the p means should be different in the population prior to conducting the mean contrasts. In this situation, using post-hoc mean contrast tests that apply an alpha-correction is necessary (see Weber, 2007). In situations where researchers indeed formulated a specific hypothesis about a concrete mean contrast prior to collecting and analyzing data, there is some agreement that considering alpha-correction is not necessary. Those hypothesis-based mean contrasts are called a-priori mean contrasts. Essentially, the difference is that in post-hoc mean contrasts researchers apply a "trial and error" search for significant contrasts while in a-priori mean contrasts researchers have a predetermined hypothesis or a prediction about which contrast should turn out as significant. Post-hoc explanations of an empirical result need to be treated differently than a-priori predictions, because it is usually much easier to explain an event that has occurred than to predict it before it has occurred.

TESTS FOR FREQUENCY DIFFERENCES (CHI-SQUARE TESTS)

So far, we've discussed tests for mean differences. Outcome variables have been quantitative, or have been measured on an interval scale. Now we want to learn about another class of statistical tests: those that are designed to test differences in frequencies or relationships between variables based on frequencies.

Tests for differences in frequencies can be applied for categorical (nominal scaled) variables and also for ordinal variables and even interval variables treated as (or converted into) categorical variables. For example, think of the variable "time spent using video games" (measured in minutes per day). This ratio-level variable can be easily converted into an ordinal level variable and treated as a nominal level variable by categorizing study participants' time spent with video game into groups (e.g., 1–15 minutes/day, 16–30 minutes/day, more than 2 hours/day). However, keep in mind that tests for differences in frequencies of a categorical variable do not use the advantage of more information in ordinal and interval variables. Therefore, these tests are usually less powerful. If possible, always apply a test designed for the highest possible measurement level of your data.

Parametric tests for frequency differences are usually based on a chi-square distributed test statistic, that is, the sampling distribution of a standardized, frequency-based test statistic follows a chi-square distribution (see Appendix A, Table 4). Therefore, tests for frequency differences are also called chi-square tests. Every chi-square test follows a five-step procedure, which we will describe below. These five steps essentially provide answers to the four basic questions we need to ask in any parametric statistical test (see under Independent Samples *t*-Test/Background).

Chi-square tests can be classified by the number of categorical variables they include (one variable: one-dimensional chi-square tests; two variables: two-dimensional chi-square tests, etc.) and whether measurements are independent (one measurement) or dependent (two or repeated measurements). In the following we will introduce the most popular (and most frequently) used variants of chi-square tests. As you will see, chi-square tests follow a five-step procedure. We begin with the simplest version of a chi-square test—the one measurement, one dimensional chi-square test with two categories.

One Measurement, One-Dimensional Chi-Square Test

Two Categories

Recently, there has been growing concern in the United States about declining enrollment in degrees related to science, technology, engineering, and mathematics (STEM). Researchers have focused on how STEM careers have tended to be more male dominated rather than equally distributed among the sexes. In line with this, communication scholars such as Buzzanell, Berkelaar, & Kisselburgh (2011) and Myers, Jahn, Galliard, & Stoltzfus (2011) have focused on anticipatory socialization into STEM careers, for both boys and girls. After reading research by the aforementioned communication researchers, your college dean appoints a task force to look at enrollment in science, technology, engineering, and mathematics departments. Specifically, you have been tasked to look at enrollment in the Chemistry Department of your university. Based on your report, the dean will determine whether to conduct greater outreach to women (including whether to have science outreach programs at the local high school). You find out that there are 869 men and 576 women who are declared majors and who are in academic good standing. If we consider the students of your institution's chemistry department as a sample of chemistry students, can this gender difference be explained by random distribution? The answer to this question depends on what we define as our null hypothesis, which leads to the first step of any chi-square test:

Step 1: Calculate Expected Frequencies According to the Null Hypothesis

In our example, researchers may have different null hypotheses. We describe here two possibilities that provide reasonable null hypotheses:

> Case 1 (H_0): The gender distribution (gender frequencies) in your chemistry department is in accordance with the general gender frequencies in the US, which is close to 50:50. The assumed population would be people in the US (H_1: The gender distribution in your chemistry department deviates from the general gender distribution in the US.)

Case 2 (H_0): The gender distribution (gender frequencies) in your chemistry department is in accordance with the general gender frequency at chemistry departments in the US, which is about 6.6(men):1(woman). The assumed population would be people at chemistry departments in the US. (H_1: The gender distribution in your chemistry department deviates from other chemistry departments in the US.)

Essentially, the difference between the first and the second case is what we consider as gender bias: that more men than women are enrolled in your chemistry department compared to US population or compared to enrollment numbers in chemistry department at other US universities. Based on what we consider to be the population we calculate what gender distribution (gender frequencies) we would expect if the null hypothesis distribution were true. Let's do this in our example.

Case 1: H_0: 50:50

Sample size: $n = 869 + 576 = 1{,}445$

Expected frequencies (f_e):

50:50 means that we would expect 50% of our sample size should be male and the other 50% female, or:

$$f_e = \frac{f_{o(1)} + f_{o(2)}}{2} = \frac{869 + 576}{2} = \frac{1{,}445}{2} = 722.5$$

Case 2: H_0: 6.6:1

$$f_{e(\text{Male})} = n \cdot \frac{6.6}{6.6 + 1} = 1445 \cdot 0.87 = 1{,}257.15$$

$$f_{e(\text{Female})} = n \cdot \frac{1}{6.6 + 1} = 1445 \cdot 0.13 = 187.85$$

As we can see, the expectations for the gender distribution are quite different if we assume different populations for our null hypothesis.

Step 2: Calculate Empirical Chi-Square Test Statistic

The next step is that we want to evaluate the deviations of the observed frequencies (f_o) from the expected frequencies (f_e). The higher the deviations, the less likely is the null hypothesis, because expected frequencies were calculated in accordance with the null hypothesis. Hence, these deviations contain information that we want to evaluate and, therefore, build the basis for our test statistic.

Because the sum of deviations ($f_o - f_e$) across all categories is always zero, we calculate the squared deviation. This also means that large deviations are weighted higher. We also divide by the expected frequency. This ensures that the deviation is standardized by the expected frequency. As an example, a deviation of 10 while expected 1,000 is certainly less important than a deviation of 10 while expected 100. In sum, to evaluate the deviations, we calculate the following chi-square test statistic:

$$\chi^2_{\text{emp}} = \sum_{j=1}^{2} \frac{(f_{o(j)} - f_{e(j)})^2}{f_{e(j)}}$$

This empirical chi-square test value can only be zero if the expected frequencies are equal to the observed frequencies. Small chi-square test values would suggest non-rejection of the null hypothesis; large chi-square test values would suggest rejection of the null hypothesis. In our example, we yield:

$$\text{Case 1:} \quad \chi^2_{\text{emp}} = \frac{(869 - 722.5)^2}{722.5} + \frac{(576 - 722.5)^2}{722.5} = 59.41$$

$$\text{Case 2:} \quad \chi^2_{\text{emp}} = \frac{(869 - 1{,}257.15)^2}{1{,}257.15} + \frac{(576 - 187.85)^2}{187.85} = 921.87$$

Just looking at the chi-square values, both suggest rejecting the null hypothesis, but both chi-square values (i.e., the frequency differences behind them) could have occurred randomly. In order to answer the question whether our frequency differences could be a random result and the null hypothesis is in fact true in the population, we need a statistical test, which is step 3 of a chi-square test procedure:

Step 3: Chi-Square Test: Finding the Critical χ^2-Test Value in a χ^2-Distribution Table

If we draw a large number of samples from our H_0 population and each time calculate a chi-square value according to the procedure above we would yield a sampling distribution of chi-square values. It can be shown that this sampling distribution follows a chi-square distribution with a specific expected value and standard error. Like in other statistical tests, with this information we are able to calculate the *p*-value of our empirical chi-square test value; that is, we calculate the probability of our sample chi-square value or even more extreme values under the assumption that the null hypothesis is true.

If you take a look at the chi-square distribution in Appendix A (Table 4), you can see that, like a *t*-distribution, the chi-square distribution requires you to know the degrees of freedom of the chi-square test statistic. To determine how many degrees of freedom our chi-square test value has, we have to count how many summands can randomly vary when calculating the chi-square value. In our example, we only have two summands, because we only have two categories (men and women). Only one of them can randomly vary because the total sum of students is fixed, that is, is given by the study's sample size (n). Thus, the chi-square test value in our example has only one degree of freedom. With the usual alpha-level of 5% we yield a critical test value of χ^2_{crit} $(df = 1, \alpha = 5\%) = 3.84$.

Step 4: Test Conclusion and Interpretation

The decision rule for the test conclusion is the same as in all other tests, but the test interpretation is of particular importance in chi-square tests as we will demonstrate now. The decision rule is:

$$\text{If } \chi^2_{emp} > \chi^2_{crit}\,(df, \alpha) \quad \Rightarrow \quad \text{Result is significant at the } \alpha\% \text{ level}$$

As we can see, in both cases, the empirical value is greater than the critical value. Hence, both results are significant. However, be aware of the interpretations. They may differ in chi-square tests!

In case 1, fewer women study chemistry (576) than we had expected (722.5). Compared with the general gender ratio in the United States (H_0), chemistry seems to be a less attractive choice for women at this university.

In case 2, the number of women studying chemistry (576) is greater than expected (187.85). Compared with the gender ratio at chemistry departments in the US (H_0), chemistry seems to be a preferred program among women at this university.

Step 5: Discussion of the Test's Assumptions

Like other parametric tests, the chi-square test is based on a sampling distribution and long-term error rates (Type-I and Type-II errors). This means that chi-square tests, too, require that certain test assumptions are met. The one measurement, one-dimensional chi-square test for two categories requires that two essential assumptions are met:

1. Every study element of the sample has to be assigned exclusively to only one category. That is, the assignment of a study element to a category must be unambiguous.
2. If a chi-square test statistic has only one degree of freedom, the *expected* frequencies in each category must be at least greater than 10. This essentially means that the sample size has to be large enough.

Usually it is not difficult to meet these assumptions in a study. However, if one, or both assumptions are violated, it would no longer be appropriate to use the chi-square sampling distribution in Appendix A as a reference. In this case, researchers would use a non-parametric test variant of the chi-square test (see Sprent & Smeeton, 2007).

In conclusion, the chi-square test for both cases in our example indicated a significant result (with different interpretations) and all assumptions were met. That is, the researcher can rely on the chi-square tests within the limitations of the usual Type-I error rate.

Multiple Categories (Goodness of Fit Chi-Square Test)

Our previous example considered only two categories (men, women) of the categorical (nominal level) variable gender. The one measurement, one-dimensional chi-square test can be easily generalized for the analysis of a categorical variable with more than two categories. In fact, the procedure is almost identical with the case of two categories, which is why we will now present the five-step procedure in a simplified, shortened version. Again, we begin with an example:

A researcher (c.f., Chory-Assad, 2004) believes that following exposure to sitcoms on television, aggressive responses in a story completion task will not be equally distributed across four categories of aggression type (character attacks, competence attacks, expression of dislike, and physical appearance attacks). A random sample of $n = 100$ participants were exposed to a popular television sitcom and afterward completed a story completion task. Content analyses revealed the following numbers (Table 7.13):

Table 7.13. Example—Aggressive responses after exposure to sitcoms

A. Character attacks	61
B. Competence attacks	24
C. Expressions of dislike	8
D. Physical appearance attacks	6
Total	100

Step 1: Calculate Expected Frequencies According to the Null Hypothesis

The researcher formulates the following hypotheses:

H_0 : There are no differences in the number of aggressive responses after exposure to a popular sitcom on television. In other words: The distribution of aggressive responses follows a univariate distribution.

H_1 : There are differences in the number of aggressive responses after exposure to a popular sitcom on television. The distribution of aggressive responses deviates from a univariate distribution.

If the null hypothesis assumes a univariate (equal) distribution of aggressive responses then we would expect the following frequency for each category (k = the number of categories):

$$f_{e(A)} = 1/k \cdot n \text{ or } n/k = {}^{1}/_{4} \cdot 100 = 25 = f_{e(B)} = f_{e(C)} = f_{e(D)}$$

Note: Because the one measurement, one-dimensional chi-square test is primarily used to test hypotheses about the distribution of discrete variables (see Chapter 3), this variant of a chi-square test is also called the Goodness of (Distribution) Fit Chi-Square Test.

Step 2: Calculate Empirical Chi-Square Test-Statistic

Instead of $k = 2$ categories, we now need to consider $k = 4$ categories:

$$\chi^2_{\text{emp}} = \sum_{j=1}^{k=4} \frac{(f_{o(j)} - f_{e(j)})^2}{f_{e(j)}} = \frac{(61 - 25)^2}{25} + \frac{(24 - 25)^2}{25} + \frac{(8 - 25)^2}{25} + \frac{(6 - 25)^2}{25} = 77.88$$

Step 3: Chi-Square Test: Finding the Critical χ^2-Test Value in a χ^2-Distribution Table

The rationale of this step is the same as in the case of two categories. That is, we need to answer the question: How many summands can randomly vary when calculating the chi-square value.

In our example, we have $k = 4$ categories or summands. Only three of them can randomly vary because the total sum of participants is fixed, that is, is given by the study's sample size (n). Thus, the chi-square test value in our example has three degrees of freedom or in general, the one measurement, one-dimensional chi-square test has $df = k - 1$ degrees of freedom. With an alpha-level of 1% for this example we yield a critical test value (see Appendix A, Table 4) of χ^2_{crit} $(df = 3; \alpha = 1\%) = 11.34$.

Step 4: Test Conclusion and Interpretation

We compare our empirical chi-square test value with the critical chi-square test value. The common decision rule is:

$$\text{If } \chi^2_{emp} > \chi^2_{crit}\,(df, \alpha) \quad \Rightarrow \quad \text{Result is significant at the } \alpha\% \text{ level}$$

As we can see, the empirical value is clearly greater than the critical value and thus "cuts" less than $\alpha\%$ from the chi-square sampling distribution in Appendix A. The researcher would reject the null hypothesis and interpret the finding in accordance with the research hypothesis, that is, the distribution of aggressive responses after exposure to a popular television sitcom deviates from a univariate distribution; while character attacks in a story completion task occur more frequently than expected, physical attractiveness attacks and expressions of dislike occur less frequently than expected.

Step 5: Discussion of the Test's Assumptions

The one measurement, one-dimensional chi-square test for multiple categories requires that two essential assumptions are met:

1. Every study element of the sample has to be assigned exclusively to one and only one category. That is, the assignment of a study element to a category must be unambiguous.
2. If a chi-square test statistic has more than one degree of freedom, the *expected* frequencies in each category must be at least greater than 5 (i.e., the sample size has to be sufficiently large).

If one or both assumptions are violated, it would be no longer appropriate to use the chi-square sampling distribution in Appendix A as a reference and researchers are advised to use a nonparametric test variant of the chi-square test (see Sprent & Smeeton, 2007).

As both assumptions are met in our example, the researcher can rely on the chi-square test. Note that the test decision can still be wrong, but the probability of a wrong decision is smaller than 1% in our example.

Two Measurements, One Dimensional Chi-Square Test

A researcher (c.f., Benoit, McKinney, & Stephenson, 2003) is interested in whether exposure to presidential candidates during a television debate leads to a change in preference for candidates.

The researcher takes voter preferences of $n = 237$ viewers before watching a debate, and after watching a debate. That is, the researcher takes two measurements of one categorical variable (voter preference), one before and one after the debate. The results of the study are in Table 7.14. (the four cells of the table are labeled with a, b, c, and d).

Table 7.14. Changes in voter preference after watching an Obama/Clinton debate

Preference	**After the debate**	
Before the debate ⇓	Clinton	Obama
Clinton	80 (cell a)	25 (cell b)
Obama	12 (cell c)	120 (cell d)

Looking at the data table it seems that Barack Obama was the winner of the debate. 25 of the 237 viewers preferred Hillary Clinton before the debate and Barack Obama after the debate. But only 12 preferred Obama before the debate and Clinton after the debate. That is, more viewers changed from Clinton to Obama than from Obama to Clinton. But did this change occur at random? This question is answered by applying a statistical test—the McNemar chi-square test (also called the significance of change chi-square test).

Step 1. Calculate Expected Frequencies According to the Null Hypothesis

The hypotheses of the McNemar test are:

H_0: Either no viewers change their preference or the number of viewers that change from Clinton to Obama corresponds to the number of viewers that change from Obama to Clinton. In other words, no real change occurred.

$H_{1(a)}$: The debate changes voter preferences (non-directional).

The research hypothesis of the McNemar test can be (and usually is) a directional hypothesis. For example: The Obama campaign team found that prior to the debate Obama's messages resonated better with the voters. Therefore, they hypothesize:

$H_{1(b)}$: The debate changes voter preferences in favor to Barack Obama (directional).

If the null hypothesis predicts that no real change has occurred then we would expect no change for the frequencies in cell b and c (the change cells):

$$f_{e(b)} = f_{e(c)} = \frac{b + c}{2} = \frac{25 + 12}{2} = 18.5$$

Step 2. Calculate Empirical Chi-Square Test Statistic

For the calculation of the empirical chi-square test statistic we use the same chi-square formula that we introduced earlier:

$$\chi^2_{emp} = \sum_{j=1}^{2} \frac{(f_{o(j)} - f_{e(j)})^2}{f_{e(j)}} = \frac{(12 - 18.5)^2}{18.5} + \frac{(25 - 18.5)^2}{18.5} = 4.57$$

With a few simple transformations the calculation of the empirical chi-square value can be simplified (the characters b and c indicate the cell of data Table 7.14):

$$\chi^2_{emp} = \frac{\left(b - \frac{b+c}{2}\right)^2 + \left(c - \frac{b+c}{2}\right)^2}{\frac{b+c}{2}} = \frac{(b-c)^2}{b+c} = \frac{(25-12)^2}{37} = 4.57$$

Step 3. Chi-Square Test: Finding the Critical χ2-Test Value in a χ2-Distribution Table

With two summands when calculating the empirical chi-square value we have one degree of freedom. From this, we can find the critical value in the chi-square distribution table in Appendix A. The value is $\chi^2_{crit}\ (df = 1, \alpha = 5\%) = 3.84$. This critical chi-square value is valid for non-directional research hypotheses because the chi-square distribution table is created for non-directional hypotheses and two-sided tests.

In step 1 we also presented a directional research hypothesis that needs to be tested with a one-sided test. In this case, we simply double α and find the critical value with $\chi^2_{crit}(df = 1, 2 \cdot \alpha = 10\%) = 2.71$.

Step 4. Test Conclusion and Interpretation

One-sided test: If $\chi^2_{emp} > \chi^2_{crit}\ (df, \alpha) \Rightarrow$ Result is significant at the $\alpha\%$ level

Two-sided test: If $\chi^2_{emp} > \chi^2_{crit}\ (df, 2 \cdot \alpha) \Rightarrow$ Result is significant at the $\alpha\%$ level

The empirical chi-square test value is greater than the critical chi-square value for both the one-sided and two-sided test. We conclude that the debate had an effect on viewers' preferences—significantly more viewers preferred Barack Obama after the debate than Hillary Clinton.

Step 5. Discussion of the Test's Assumptions

The McNemar test requires that two essential assumptions are met:

1. All respondents from the first measurement (i.e., before the debate) are also included in the second measurement (i.e., after the debate). That is, no respondents (viewers) were lost between the first and second measurement.

2. As the McNemar test statistic has one degree of freedom, the *expected* frequencies in each of the two cells b and c must be greater than 10.

In our example both assumptions are met and we can rely on the test's decision. If one or both assumptions were violated we would need to be careful with our interpretations and consider an alternative non-parametric test (see Sprent & Smeeton, 2007).

One Measurement, Two-Dimensional Chi-Square Test

The previous sections discussed one-dimensional chi-square tests, that is, chi-square tests that consider the distribution of only one categorical (nominal level) variable. We will now present two-dimensional chi-square tests, which can analyze the frequency distributions of two categorical variables simultaneously. You will see that, as with all chi-square tests, two-dimensional chi-square tests follow the same logic and five-step procedure that we have already introduced. We begin with the one measurement, two-dimensional chi-square test for two categories, which is also called the "2 by 2" Chi-Square Test.

Two Categories

Prior research has shown that bloggers may be conscious about what they disclose due to concern for possible negative consequences. A researcher (c.f., Qian & Scott, 2007) is interested in whether the frequency of identity disclosure (e.g., use of real name, real photos) differs between different types of intended audience (e.g., online and unknown other bloggers; offline and known others such as friends and family).

Assumptions about frequency distributions across two variables simultaneously can also be expressed as a relationship between the two variables. That is, we can also say that the researcher in our example claims a relationship between the variables "intended audience" and "identity disclosure" among bloggers. The researcher draws a random sample of $n = 200$ bloggers from blogging communities (e.g., LiveJournal, Blogger) and asks them a series of questions about their intended audience. The researcher follows up with a content analysis to determine the extent to which bloggers provide identifying or anonymous information (e.g., use of real name, real photos) on their web pages. The frequencies are as appears in Table 7.15.

Table 7.15. Example: Intended audience and identity disclosure

		Identity Disclosure		
		Anonymous	**Identifiable**	**Total**
Audience	**Online**	65	35	100
	Other	37	63	100
	Total	102	98	$n = 200$

As in all other chi-square tests, the test procedure begins with the hypotheses and the calculations of expected frequencies:

Step 1: Calculate Expected Frequencies According to the Null Hypothesis

The researcher formulates the following hypotheses:

H_0: Bloggers whose audiences are online compared to bloggers whose audiences are other-based do not differ in their frequency to disclose their identity. Alternatively, the frequency distributions of "intended audience" and "identity disclosure" in the population of bloggers are unrelated, that is, the two categorical variables are independent.

H_1: Bloggers whose audiences are online compared to bloggers whose audiences are other-based differ in their frequency to disclose their identity. Alternatively, the frequency distributions of "intended audience" and "identity disclosure" in the population of bloggers are related, that is, the two categorical variables are dependent.

The question in step 1 is what frequencies would we expect if we assume the null hypothesis (independence) is true in the population of bloggers? Recall Chapter 3, particularly Table 3.7, "Rule of Multiplication for Independent Events." The multiplication rule in probability theory tells us how to calculate the probability of a combined event when the individual events are either dependent or independent. In the case of independence, the multiplication rule says that if the events *A* and *B* come from the same sample space, the probability that both *A* and *B* occur equals the probability that event *A* occurs times the probability that *B* occurs if the two events are independent.

The null hypothesis of our example stated independence of both variables. Thus, if events (categories) of the variables "intended audience" and "identity disclosure" are indeed independent in the population of bloggers, we would expect that frequencies distribute in accordance with the rule of multiplication for independent events. For example, for the probability of the combined event "audience-online" (event *A)* and "identity disclosure-anonymity" (event B) would be given by:

$$P(A \cap B) = P(A) \cdot P(B)$$

As we have learned in Chapter 3, probabilities are estimated with relative frequencies:

$$P(A \cap B) = \frac{f_e(A \cap B)}{n} \quad \Rightarrow \quad f_e(A \cap B) = P(A \cap B) \cdot n = P(A) \cdot P(B) \cdot n$$

Given the data in Table 7.15, what is the probability that a blogger's intended audience is online (*A*) and she does not reveal her identity (*B*)? For *P*(*A*) we would estimate 100/200 and for *P*(*B*) 102/200, thus for $f_e(A \cap B)$ we yield:

$$f_e(A \cap B) = P(A) \cdot P(B) \cdot n = \frac{100}{200} \cdot \frac{102}{200} \cdot 200 = \frac{100 \cdot 102}{200} = 51$$

The rule of multiplication for independent events applied to the 2 × 2 frequency table in Table 7.14 can be generalized to:

$$f_{e(i,j)} = \sum_{i=1}^{2}\sum_{j=1}^{2} \frac{\text{Sum of frequencies in row i} \cdot \text{Sum of frequencies in column j}}{n}$$

Using this formula, Table 7.16 provides the observed and expected frequencies for each cell.

Table 7.16. Observed and expected frequencies of audience and identity disclosure

		Identity Disclosure		
		Anonymous	**Identifiable**	**Total**
Audience	**Online**	Cell a: 65 ($f_e = 51$)	Cell b: 35 ($f_e = 49$)	100
	Other	Cell c: 37 ($f_e = 51$)	Cell d: 63 ($f_e = 49$)	100
	Total	102	98	$n = 200$

Step 2: Calculate Empirical Chi-Square Test-Statistic

The calculation of the empirical chi-square test value follows the same rationale and procedure like in all chi-square tests; instead of just one variable's frequency distribution, however, we now have to consider two variables' frequency distribution. This means for our chi-square formula that we have to sum the deviations of observed frequencies from the expected frequencies in step 1 across two dimensions (that is, across the rows and columns of Table 7.15). Arithmetically this means we need two sum signs in our formula:

$$\chi^2_{\text{emp}} = \sum_{i=1}^{2}\sum_{j=1}^{2} \frac{(f_{o(i,j)} - f_{e(i,j)})^2}{f_{e(i,j)}}$$

$$\chi^2_{\text{emp}} = \frac{(65-51)^2}{51} + \frac{(35-49)^2}{49} + \frac{(37-51)^2}{51} + \frac{(63-49)^2}{49} = 15.68$$

Or, alternatively, with a few mathematical transformations we can combine step 1 and step 2 and directly calculate the empirical chi-square value using the frequencies in cell a to d in Table 7.15:

$$\chi^2_{\text{emp}} = \frac{n \cdot (a \cdot d - b \cdot c)^2}{(a+b)(c+d)(a+c)(b+d)} = \frac{200 \cdot (65 \cdot 63 - 35 \cdot 37)^2}{(65+35)(37+63)(65+37)(35+63)} = 15.68$$

Step 3: Chi-Square Test: Finding the Critical χ^2-Test Value in a χ^2-Distribution Table

As in all other chi-square tests, the main question in step 3 is how many degrees of freedom does our chi-square test value have? We answer this question by considering how many summands the chi-square test value has and how many of them can randomly vary given a sample size of *n,* which is determined by the researcher. Each of the two variables has two categories, that is, for each variable we obtain one degree of freedom or combined we yield $1 \cdot 1 = 1$ degree of freedom. Alternatively, we could think about how many frequencies in Table 7.15 can randomly vary if the sample size and the marginal distribution are given? The answer is one, only one of the four frequencies can be modified, all other frequencies have to add up to the row or column sum and finally to the sample size. In general, the 2×2 Chi-Square Test has $df = (k - 1) \cdot (l - 1)$ degrees of freedom with $k = l = 2$. With an alpha-level of 1% for this example we yield a critical test value (see Appendix A, Table 4) of χ^2_{crit} $(df = 1; \alpha = 1\%) = 6.63$.

Step 4: Test Conclusion and Interpretation

We compare our empirical chi-square test value with the critical chi-square test value:

$$\text{If } \chi^2_{emp} > \chi^2_{crit}\,(df, \alpha) \quad \Rightarrow \quad \text{Result is significant at the } \alpha\% \text{ level}$$

The empirical test value is greater than the critical test value in our example. Hence, we reject the null hypothesis and interpret the research hypothesis as our test result: The study found evidence that bloggers whose audiences are online or other-based differ in their frequencies to disclose their identity. Alternatively, we could say that the researcher found evidence of a relationship between the variables "intended audience" and "identity disclosure."

As always, it is of particular importance in chi-square tests that we inspect the data in order to interpret our findings. Looking at Table 7.16 it seems that bloggers whose intended audiences are online blog anonymously more frequently than expected, whereas bloggers whose audiences are offline reveal their identity more frequently than expected.

Step 5. Discussion of the Test's Assumptions

With one degree of freedom, the one measurement, two-dimensional chi-square test for two categories requires that the following assumptions are met:

1. Every study element of the sample has to be assigned exclusively to one and only one category for each variable.
2. The *expected* frequencies in each category must be at least greater than 10.

Both assumptions are met in our example; that is, the researcher can rely on the chi-square test.

If the assumptions were violated, it would no longer be appropriate to use the chi-square sampling distribution in Appendix A as a reference for null hypothesis expectations and researchers would be advised to use a non-parametric test variant of the chi-square test such as Fisher's Exact Test for 2x2 table (see Sprent & Smeeton, 2007).

Multiple ($k \times l$) Categories

The one measurement, two-dimensional chi-square test for multiple ($k \times l$) categories follows the exact same procedure as the 2 × 2 Chi-Square Test. Therefore, we only briefly demonstrate the calculations for the $k \times l$ Chi-Square Test using our five-step procedure and the following example:

Mass communication researchers often emphasize the role of news media as a source of stereotypical images. An underlying assumption is that these stereotypes produce or reinforce prejudice and discrimination by audiences. A researcher (c.f., Gorham, 2006) is interested in how the race of a subject (white, black) in crime news story influences the abstractness of the language viewers use to describe the subject (from most concrete to most abstract: descriptive action verbs(DAVs), interpretive action verbs (IAVs), state verbs (SVs), and adjectives (Adjs)). Language abstractness is thought to reveal the extent to which observed behaviors are tied to circumstances or are inherent traits of a particular group. The researcher conducts an experiment in which he randomly assigns $n = 199$ to either view a news story about a black crime suspect or a white crime suspect. The dependent variable in this case is the choice of language the viewer chooses to describe the crime suspect. The frequencies of the research participants' choices across the two variables can be found in Table 7.17.

Table 7.17. Example: Language abstractness chosen to describe a crime suspect

		Language Descriptors				
		DAVs	**IAVs**	**SVs**	**ADJs**	**Total**
Race	**White**	6 (5.97)	63 (46.76)	10 (8.46)	20 (40.79)	99
	Black	6 (6.03)	28 (47.24)	4 (8.54)	62 (41.21)	100
	Total	12	94	17	82	***n* = 199**

Step 1: Calculate Expected Frequencies According to the Null Hypothesis

The researcher formulates the following hypotheses:

H_0: The frequency distributions of "race" and "language descriptors" in the population of crime news stories are unrelated, that is, the two categorical variables are independent.

H_0: The frequency distributions of "race" and "language descriptors" in the population of crime news stories are related; that is, the two categorical variables are dependent.

We apply the same general rule as in the 2 × 2 Chi-Square Test but this time with $k = 2$ categories for "race" and $l = 4$ categories for "language descriptors":

$$f_{e(i,j)} = \sum_{i=1}^{k=2} \sum_{j=1}^{l=4} \frac{\text{Sum of frequencies in row i} \cdot \text{Sum of frequencies in column j}}{n}$$

For example, for the first row and first column of Table 7.17 we obtain: $f_{e(1,1)} = (99 \cdot 12)/199 = 5.97$. Table 7.17 provides the observed and expected frequencies (in parentheses) for each cell.

Step 2: Calculate Empirical Chi-Square Test-Statistic

We use the same rationale and formula as for the 2 × 2 Chi-Square Test:

$$\chi^2_{\text{emp}} = \sum_{i=1}^{k=2}\sum_{j=1}^{l=4}\frac{(f_{o(i,j)} - f_{e(i,j)})^2}{f_{e(i,j)}}$$

$$\chi^2_{\text{emp}} = \frac{(6-5.97)^2}{5.97} + \frac{(63-46.74)^2}{46.74} + \ldots + \frac{(4-8.54)^2}{8.54} + \frac{(62-41.21)^2}{41.21} = 37.25$$

Step 3: Chi-Square Test: Finding the Critical χ^2-Test Value in a χ^2-Distribution Table

The same rationale and rule discussed for 2 × 2 Chi-Square Tests applies here, that is, $df = (k - 1) \cdot (l - 1)$ degrees of freedom with $k = 2$ and $l = 4$. Hence, the chi-square test value in our example has $1 \times 3 = 3$ degrees of freedom. With an α-level of 1% we can find in Table 4 of Appendix A a critical chi-square value of $\chi^2_{\text{crit}}(df = 3; \alpha = 1\%) = 11.34$.

Step 4. Test Conclusion and Interpretation

Our empirical chi-square value is greater than the critical chi-square value, that is, the test result is significant at the $\alpha = 1\%$ level—language descriptors depend on the race of the crime news subject. A more detailed interpretation of our results requires a closer inspection of Table 7.17 and particularly the comparison of the observed with the expected frequencies. For example, it seems that readers are more likely to employ adjectives for black crime suspects.

Step 5. Discussion of the Test's Assumptions

With more than one degree of freedom, the one measurement, two-dimensional chi-square test for multiple categories requires that the following assumptions are met:

1. Every study element of the sample has to be assigned exclusively to one and only one category for each variable.
2. The *expected* frequencies in each category must be at least greater than 5.

Both assumptions are met in our example and the researcher can rely on the chi-square test result, that is, can rely on the result's *p*-value.

As a final note: Even if the requirement for the expected frequencies ($f_e = 10$ for $df = 1$; $f_e = 5$ for $df > 1$) is violated, chi-square tests still decide robustly if at least 80% of all categories or cells have an expected frequency of 5 or higher. Like *t*-tests, chi-square tests that are based on a reasonably large sample size can be considered as extremely robust tests.

SUMMARY

In this chapter, we focused on specific hypothesis tests for differences, which include mean differences (interval and ratio-level data) and frequency differences (nominal level data). For mean differences we presented the independent samples *t*-test and the dependent (paired) samples *t*-test. For problems with more than two groups we introduced Analysis of Variance (ANOVA). For frequency differences we discussed different variants of the chi-square tests and suggested a five-step procedure for executing (and understanding) chi-square tests.

We explained the background of each test including the rationale for a test's ability to provide statistical inference. Various examples from the communication literature illustrated the calculations and the interpretations of test results.

TERMS

- **Independent samples *t*-test** is a statistical test that tests a hypothesis about a (mean) difference between two groups. The two groups are assumed to be independent, meaning that the two groups we are interested in are sampled such that the selection of elements (research participants) in one group does not influence the selection of elements in the other group.
- **Dependent samples *t*-test** is a statistical test that tests a hypothesis about a (mean) difference between two groups. The two groups are assumed to be dependent, meaning that there is a relationship between study elements (research participants) of the two groups.
- **Analysis of Variance (ANOVA)** is a statistical test procedure that tests hypotheses about (mean) differences among groups of one or many independent variables (factors).
- **Chi-square tests** are a class of statistical tests that are designed to test differences in frequencies of variables that are measured on a nominal scale.

CHAPTER 7 PROBLEM SET

1. For each of the following calculated *t*-values and sample sizes, indicate the degrees of freedom and whether you should reject or fail to reject the null hypothesis of a *t*-test for small and independent samples (indicate whether it is at the .05 or .01 significance level). Presume a non-directional hypothesis (i.e., a two-sided test).
   a. $t = 2.18$ $n1 = 5$ $n2 = 5$
   b. $t = -2.05$ $n1 = 12$ $n2 = 10$
   c. $t = 2.18$ $n1 = 15$ $n2 = 15$
   d. $t = -2.05$ $n1 = 16$ $n2 = 16$

2. An instructor is teaching an introductory communication course and wants to develop two separate tests, one that is more challenging for advanced students and one that is slightly easier. In order to determine whether the two tests forms are different, both tests will be given to two groups of students with similar skills. Test whether the test scores are significantly different. The data are shown below:

a. State the hypotheses

b. Conduct a t-test with $\alpha = .05$

c. State your statistical decision

Subject	Test A/Group A	Subject	Test B/Group B
1	73	11	99
2	82	12	90
3	91	13	95
4	78	14	91
5	75	15	70
6	81	16	80
7	66	17	75
8	84	18	86
9	77	19	88
10	63	20	90

3. A researcher decides to test the adage, "Blondes have more fun/dates," by conducting a random sample of single college female students. The sample consists of fifteen females who have brunette hair at the start of the study. The dependent variable is the number of dates from the previous month. After measuring these women once, the participants bleach their hair blonde and the number of dates that they go on is tallied. Thus, the subjects started with dark hair for the first measurement and then went to blonde hair for the second measurement.

a. State the null and research hypothesis

b. Conduct a t-test with $\alpha = .05$ and the following data

c. State your statistical decision

Sample Time 1 (Dark hair)	Sample Time 2 (Blonde hair)
4	1
8	8
7	5
6	6
10	3
6	9
5	4
8	5
7	2
3	7
4	4
11	8
9	9
4	3
1	7

4. A psychologist was working on a study and compiles data on male and female attitudes about a political candidate. He wanted to know if they have different attitudes. Unfortunately after making a table of the data, he spilled coffee on the table and made some of the values unreadable. Here is what remains:

	Attitude		
Gender	**Positive**	**Negative**	**Totals**
Male	48		60
Female			
		15	$N = 100$

a. Is it possible to re-create the data? Fill in the missing values
b. State the hypotheses
c. Apply the appropriate chi-square test at $\alpha = .01$ and make sure to conduct (and label) all five steps.

5. A researcher employs an injunctive norm campaign directed at sexually active college students encouraging them to use condoms. The results of the campaign are listed below:

	After campaign	
Before campaign	**Uses condoms**	**Doesn't use condoms**
Uses condoms	70	10
Doesn't use condoms	50	20

   a. State the hypotheses
   b. Conduct the appropriate chi-square test at $\alpha = 0.05$ and make sure you conduct (and label) all 5 steps
6. In a *t*-test, which of the following statements are true:
   a. When it is very unlikely ($p \leq 5\%$) that 2 groups represent the same population, the conclusion is that the 2 groups are significantly different from one another at the $\alpha < 0.05$ level.
   b. The larger the sample size *(n)* the greater your ability to reject the t-test's H_0.
   c. A *t*-test is more appropriate when population parameters are known, compared to a *z*-test.
   d. A paired-difference *t*-test should be conducted when the two groups you are comparing are independent from one another.
7. Indicate which of the following statements are true:
   a. In a 2×2 contingency table, when all the expected frequencies equal the observed frequencies, the calculated χ^2-statistic equals zero.
   b. With chi-square tests, the degrees of freedom are determined by the sample size.
   c. The chi-square distribution is symmetrical around the mean.
   d. In a chi-square test, expected frequencies are always calculated according to the null hypothesis.
   e. Both large negative and large positive chi-square values suggest evidence for to the research hypothesis.

REFERENCES

Benoit, W. L., McKinney, M. S., & Stephenson, M. T. (2003). Effects of watching primary debates in the 2000 US presidential campaign. *Journal of Communication, 52*(2), 316–332.

Buzzanell, P. M., Berkelaar, B. L., & Kisselburgh, L. (2011). From the mouths of babes: Exploring families' career socialization of young children in China, Lebanon, Belgium, and the United States. *Journal of Family Communication, 11*(2), 148–164.

Chory-Assad, R. M. (2004). Effects of television sitcom exposure on the accessibility of verbally aggressive thoughts. *Western Journal of Communication,* 68 (4), 431–453.

Goodboy, A. K., & Myers, S. A. (2008). The effect of teacher communication on student communication and learning outcomes. *Communication Education, 57*(2), 153–179.

Gorham, B. W. (2006). News media's relationship with stereotyping: The linguistic intergroup bias in response to crime news. *Journal of Communication, 56,* 289–308.

Hancock, J. T., & Toma, C. T. (2009) Putting your best face forward: The accuracy of online dating photographs. *Journal of Communication,* 59, 367–386.

Keppel, G., & Wickens, T. D. (2004). *Design and analysis. A researcher's handbook.* Upper Saddle River: Pearson.

Myers, K. K., Jahn, J. L. S., Galliard, B. M., & Stoltzfus, K. (2011). Vocational anticipatory socialization: A communicative model of adolescents' interest in STEM. *Management Communication Quarterly, 25*(1), 87–120.

Qian, H., & Scott, C. R. (2007). Anonymity and self-disclosure on weblogs. *Journal of Computer-Mediated Communication, 12,* 1428–1451.

Rutherford, A. (2001). *Introducing ANOVA and ANCOVA. A Glm approach.* Thousand Oaks: Sage.

Sprent, P., & Smeeton, N. C. (2007). *Applied nonparametric statistical test methods.* Boca Raton, FL: Taylor & Francis.

Turner, R., & Thayer, J. (2001). *Introduction to analysis of variance.* Thousand Oaks: Sage.

Weber, R. (2007). To adjust, or not to adjust alpha in multiple testing: That is the question. Guidelines for alpha adjustment as response to O'Keefe's and Matsunaga's critiques. *Communication Methods and Measures,* 1(4), 281–289.

CHAPTER 8

SPECIFIC SIGNIFICANCE TESTS FOR DEPENDENCIES

Dependencies among variables are of great interest to both communication professionals and researchers. For example, as the marketing manager for an online music retailer, you may be interested in exploring the relationship between time spent on your online store and number of purchases. Or, you may be interested in the number of monthly visits to a site and items that are "orphaned" in an online shopping cart. That is, when a user does not complete a purchase. A better understanding of these dependencies may help you to answer questions about customer loyalty or whether increasing online traffic through targeted marketing is a decision that may pay off. As a communication researcher, you may be interested in studying music consumption via new media. You have noticed (and so has the music recording industry) a growing trend in illegal downloading activity. It would be interesting to learn more about the relationship between music fans' current downloading activity, the expected outcomes of this activity, and future illegal downloading intentions. If you find a dependency among these variables it may be possible to predict one variable (e.g., future downloading activity) with the information provided by other variables.

In Chapter 7, we focused on specific tests for differences. This chapter will focus on specific tests for dependencies among variables. Such tests are very important in empirical research. We can use one variable to predict another only if we know that two or more variables are dependent on each other. Needless to say that predicting variables is one of the most important goals of empirical research. However, there is an important difference between the social sciences, which includes the field of communication, and the natural or physical sciences: In the natural and physical sciences most dependencies between variables are deterministic by nature. For example, if we know the velocity with which an object was moving, and the amount of time it was in motion, we can precisely predict the distance traveled—we simply multiply velocity and time. We can do this because the relationship between distance and time defines a law and the relationship between distance and time is a deterministic dependency. In the social sciences we usually do not know these types of deterministic dependencies. There are no laws. We still can observe dependencies among variables, but those dependencies are all probabilistic in nature, which means that a relationship between variables is never perfect in the social sciences. For example, the time students spent preparing for exams and their exam scores are positively related such that with increasing preparation time, students' exam scores tend to be better. Notice that we wrote "tend to be better." If we collected data on exam preparation times and exam scores we would certainly see unusual cases in which a high exam scores corresponds to very little preparation and vice versa. Why? Because there are many other variables that have an impact on exam scores (e.g., general ability, motivation, test anxiety) and because there is usually also randomness involved.

In the following sections, we will introduce regression and correlation analysis. Regression and correlation analysis are statistical procedures that enable social science researchers (and professionals) to explore and test probabilistic dependencies among variables. Regression and correlation are among the most important statistical procedures for communication scholars. Without the ability to understand and correctly interpret the results of a regression and correlation analysis it may be very difficult to follow the research findings of scholarly articles or, as a communication professional, to draw accurate conclusions and make the right decisions.

REGRESSION ANALYSIS

Let us begin with an example: A communication researcher wants to determine whether students' school performance and playing time of a vocabulary-building video game are related. The researcher defines school performance as the dependent variable with the scores 0 to 10 for correct answers on a vocabulary exam. Playing time is defined as the independent variable and measured in minutes per day. For illustration purposes, we assume that the researcher collected data from only ten children. Figure 8.1 visualizes the data. The coordinates of each dot in the figure stands for a measurement on both variables—on the y-axis the measurement for the vocabulary exam score, and on the x-axis the measurement for playing time. For example, the thick black dot in Figure 8.1 stands for a child with a playing time of 40 minutes a day and an exam score of 6 correct answers. Figure 8.1 defines a so-called two dimensional scatter plot.

As we can see, with increasing playing time school performance tends to increase, but the relationship is not perfect. Why? It may be that the children have different skills, motivation, or support from their parents, all of which may also affect their performance on a vocabulary exam. In situations like this the question is: How can we best represent this probabilistic dependency knowing that the relationship is not perfect? If we asked you to look at Figure 8.1 and think about a graphical representation of the data you would probably intuitively draw a line through the dots such that most dots show about an equal distance to the line. This intuitive solution is similar to what a regression analysis does:

In the language of mathematics, drawing a line means fitting a linear equation to the data. Thus, a regression analysis fits a linear equation to the data that best describes the relationship between the dependent and the independent variable. A linear equation is defined as:

$$\hat{y}_i = b \cdot x_i + a$$

$\hat{y}_i$ stands for the measurements of the dependent variable and x_i for the measurements of the independent variable. b represents the slope of the regression line, and a its y-intercept.

Figure 8.1. Scatterplot of educational video game playing and vocabulary performance

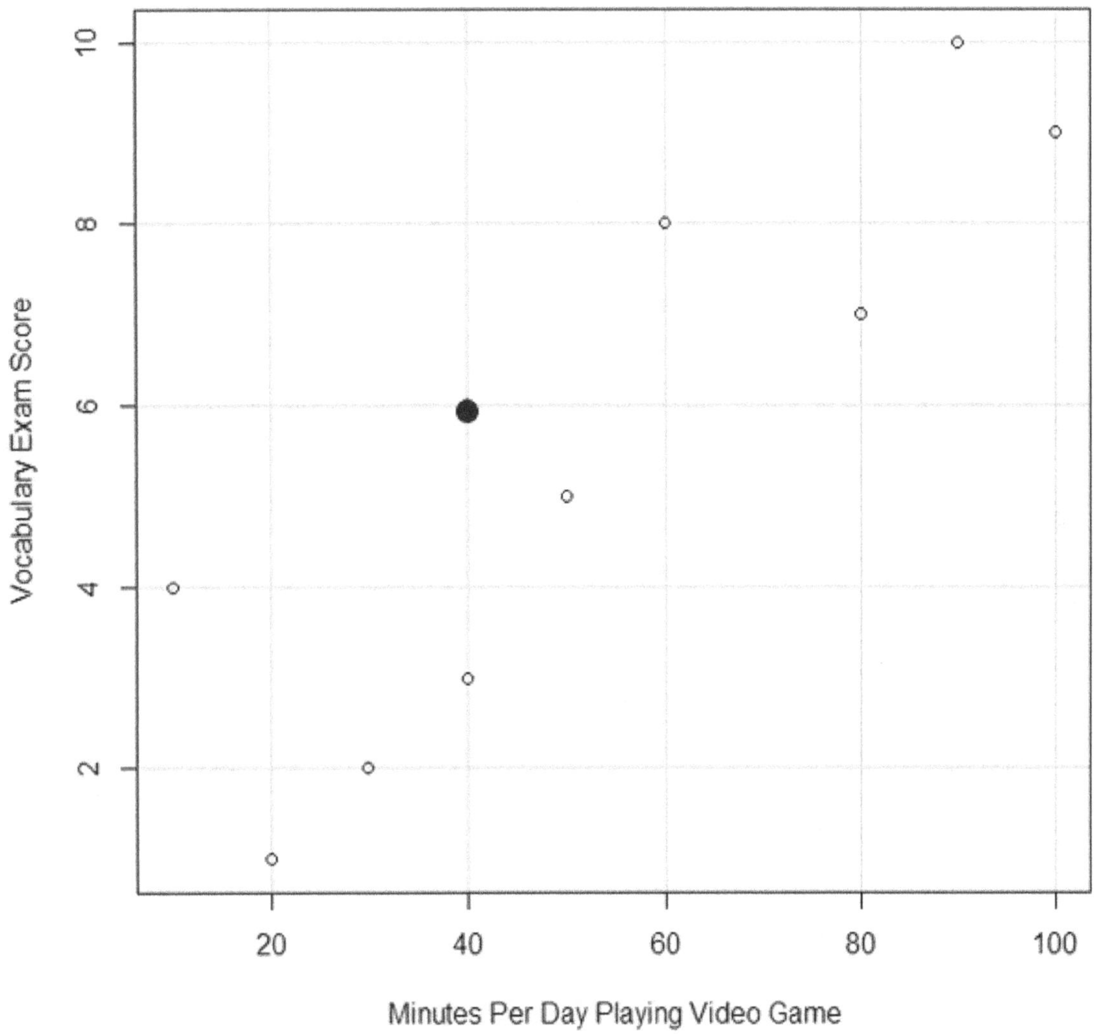

b and *a* are called the regression coefficients. Regression analyses that consider only one dependent and one independent variable (and therefore only need two regression coefficients) are called bivariate regressions. Regression analyses that describe the relationship between one dependent and more than one independent variable are called multiple regression analyses and require more than two regression coefficients (one for each independent variable plus a coefficient for the *y*-intercept). In this introduction we focus our explanations on the bivariate case. Figure 8.2 shows the scatterplot of our example together with a fitted (linear) regression line.

Figure 8.2. Scatterplot with fitted linear regression line

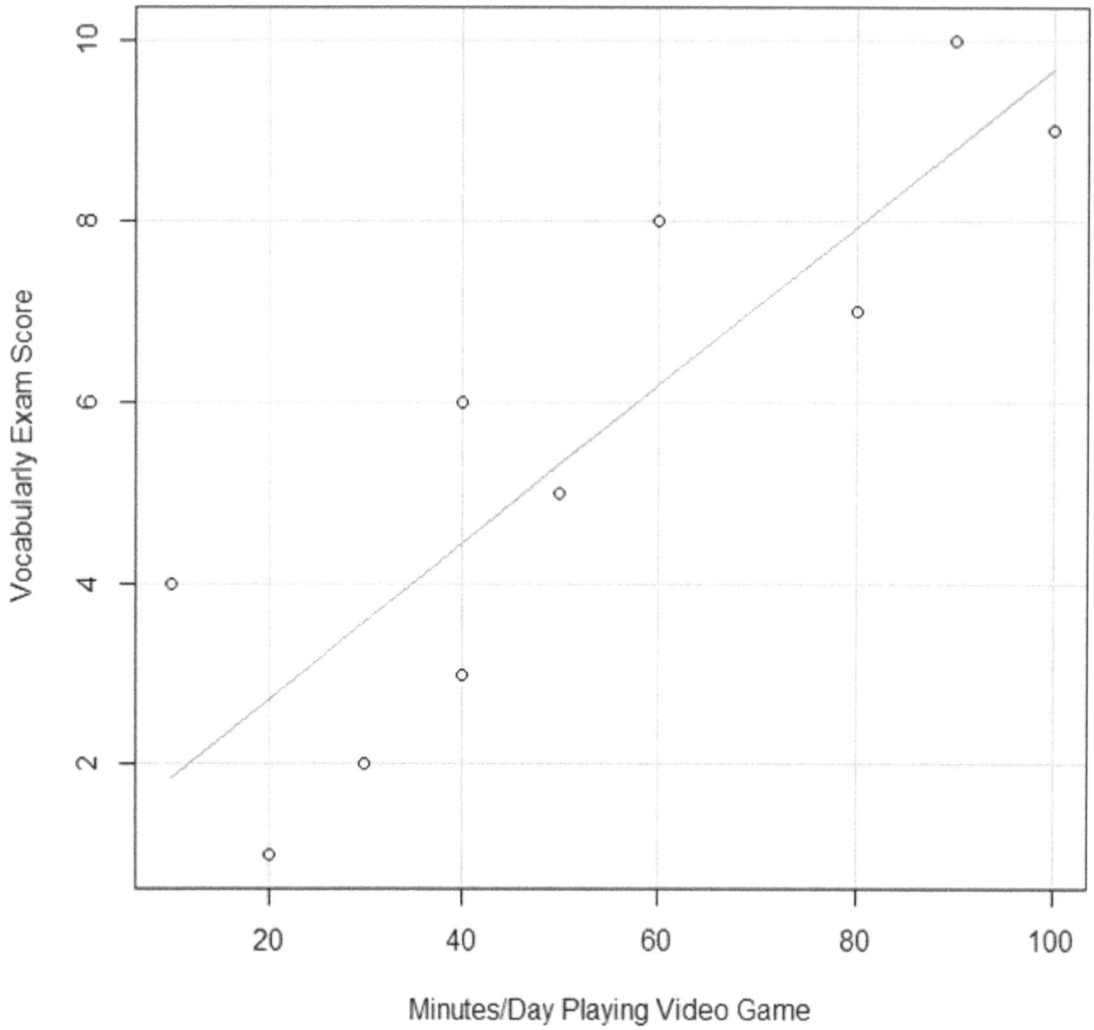

Regression Coefficients and the Least-Squares Criterion

Until now it remained unanswered how exactly a regression analysis fits a linear equation to the data. In other words, which of the many possible linear equations (each with a different slope and y-intercept) best represents the data? The criterion that regression analysis uses to determine the best (fitting) linear equation is based on a squared residual error $(y_i - \hat{y}_i)^2$ when predicting the dependent variable. Figure 8.3 visualizes all squared errors that we commit when using the figure's regression line as the line that best represents the data.

As you can imagine, if the slope and y-intercept of the regression line change, all the squared residual errors and the sum of all squared residuals will change. When we conduct a standard regression analysis, then the regression line (with slope and y-intercept as its coeffi-

Figure 8.3. Regression line and squared residuals

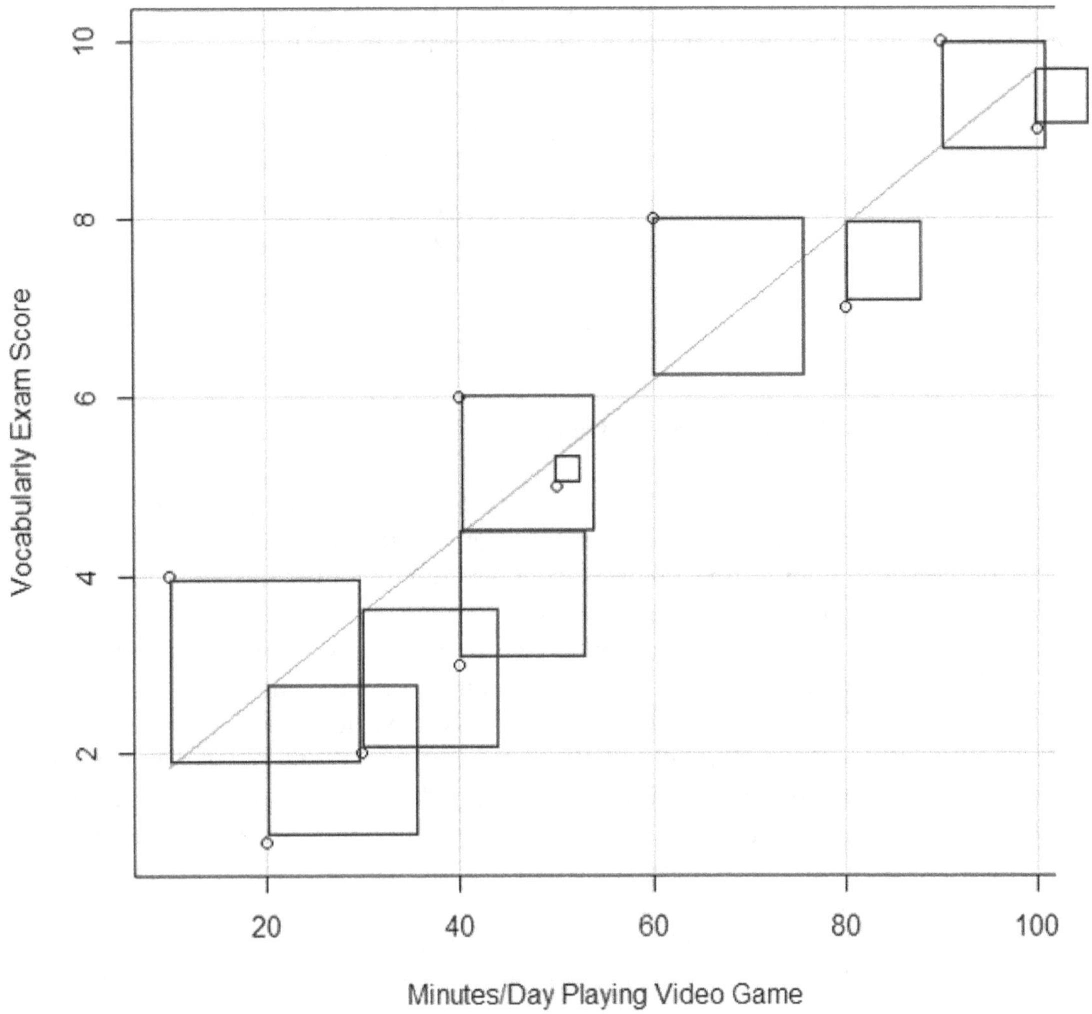

cients) that *minimizes* the sum of all squared residuals is considered as the line that best represents the data. That is:

$$\sum_{i=1}^{n}(y_i - \hat{y}_i)^2 = \sum_{i=1}^{n}(y_i - b \cdot x_i + a)^2 \rightarrow \text{Min}$$

This criterion is defined as the least-squares criterion. The method is called the method of least squares, the estimates are called least squares estimates, and the line that meets the least squares criterion is called the least squares line. The lines in Figure 8.2 are drawn based on equations that meet the least squares criterion. We can see that some values for children's performance can be predicted quite well, whereas others are predicted quite poorly.

The coefficients for the regression line that meet the least squares criterion can be derived through differential calculus (not shown here). The equations for the least squares regression coefficients (slope b, and y-intercept a) are given below (x stands for measurements of the independent variable and y for measurements of the dependent variable):

$$b = \frac{n \cdot \sum_{i=1}^{n} x_i \cdot y_i - \sum_{i=1}^{n} x_i \cdot \sum_{i=1}^{n} y_i}{n \cdot \sum_{i=1}^{n} x_i^2 - \left(\sum_{i=1}^{n} x_i\right)^2}$$

$$a = \bar{y} - b \cdot \bar{x}$$

Significance Test for Regression Coefficients

So far, we've discussed the concept of regression from a descriptive perspective only. Usually an inferential perspective on regression is more interesting; that is, generalizing a regression equation for a population or testing whether hypotheses about regression coefficients withstand the confrontation with empirical data in a sample.

In a deterministic regression model with one dependent and one independent variable each value of x corresponds to a single value of y and all measurements define a perfect regression line. In practical communication research this scenario is not realistic—for the same values of x we will see a variation of y. Why? Because there are no simple laws that explain communication phenomena. Variation in y has usually different reasons, some accounted for by x and some unaccounted for. For example, even for students who play exactly the same amount (in minutes) of an educational video game, there will be a variation in school performance. This is because school performance (here scores on a vocabulary test) depends on many other variables as well (e.g., cognitive skills, motivation, parental support). From an inferential perspective there is also another source of variation in y values. If we draw repeatedly random samples out of the same population, we would expect y values to randomly vary for fixed x values.

The distribution of y given a fixed x value is called the conditional y-distribution (see Figure 8.4). The conditional y-distribution has a mean and a standard deviation, which is called the conditional mean and the conditional standard deviation. Similarly, for different values of x we expect another conditional distribution for y with another mean (and maybe another standard deviation).

The conditional standard deviations provide information on how accurate we are able to predict y values based on x values and can be estimated with the sample data through:

$$\hat{\sigma}_{y|x} = \sqrt{\frac{SSE}{n-2}} = \sqrt{\frac{\sum (y - \hat{y})^2}{n-2}}$$

Figure 8.4. Conditional distributions

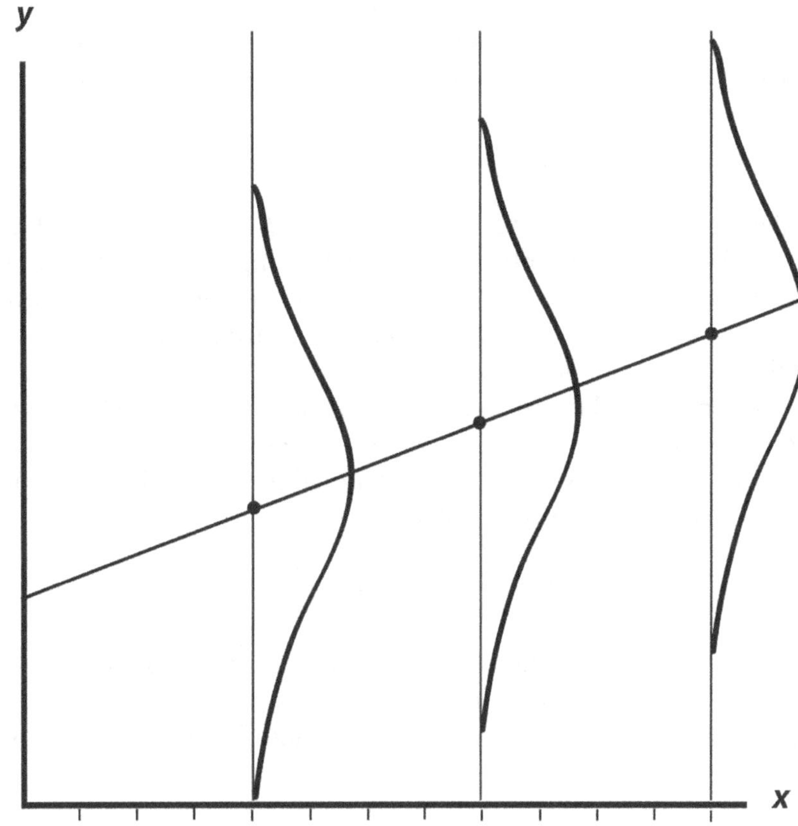

If we know that there are other variables in place that also (systematically) explain variation in y values, then our regression model is underspecified, that is, other independent variables besides x are missing in the regression equation, and the mean, standard deviation, and shape of the conditional distributions are affected by other explanatory variables. In this case, a multiple regression analysis would be required (not discussed here). If, however, we assume (or know) that there are no other explanatory variables besides x, then our bivariate model would be correctly specified, the mean and standard deviation of y would be a function of x, and the shape of the conditional distribution would be the result of many unsystematic effects from sample to sample, which we also call random effects or random variation. In this case we would expect the conditional distributions to follow a normal distribution. Thus, in the context of an inferential perspective, a bivariate regression equation is nothing more than a linear function that describes how the *mean* (and sometimes the standard deviation) of a response variable, y, changes according to the values of an explanatory variable, x, in a correctly specified regression model. However, the parameters of this function are unknown; that is, we have to estimate them based on our sample data by using our sample regression coefficients introduced above. To recall, we would use the sample's slope coefficient, b, to estimate the population's slope coefficient, β, and the sample's y-intercept, a, as estimate of the population's y-intercept, α. In other words, we would use the sample regression coefficients b and a to estimate the population's regression equation:

$$\hat{y}_i = \beta \cdot x_i + \alpha$$

Thus, as with any other parameters we have discussed so far, β and α are estimated by sample statistics. Further, if y-values vary randomly from sample to sample and form a normal distribution in case of a correctly specified model, the regression coefficients b and a will vary randomly from sample to sample with a specified sampling distribution. Usually, one is more interested in a statistical test for the slope coefficient, because this coefficient defines the relationship between the dependent and the independent variable. If there is a slope of 0 then there is no relationship between the variables, because for any given value of x we would predict the same value for y. Thus, a statistical test in its non-directional form would test the following hypothesis (of course, a one sided test is also possible):

$$H_0: \beta = 0 \qquad H_1: \beta \neq 0$$

The slope coefficient's sampling distributions for $n < 30$ is a t-distribution with $(n - 2)$ degrees of freedom and for $n > 30$ a normal distribution. The expected values (means) of both sampling distributions are the true regression coefficients in a population. The standard error of the slope can be calculated as:

$$\hat{\sigma}_b = \frac{\hat{\sigma}_{y|x}}{\sqrt{\sum (x - \overline{x})^2}} = \frac{\hat{\sigma}_{y|x}}{\hat{\sigma}_x \sqrt{n - 1}}$$

Once we have the estimate for the slope parameter, the slope parameter's expected value in the population, and the estimated standard error, we can easily calculate a standardized empirical value, which is computed as always: (statistic—mean/expected value of statistic)/(standard error of statistic) or with the following formula:

$$t = \frac{b - E(b)}{\hat{\sigma}_b} = \frac{b - \beta}{\hat{\sigma}_b}$$

As according to H_0 the expectation for β is 0, the calculation of the standardized empirical test value simplifies to:

$$t = \frac{b}{\hat{\sigma}_b}$$

As usual (see Chapter 6), we compare this empirical value with a critical test value of the appropriate sampling distribution that corresponds to a pre-determined maximum Type-I error α (e.g., 5%). For $n < 30$ we look at the t-distribution table ($df = n - 2$), and for $n > 30$ we look at the standard normal distribution table (see Appendix A, Table 1).

Confidence Interval for Regression Coefficients

If we are more concerned about the size of the slope and the precision of its estimation than in just testing whether the slope is 0, then we calculate a confidence interval. In line with what we learned before (see Chapter 6) and with the information about the regression slope's sampling distribution, this confidence interval can be easily calculated with:

$$P(b \pm t_{(^{\alpha}/_{2})} \cdot \hat{\sigma}_b) = P\left(b \pm t_{(^{\alpha}/_{2})} \cdot \frac{\hat{\sigma}_{y|x}}{\hat{\sigma}_x \sqrt{n-1}}\right) = 1 - \alpha$$

Looking at the formula, what factors affect the width of the regression slope coefficient's confidence interval? (a) As in all confidence intervals, a higher confidence coefficient (e.g., 99%) leads to a larger confidence interval, as a smaller α increases the t (or z) value of the corresponding standardized sampling distribution. By extension, a lower confidence coefficient (e.g., 90%) leads to a smaller confidence interval. (b) A higher conditional standard deviation leads to a larger confidence interval width. More random variation in y-values increases the uncertainty in predicting the dependent variable. By extension, a smaller conditional standard deviation leads to a smaller confidence interval width. (c) A larger sample size leads to a smaller confidence interval width. More data means more information, which pays off in the form of smaller standard errors. By extension, a smaller sample size leads to a wider confidence interval. And finally, (d) the higher the variance of the x values, the smaller the confidence interval width. Exhausting the entire variation of x-values and observing the responses in y-values rather than restricting observations to a small interval of x-values leads to more reliable estimation of y-values and to a smaller confidence interval. By extension, the smaller the variance of x values, the larger the confidence interval width. In sum, the most accurate predictions (i.e., predictions with a small confidence interval) are based on large samples ($n > 30$), high variance in x values, low variance in y values, and on the smallest confidence coefficients that make sense in a particular situation (usually 99%, 95%, or 90%).

Assumptions and Issues

Significance tests and confidence intervals rest on specific assumptions. If assumptions are violated, then the results of statistical tests and confidence intervals are biased (see Chapter 6). Unbiased results of significance tests and confidence intervals for regression coefficients can be expected if the following assumptions are met.

Random Sample and Valid Model Specification

Unbiased statistical inference assumes that the data collection is based on a sampling technique that provides a true sample (see Chapter 4). Usually, a classical and well-conducted random sampling technique meets this requirement. In addition, statistical inference can only be unbiased if the underlying regression model is valid, that is, correctly specified. A valid specification includes two aspects: First, does the model contain all relevant variables or are important explanatory/independent variables missing? Second, are the variables indeed linearly related to the outcome/dependent variable? If the answer to the first question is uncertain, then it is likely that the regression coefficient of the independent variable(s) in the equation does not reflect the variable's real impact on the dependent variable and, as a consequence, inferential interpretations would be flawed. Even more important, if the answer to the second question is "no," that is, there is a chance that the variables stand in a non-linear relationship (see Figure 8.5), then—again—any inferential claim is likely to be flawed.

Take the right plot in Figure 8.5 as an example. Here we deal with a reversed u-shaped function. As independent variable we could imagine the information density of a television ad and as dependent variable the recall of this ad. Too little information per time unit typically leads to a boring ad and subsequently to bad results in recall tests. Too much information per time unit usually leads to cognitive overload, a reassignment of cognitive resources to other tasks (distraction), and—again—subsequently to bad results in a recall test. Ads with an information density in the middle tend to lead to the best results in recall tests. If we fitted a linear regression line as demonstrated above to the right plot in Figure 8.5, we would obtain an estimated regression slope coefficient b of 0 in this ideal example, indicating no relationship between the independent variable x and dependent variable y. Obviously, interpreting and generalizing the relationship in our example as "no relationship" would be incorrect. In reality, there is a strong relationship between the two variables, just not a linear relationship.

Thus, in cases like in this example, it would be inappropriate to use a linear regression analysis to describe and test the relationship between the two variables. There are other regression techniques that can deal with nonlinear relationships (not discussed here).

Distribution Assumptions

As mentioned above, the statistical test and confidence intervals assume that the conditional distributions are normal and that the conditional standard deviations are identical at each x-value. Actually, distribution assumptions are even stricter as we need to assume a bivariate normal distribution (like a bell) between y and x. However, if one can show that the conditional distributions $y|x$ (y for fixed x) are normal then one can assume that the bivariate distribution is normal, too. Also true is that this assumption is only important if one deals with small samples. For large samples, the Central Limit Theorem (see Chapter 5) assures that the sampling distribution for the slope coefficient is normal even if the conditional distributions deviate from a normal distribution.

If the assumption of identical conditional standard deviations (also called "Homoscedasticity") is violated (which would be called "Heteroscedasticity") then one can show that the least-square criterion no longer provides a regression model that best represents the data. Other criteria would

Figure 8.5. Examples of nonlinear relationships

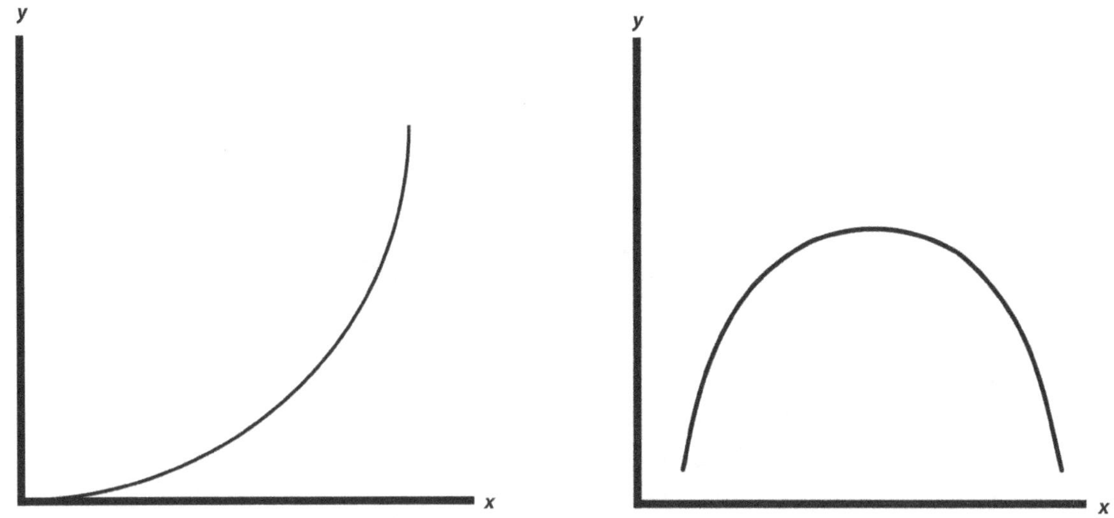

lead to a lower sum of squared residual errors in this case and, as a consequence, the estimation for the slope coefficient including its *p*-value would be slightly different.

Finally, researchers have to assume that all measurements are independent; that is, are based on unique units of analyses. If, for example, all measurements in Figure 8.2 were collected from the same participant (e.g., a student's performance in a vocabulary test over time with repeated exposure to an educational video game), then this assumption would be violated and the statistical test would provide biased results. It is unlikely that researchers violate this assumption if they work with random samples as discussed above and do not apply a repeated-measurement or time-series design.

For violations of all three distribution assumptions there are statistical techniques and modifications of the basic regression models that can deal with these violations and correct biases in the results. For a more advanced discussion of how to deal with violations of assumptions we recommend Mendenhall and Sincich (2011).

Other Issues

The results of a regression analysis are also sensitive to so-called "data problems" which—technically—do not count as violations of statistical assumptions, but nevertheless lead to a biased estimation of regression coefficients. One of the most important data problems is the existence of outliers; that is, observations that clearly stand apart from all the other measurements. Figure 8.6 visualizes an outlier observation. Look at the squared residual of the outlier with the line of best fit for the rest of the data points.

Due to the least-squares criterion, outliers are characterized by very large squared residuals and thus become very influential for the estimation of the slope coefficient. Graphically, outliers

Figure 8.6. Effect of an outlier observation

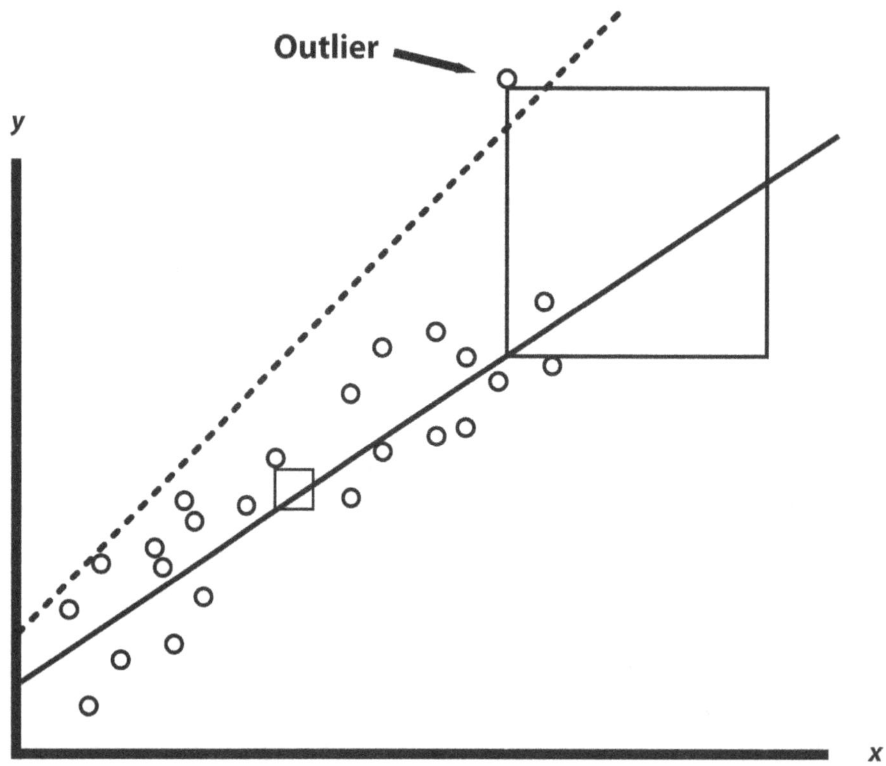

"pull" the regression line in their direction and the resulting regression line (with a biased slope coefficient) no longer represents the data well.

Another issue deals with the fact that it is difficult to apply an estimated regression equation to values of x outside the range of observed values. For example, if most of the observed x values in Figure 8.2 fall between 20 and 100 minutes playing time per day, then it is recommended that one does not use the regression equation for predicting the performance in a vocabulary exam for students that played an educational video game for 180 minutes a day. Usually, a regression equation can only be reliably used for predictions within the range of the actual observations. Predictions outside this range can be greatly inaccurate and even impossible. Other issues that are related to the range of observations of either the dependent or the independent variable are discussed below in the context of a correlation analysis.

In multiple regression analysis, that is, in regression analyses with more than one independent variable, x, there are even more complex data problems to consider before interpreting the results. The most important data problems are labeled "multicollinearity" and "suppression." Both problems occur when there are strong relationships (correlations) among the independent variables. An advanced discussion of multicollinearity and suppression problems and how to deal with these problems in practical research can be found in Mendenhall and Sincich (2011).

Recommendations

With the exception of the independence assumption and the data problems outlined above, violations of assumptions in regression analysis usually result in conservative test decisions. That is, they result in increased Type-I and Type-II error probabilities. Simulations show that significance tests for regression slope coefficients are more or less robust if the sample size is large. We recommend that you always inspect the scatterplot of two related variables. Violations such as Heteroscedasticity, non-independence, non-linearity, or outliers, can be visually identified based on a simple scatter plot.

Example: Regression Analysis of Perceived Reader Support and Blogger Personal Growth

A health communication researcher (c.f., Raines & Keating, 2011) is interested in the relationship between bloggers' perceptions of reader support and bloggers' personal growth. Specifically, the researcher believes that perceived reader support is consequential for personal growth for individuals who are dealing with a health condition. The researcher surveys bloggers about perceived reader support and personal growth (both variables on a scale of 1 to 7; 1 = "very little support/growth" to 7 = "very much support/growth"). Personal growth refers to respondents' beliefs that they are evolving and realizing their potential. The researcher is pilot testing her measurement, and got the following data from five bloggers, displayed in Table 8.1.

Table 8.1. Perceived reader support and personal growth—raw data

Participant	x (perceived support)	y (personal growth)
1	1	2
2	2	1
3	6	7
4	4	5
5	2	3

We want to know, can we predict personal growth based on bloggers' perceived reader support? We know from the formula to calculate the regression coefficients that we need the sum of all x-values, the sum of squared x-values, and the sum of the product of x and y. Table 8.2 provides the computed values for each of these components.

Table 8.2. Perceived reader support and personal growth—formula components

Participant	x	y	x^2	$x \cdot y$
1	1	2	1	2
2	2	1	4	2
3	6	7	36	42
4	4	5	16	20
5	2	3	4	6
Σ	15	18	61	72
Mean	3	3.6	—	—

With the information provided in Table 8.2, we can calculate the slope and y-intercept of the regression line that best represents the data. Table 8.3 provides the computations.

Table 8.3. Perceived reader support and personal growth—regression coefficients

Step 1. Compute the value of the slope

$$b = \frac{n \cdot \sum_{i=1}^{n} x_i \cdot y_i - \sum_{i=1}^{n} x_i \cdot \sum_{i=1}^{n} y_i}{n \cdot \sum_{i=1}^{n} x_i^2 - \left(\sum_{i=1}^{n} x_i\right)^2} = \frac{5 \cdot 72 - 15 \cdot 18}{5 \cdot 61 - 225} = \frac{90}{80} = 1.125$$

Step 2. Compute the intercept

$$a = \bar{y} - b \cdot \bar{x} = 3.6 - 1.125 \cdot 3 = 0.225$$

Step 3. State the linear regression equation

$$\hat{y}_i = 1.125 \cdot x_i + 0.225$$

As we can see, perceived reader support and personal growth are positively related with a slope coefficient of 1.125. This means, that if perceived reader support increases one scale point, self-reported personal growth increases 1.125 scale points. If a participant reports no perceived reader support, personal growth would still be 0.225 and not zero.

The next step, with a known regression equation, we can compute the estimations for y ($\hat{y}$), the squared residuals $(y - \hat{y})^2$, and the squared deviation of x-values from their mean $(x - \bar{x})^2$. We need this information in order to conduct the significance test for the slope coefficient of the regression. Table 8.4 provides the computations.

Table 8.4. Perceived reader support and personal growth

Participant	x	y	x^2	$x \cdot y$	$\hat{y}$	$(y - \hat{y})$	$(y - \hat{y})^2$	$(x - \bar{x})^2$
1	1	2	1	2	1.35	0.65	0.4225	4
2	2	1	4	2	2.475	−1.475	2.1750	1
3	6	7	36	42	6.975	0.025	0.0006	9
4	4	5	16	20	4.725	0.275	0.0756	1
5	2	3	4	6	2.475	0.525	0.2756	1
Σ	15	18	61	72	15	0	2.9500	16
Mean	3	3.6						

The researchers hypothesized that with increasing perceived reader support the self-reported personal growth should also increase. Thus, the hypotheses for our example are:

$$H_0 : \beta \leq 0 \qquad H_1 : \beta > 0$$

Table 8.5 provides the step-by-step computations.

COVARIANCE

The covariance statistic is another measure to describe how two variables vary together. The covariance statistic is closely related to regression analysis. The covariance statistic is defined as:

$$\text{cov}(x, y) = \frac{\sum_{i=1}^{n} (x_i - \bar{x}) \cdot (y_i - \bar{y})}{n}$$

Stated in words, covariance is determined by subtracting the mean of x from each x value, subtracting the mean of y value from each y value, multiplying these values (calculating the cross-product), summing the cross-product, and dividing by the sample size. The result of these computations is a number that can be either negative, zero, or positive. Looking at the formula, what does a positive covariance mean? That means that x values above average occur more frequently with y-values above average, and x-values below average occur more frequently with y-values below average. Correspondingly, a negative covariance means that x-values above average occur more frequently with y-values below average and x-values below average occur more frequently with y-values above average. And finally, a covariance of zero would suggest that for x-values below and above their average we see about the same number of y-values below and above average. Figure 8.7 illustrates positive, negative, and zero covariance.

Table 8.5. Perceived reader support and personal growth—data for the significance test

Step 1: Compute the estimated conditional standard deviation

$$\hat{\sigma}_{y|x} = \sqrt{\frac{SSE}{n-2}} = \sqrt{\frac{\sum (y-\hat{y})^2}{n-2}} = \sqrt{\frac{2.95}{3}} = 0.991$$

Step 2: Compute the estimated standard error of the slope b

$$\hat{\sigma}_b = \frac{\hat{\sigma}_{y|x}}{\sqrt{\sum (x-\bar{x})^2}} = \frac{0.991}{\sqrt{16}} = 0.247 \quad \left[= \frac{\hat{\sigma}_{y|x}}{\hat{\sigma}_x \sqrt{n-1}} \right]$$

Step 3: Compute the standardized empirical test value

$$t_{emp} = \frac{b - E(b)}{\hat{\sigma}_b} = \frac{b-\beta}{\hat{\sigma}_b} = \frac{b}{\hat{\sigma}_b} = \frac{1.125}{0.247} = 4.55$$

Step 4: Compare with critical t-value
See t-table: $df = 3$; $\alpha = 5\%$ (one-sided test) $\Rightarrow t_{crit} = 2.353$; $t_{emp} > t_{crit}$

Step 5: Summarize and interpret test outcome
From the above, we can say that perceived reader support was a significant predictor of bloggers' perceptions of personal growth ($\beta = 1.125$, $t = 4.55$, $df = 3$, $p < 0.05$).

The covariance statistic can be understood as the product of corresponding deviations. High positive products of corresponding deviations lead to high positive covariance. High negative products lead to high negative covariance. Thus, a positive covariance indicates a positive linear relationship among the variables, that is, as x increases, y increases. A negative covariance indicates a negative linear relationship among the variables, that is, as x increases, y decreases. A covariance of zero (or close to zero) indicates no linear relationship.

Covariance and Regression

As mentioned, the concepts of covariance and regression are closely related. If you have already calculated the covariance of two variables, you can easily calculate the regression line. Recall that covariance is defined by:

$$\text{cov}(x, y) = \frac{\sum_{i=1}^{n} (x_i - \bar{x}) \cdot (y_i - \bar{y})}{n}$$

Figure 8.7. Positive, negative, and zero covariance

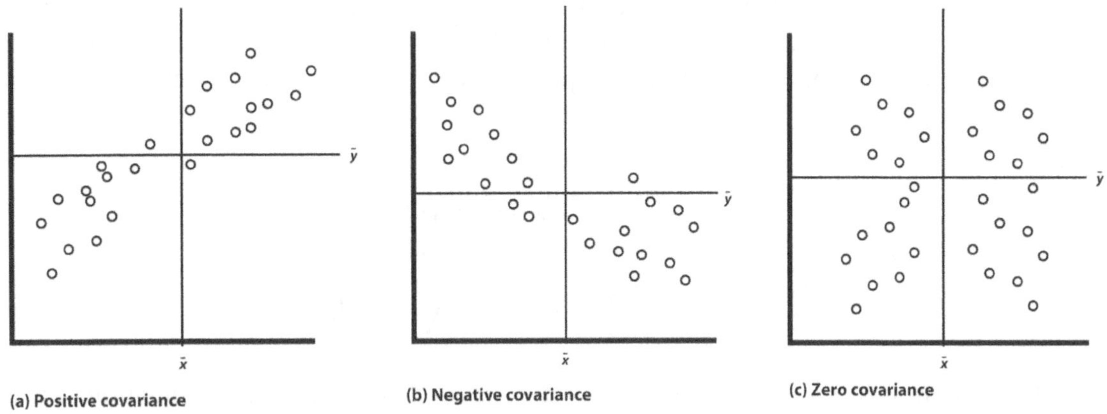

The regression slope coefficient is nothing else than the covariance between the dependent variable y and the independent variable x divided by the variance of the independent variable x:

$$b = \frac{\text{cov}(x, y)}{s_x^2} = \frac{\frac{1}{n}\sum (x_i - \bar{x})(y_i - \bar{y})}{\frac{1}{n}\sum (x_i - \bar{x})^2}$$

Thought question: Looking at this formula and Figure 8.5, what would be the best prediction for the dependent variable y if we have a covariance of 0 (which means that variable x has no impact on y)? In this case, the best prediction for y would simply be the mean of y (or in a more complicated language, it would be the mean of all conditional distributions, which does not differ for any given x value). The intercept coefficient a of the regression equation stands for the dependent variable's mean. Thus, if there is no relationship between x and y, both the covariance and regression slope coefficient is zero, and the best prediction would be a, the intercept coefficient.

Unfortunately, the covariance has a major disadvantage. The covariance is dependent on the scales or units of the variables. For example, if we multiplied the independent variable x with factor k, and the dependent variable y with factor l, the covariance would change with factor $(k \cdot l)$. This is unfortunate because scales or units of variables are often arbitrary in social sciences. In Chapter 2 we learned that scores on different scales can be compared if we z-standardize the variables' scores. If we z-standardize both the independent and the dependent variable and then calculate the covariance of the z-standardized variables, we obtain a statistic providing information about the relationship between the two variables that is independent from the variables' scale units. This statistic is called the Pearson Correlation Coefficient r.

CORRELATION COEFFICIENT R (PEARSON CORRELATION COEFFICIENT)

In general, a correlation provides information on the strength and the direction of a linear relationship between two metric variables. We use the character r for a correlation in a sample and the Greek letter ρ (rho) for a correlation in a population. Correlations are calculated using the following formula:

$$r = \frac{\text{cov}(x, y)}{s_x \cdot s_y} = \frac{\frac{1}{n}\sum (x_i - \bar{x})(y_i - \bar{y})}{\sqrt{\frac{1}{n}\sum (x_i - \bar{x})^2} \cdot \sqrt{\frac{1}{n}\sum (y_i - \bar{y})^2}}$$

Correlations are insensitive to scales or units of measurement because the computation of a correlation coefficient includes a z-standardization. The formula illustrates the standardization. The means of x and y are subtracted from each x and y value, respectively, divided by the standard deviations of x and y. These values are then summed, and divided by the sample size. The formula below conveys another way to represent correlation.

$$r = \frac{\sum_{i=1}^{n} (x_i - \bar{x}) \cdot (y_i - \bar{y})}{n \cdot s_x \cdot s_y} = \frac{1}{n} \cdot \sum_{i=1}^{n} \left(\frac{x_i - \bar{x}}{s_x} \cdot \frac{y_i - \bar{y}}{s_y} \right) = \frac{1}{n} \cdot \sum_{i=1}^{n} z_{xi} \cdot z_{yi}$$

Thus, a correlation coefficient can also be understood as the average cross-product of corresponding, z-standardized scores. With a few simple mathematical transformations, r can be calculated with another formula that only requires the sum of values and the sum of squared values. The formula looks more complicated, but the calculations (especially if done in a table) are easier with this formula:

$$r = \frac{n \cdot \sum_{i=1}^{n} x_i \cdot y_i - \left(\sum_{i=1}^{n} x_i\right) \cdot \left(\sum_{i=1}^{n} y_i\right)}{\sqrt{n \cdot \sum_{i=1}^{n} x_i^2 - \left(\sum_{i=1}^{n} x_i\right)^2} \cdot \sqrt{n \cdot \sum_{i=1}^{n} y_i^2 - \left(\sum_{i=1}^{n} y_i\right)^2}}$$

Example: Correlation Analysis of Perceived Reader Support and Blogger Personal Growth

We are using the problem and data as in our example for a regression analysis (see above). Tables 8.6 and 8.7 contain all necessary information for the calculation of a Pearson correlation coefficient.

Table 8.6. Perceived reader support and personal growth—data for correlation analysis

Participant	x	y	x^2	y^2	$x \cdot y$
1	1	2	1	4	2
2	2	1	4	1	2
3	6	7	36	49	42
4	4	5	16	25	20
5	2	3	4	9	6
Σ	15	18	61	88	72
Mean	3	3.6			

Table 8.7. Perceived reader support and personal growth—calculation of a correlation

$$r = \frac{n \cdot \sum_{i=1}^{n} x_i \cdot y_i - \left(\sum_{i=1}^{n} x_i\right) \cdot \left(\sum_{i=1}^{n} y_i\right)}{\sqrt{n \cdot \sum_{i=1}^{n} x_i^2 - \left(\sum_{i=1}^{n} x_i\right)^2} \cdot \sqrt{n \cdot \sum_{i=1}^{n} y_i^2 - \left(\sum_{i=1}^{n} y_i\right)^2}}$$

$$= \frac{5 \cdot 72 - 15 \cdot 18}{\sqrt{5 \cdot 61 - 225} \cdot \sqrt{5 \cdot 88 - 324}}$$

$$= 0.9342$$

Because a correlation coefficient is based on standardized scores, correlations fall in an interval $-1 \leq r \leq +1$. The interpretation of correlation coefficients is as follows:

- $r = +1$: Defines a perfect positive linear relationship. All observations lie on a linear regression line with a positive slope.
- $r = -1$: Defines a perfect negative linear relationship or dependency. All observations lie on a linear regression line with a negative slope.
- $r = 0$: Defines no linear relationship or no linear dependency.

In our example in Table 8.7 we found a correlation of 0.9342, which indicates an almost perfect positive linear relationship between the variables. The example was constructed for illustration purposes. In ordinary communication research we'll never see Pearson correlations of +1 or −1 and likely not a correlation coefficient of greater than 0.8. Correlations in communication research usually vary between ± 0.5 (c.f., Weber & Popova, 2012, for a review of communication meta-analyses). A well-known and in the social sciences frequently used categorization of effect sizes stems from Cohen (1988). Cohen suggests labeling correlations of 0.1 as small, 0.3 as medium, and 0.5 as large.

Significance Test and Confidence Interval for Pearson Correlation Coefficients

Again, the correlation coefficient, r, like the regression slope coefficient, the difference of means, a frequency ratio, etc., is just another statistic with a known sampling distribution if we were to repeatedly draw samples from the same population and calculated the correlation between an independent variable x and an dependent variable y. As we have with other statistical tests, we first formulate both our null and research hypothesis. For a non-directional prediction research hypothesis our pair of hypotheses would look like the following:

$$H_0: \rho = 0 \qquad H_1: \rho \neq 0$$

For a directional research hypothesis, we conduct a one-sided test and our pair of hypotheses would look like the following:

$$H_0: \rho \leq 0 \qquad H_1: \rho > 0 \qquad \text{OR} \qquad H_0: \rho \geq 0; H_1: \rho < 0$$

As before, for a statistical test we need information on the expected value (mean) of the sampling distribution, the standard deviation of the sampling distribution (standard error), and the shape of the sampling distribution. Like in regression analysis, the correlation coefficient's sampling distribution for small samples ($n < 30$) is a t-distribution with ($n - 2$) degrees of freedom and for large samples ($n > 30$) a normal distribution. The standard error of r can be easily calculated with the following formula:

$$\hat{\sigma}_r = \sqrt{\frac{1 - r^2}{n - 2}}$$

The expected value of the correlation coefficient's sampling distribution is the true correlation coefficient ρ in the population which is, according to the null hypotheses stated, expected to be zero. Thus, the standardized empirical test values can be computed as follows:

$$t_{\text{emp}} = \frac{r - \rho}{\hat{\sigma}_r} = \frac{r}{\sqrt{\frac{1 - r^2}{n - 2}}} = \frac{r \cdot \sqrt{n - 2}}{\sqrt{1 - r^2}}$$

As always, we compare this empirical value with a critical test value that corresponds to a predetermined α (e.g., 5%). For $n < 30$ we look at the t-distribution table ($df = n - 2$), and for $n > 30$ we look at the standard normal distribution table (see Appendix A). The result is statistically significant if the absolute empirical value is greater than the critical value, because in this case the corresponding p-value of the empirical result must be smaller than α (c.f., Chapter 6)

If we are more concerned about the size of the correlation and its generalization to the population than knowing it is not zero, we calculate a confidence interval. In line with what we learned before (see Chapter 6) this confidence interval can be easily calculated with:

$$P(r \pm t_{(\alpha/2)} \cdot \hat{\sigma}_{r)} = 1 - \alpha$$

In case of large samples ($n > 30$) we replace the t-value with a z-value of a standard normal distribution (see Appendix A).

Example: Confirmation in Communication about Weight Management

A health communication researcher hypothesizes that the more couples are satisfied with communicating about dieting the higher is their mutual body weight acceptance. In a study of eighteen students a correlation of $r = 0.62$ between weight management conversation satisfaction and partner weight acceptance was found (c.f., Dailey, Romo, & Thompson, 2011). Is this correlation significant on the $\alpha = 1\%$ level? Table 8.8 provides the worked out example.

Table 8.8. Weight management conversation satisfaction and partner weight acceptance

The research hypothesis is a directional hypothesis:

$$H_0: \rho \leq 0 \qquad H_1: \rho > 0$$

Step 1: Compute the standardized empirical test value

$$t_{emp} = \frac{r \cdot \sqrt{n-2}}{\sqrt{1-r^2}} = \frac{0.62 \cdot \sqrt{18-2}}{\sqrt{1-0.62^2}} = 3.16$$

Step 2: Compare with critical t-value

See t-table: $df = 16$; $\alpha = 1\%$ (one-sided test) $\Rightarrow t_{crit} = 2.58$; $t_{emp} > t_{crit}$

Step 3: Summarize and interpret test outcome

Because $t_{emp} > t_{crit}$ the correlation is significant with $\alpha = 1\%$. We reject the H_0 and assume that the found correlation of $r = 0.62$ is not just a random result.

Correlation and Regression

Because correlation is closely related to covariance, and covariance is related to regression, correlation is related to regression analysis. Essentially, regression, covariance, and correlations are based on the same conceptual background. Both regression and correlation (and covariance) provide information on whether or not two variables stand in a linear relationship and whether the direction of this relationship is positive or negative. While regression analysis also enables us to predict values of the dependent variable when knowing the value of an independent variable (via a regression equation), correlation analysis provides us with one simple coefficient that informs us about the strength of a relationship independent from the definition of scales and units of measurement. As can be expected, correlation coefficients and regression slope coefficients are related to each other. If r is known, the slope of a regression function can be calculated as:

$$b = \frac{\text{cov}(x, y)}{s_x^2} = \frac{\text{cov}(x, y) \cdot s_y}{s_x \cdot s_y \cdot s_x} = r \cdot \frac{s_y}{s_x}$$

Thus, the regression slope coefficient is nothing else than the Pearson correlation coefficient multiplied by the ratio of the variables standard deviations. Conversely:

$$r = b \cdot \frac{s_x}{s_y}$$

From the two formulas above it follows that if x and y are z-standardized scores r equals b. Or in other words, if the regression and correlation analysis is based on z-standardized values, the regression slope coefficient b is identical with the correlation coefficient r.

Coefficient of Determination

In practical research, we are usually interested in the following question: How much of the variance of a dependent variable, y, can be explained if we know the values of an independent and related variable, x? If the extent of explained variance were 100%, then the scores of the dependent variable would be completely redundant and could be perfectly predicted with the values of the independent variable, x. As we have discussed, this “law-like” situation will never occur in the social sciences. Variance explanation is always below 100%. In communication research we usually obtain a variance explanation between 0 and 25%. A measure for variance explanation—the coefficient of determination—can be easily derived from the Pearson correlation coefficient. But first we should be able to understand what sources of variance in a given dataset exist and how regression and correlation analysis decomposes the total variance in a dataset in its sources (c.f., Analysis of Variance in Chapter 7). Table 8.9 provides an overview of the decomposition of variance when conducting a regression or correlation analysis.

The total variance of y can be decomposed in variance that is explained by x (explained variance) and variance that remains unexplained. In a regression analysis, the y-values that we predict via a regression equation ($\hat{y}$) are to 100% dependent on x. Thus, the variance of predicted values captures the variance explained by x. As discussed, the actual y-values vary for given x-values due to the impact of other variables and due to random variation (see above). Thus, the variance of the residuals or the variance of the conditional distributions captures the variance that is not explained by x. The ratio of explained variance to the total variance is defined as

Table 8.9. Sources of variance

<table>
<tr><td colspan="2">Total variance of y
σ_y^2</td></tr>
<tr><td>$\sigma_{\hat{y}}^2$
Variance of y explained by x
=
Variance of predicted values $\hat{y}$</td><td>$\sigma_{(y|x)}^2$
Variance of y unexplained by x
=
Variance of the residuals
(Variance of the conditional distributions)</td></tr>
<tr><td colspan="2">$$r^2 = \frac{\text{Variance of } y \text{ explained by } x}{\text{Total variance of } y} = \frac{\sigma_{\hat{y}}^2}{\sigma_y^2}$$ $$\frac{\sigma_{\hat{y}}^2}{\sigma_y^2} = \frac{\sigma_y^2 - \sigma_{(y|x)}^2}{\sigma_y^2} = \frac{TSS - SSE}{TSS}$$ $$= \frac{\sum (y - \bar{y})^2 - \sum (y - \hat{y})^2}{\sum (y - \bar{y})^2}$$</td></tr>
</table>

"variance explanation" or "explained variance." It can be shown that this ratio is simply the Pearson correlation coefficient squared. This squared correlation coefficient r^2 has been labeled as the coefficient of determination. So if you've computed an r value, you can easily compute the r^2 value. Needless to say that r^2 always falls between 0 and 1 (0% and 100%). In other words, you cannot have negative variance explained. That would not make sense.

Assumptions and Issues

Regression analysis and correlation analysis are closely related and on a conceptual level essentially the same. Therefore, all assumptions and issues discussed in the context of regression analysis (see above) apply here as well. However, with the knowledge of covariance and correlation analysis, some additional and, in practical research, frequently occurring issues can now be understood. In the following we will discuss these additional issues which are commonly known as selection errors in correlation analysis.

As we mentioned earlier, we assume a perfect random sample when conducting a significance test for regression slope coefficients or for correlations. In practical research, however, we are often unable to realize a perfect random sample and, as we will see, correlations (and regression coefficients) are particularly sensitive to an imperfect selection of study elements. We will discuss three of the most frequently occurring selection errors.

The first such selection error that we discuss is one where the sample does not cover the variables' of interest entire variation/range in the population. This selection error is illustrated in Figure 8.8.

Figure 8.8. Sample doesn't cover entire range

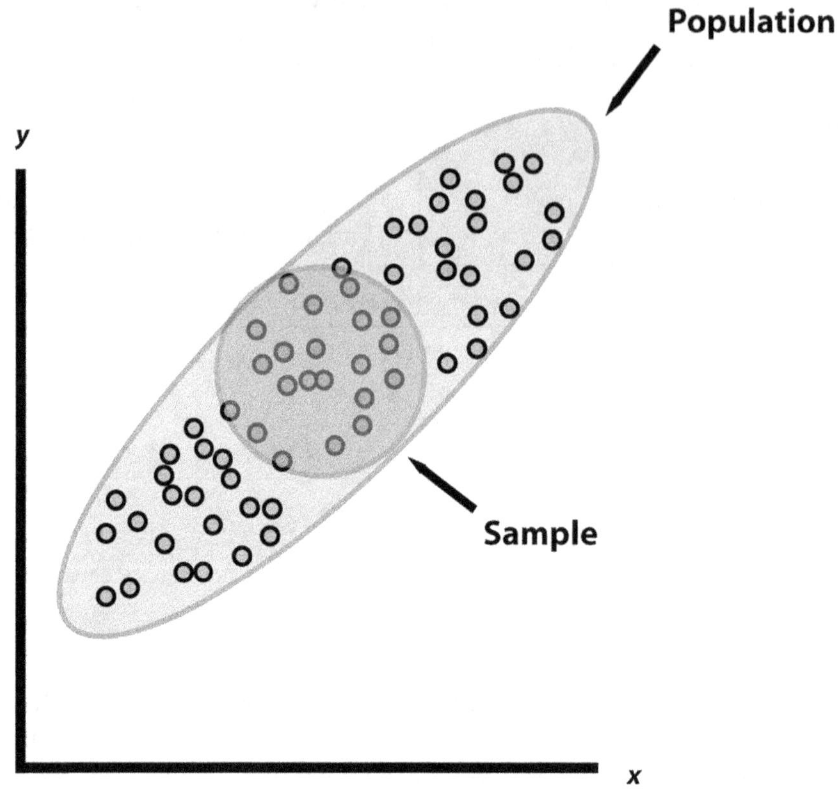

As we can see in Figure 8.8 the applied sampling procedure led to a restriction in variability of the dependent variable, *y*, and the independent variable, *x*. In the population, the variability of *y*- and *x*-values is clearly larger than in the sample. This situation frequently occurs when researchers work with undergraduate student samples, which is common in academic research. Undergraduates usually do not represent the entire variability of a given variable in the population (variables such as household income or cognitive abilities may serve as an example). As a consequence, researchers usually underestimate the true correlation in the population. Why? As Figure 8.8 illustrates, due to the restriction of variability the relationship between the *y* and *x* becomes less pronounced compared to the population.

The second such selection error that we discuss is one where the sample covers extreme groups of the population. This situation is rather common in pre-testing during research projects. Before researchers spend a lot of time and money on studying a phenomenon in all its details, they typically make sure that they can expect some relationship among the variables of interest. As a strategy, researchers sometimes select extreme groups for pre-testing. This selection error is illustrated in Figure 8.9.

For example a media researcher may compare heavy television viewers (> 8 hours per day) with non-viewers (e.g., Amish people). The rationale behind this strategy is the belief that if no

Figure 8.9. Sample covers extreme groups of the population

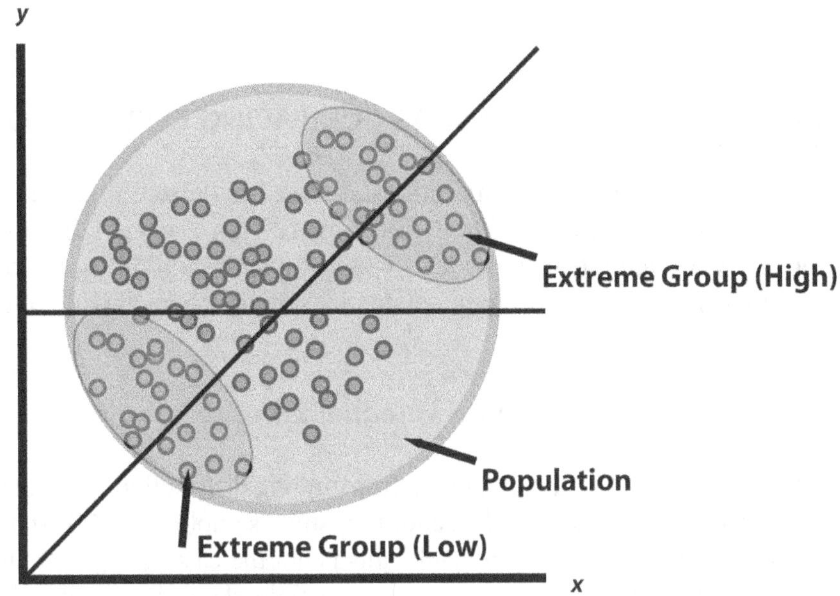

difference with regard to some meaningful dependent variable (e.g., the perceived prevalence of violent crime) can be observed using extreme groups, then it is likely that no difference can be found in less extreme groups. This strategy can prove to be dangerous on multiple levels. As Figure 8.9 illustrates, using extreme groups may suggest a relationship that is actually not existent in the population. Thus, committing a selection error by using extreme groups typically leads to an overestimation of the true correlation in the population. A real problem arises when in light of the convincing results in pre-test operating with extreme groups those results are published and the main study is dropped.

Due to the first two potential selection errors it is very important for a researcher to describe the sample and sample procedure precisely in research reports. Or for communication professionals it is very important to read the method section of research reports carefully. If a regression or correlation analysis is based on selection error samples, the results of the analyses may be misleading and for sure the results of significance tests are worthless.

A third selection error—known as the outlier problem—has already been discussed in the context of regression analysis (see Figure 8.6). Correlation and regression coefficients are very sensitive to outliers. One outlier can bias a regression line, especially in small samples. Figure 8.6 illustrates the effect of an outlier on a regression line and on the corresponding correlation. The dotted line represents the regression line with the outlier included, whereas the solid line represents the regression line with the outlier removed. Clearly, even just one outlier has the potential to bias the regression coefficients and correlation. In practical research we recommend that you inspect your sample for outliers (e.g., by inspecting a scatterplot of dependent and independent variables). Report outliers, but do not include them in a regression or correlation analyses.

FISHER'S Z-TRANSFORMATION: TESTS FOR COMPARING CORRELATIONS

So far, we have only discussed a test for correlations assumed to be zero in the population. That is, $H_0: \rho = 0$ (non-directional) or $H_0: \rho \leq | \geq 0$ (directional). This test is not the only one that we can perform. Frequently, we are interested in tests where we assume $\rho_0 \neq 0$, 0.4 for instance. In order to do a test for H_0: $\rho_0 \neq 0$, we have to understand a specific characteristic of correlations.

Sampling Distributions for Correlations Unequal Zero

As mentioned, if we assume $\rho = 0$ in a population, the sampling distribution of sample correlations follows a *t*-distribution with $n - 2$ degrees of freedom for small samples ($n < 30$) and a normal distribution for large samples ($n \geq 30$). Unfortunately, this is not the case for $\rho_0 \neq 0$. In this case we get a left-skewed sampling distribution for $\rho_0 > 0$ and a right-skewed sampling distribution for $\rho_0 < 0$. However, the famous British statistician Ronald Aylmer Fisher could show that sample correlation coefficients can be transformed so that their sampling distribution is approximately normally distributed even if $\rho_0 \neq 0$ in the population. This famous and important transformation is called Fisher's Z-transformation (capital Z) and has nothing to do with the *z*-transformation that we already know (to standardize variables). Fisher's Z can be calculated with the following formula:

$$Z = \frac{1}{2} \cdot \ln \frac{1 + r}{1 - r}$$

$$r = \frac{e^{2Z} - 1}{e^{2Z} + 1}$$ *e* stands for the Euler number ($e \approx 2.718$)

Instead of manually calculating Fisher's Z-values (or correlation coefficients *r* when Z is given), we can look them up in a Table 5 of Appendix A. Each correlation *r* has a corresponding Z value and vice versa. Note that the Table 5 in Appendix A lists only positive correlations. For $r < 0$ we select the corresponding Z-value for $r > 0$ and multiply by -1.That is, we put a minus sign in front of the Z-value.

Succinctly put, we need Fisher's Z-scores to test null hypotheses where $\rho_0 \neq 0$, and to compare and average correlations. We need Fisher's Z-scores because sampling distributions of correlations that stem from a population where $\rho_0 \neq 0$ are skewed (i.e., are not *t*- or normally distributed) and correlations do not provide a measure on an interval scale. This means that we cannot simply average and compare correlations or test correlations for significance if a population correlation is assumed to be unequal zero. Let's first discuss the problem of comparing and averaging correlation by considering an example.

Averaging and Comparing Correlations Using Fisher's *Z*-Scores

A researcher is interested in studying parents' communication skills from the perspective of the child. In study 1, a researcher found a correlation of 0.4 between communication competence and perceived social support. In study 2, a researcher found a correlation of 0.8. Although $0.8 = 2 \cdot 0.4$, the correlation in study 2 is—unfortunately—not twice as high as in study 1. This is the meaning of saying "correlations do not provide a measure on an interval scale." In order to correctly average correlations, we have to transform the correlations into Fisher's Z-scores. Fisher's Z-scores do provide a measure on an interval scale and as such can be averaged.

By looking at Table 5 in Appendix A, we can see that an r of 0.4 corresponds to a Fisher's Z-value of 0.424, and an r of 0.8 corresponds to a Fisher's Z-value of 1.099. What does this mean? 1.099 is about 2.6 times 0.424. Thus, a correlation of $r = 0.8$ indicates a dependency that is approximately 2.6 (and not 2) times stronger than a correlation of $r = 0.4$. As we can see by looking at the Fisher's Z-scores table, high correlations ($r \rightarrow \pm 1$) are more important (i.e., converted to higher Z-scores) than low correlations ($r \approx 0$). We've shown that we cannot simply compare correlations to each other. We also cannot simply average correlations or calculate variances of correlations. We will show how to use Fisher's Z-transformation to compute an average correlation.

What is the average correlation when $r_1 = 0.2$, $r_2 = 0.5$, and $r_3 = 0.9$ (the sample size n is the same in each case)? We first transform the correlations into Fisher's Z-scores. Again, we look at Table 5 in Appendix A. We get $Z_1 = 0.203$, $Z_2 = 0.549$, and $Z_3 = 1.472$. Now we can average these Z-scores. We get a mean Z-score of $\bar{Z} = 0.741$.

We can now transform this mean Fisher's Z-score back into a correlation, and we get $\bar{r} = 0.629$. The average correlation is $\bar{r} = 0.629$ and not $\bar{r} = 0.53$, which is the simple average of the three correlations without using the correct Fisher's Z-transformation. For different sample sizes, we simply calculate a weighted average (see Chapter 2) Fisher's Z-score using the sample sizes as weights. For example, using the same correlations and Fisher's Z-scores as aforementioned, but different sample sizes of $n_1 = 20$, $n_2 = 30$, and $n_3 = 50$, we obtain as average Z-score $(20 \cdot 0.203 + 30 \cdot 0.549 + 50 \cdot 1.472)/100 = 0.94$ and as corresponding average correlation coefficient 0.735.

Testing the Null Hypothesis $\rho = \rho_0$ ($\rho_0 \neq 0$) Using Fisher's *Z*-Scores

We want to test if a sample correlation, r, belongs to a population in which we expect a correlation of ρ_0 but this null hypothesis population correlation is assumed to be unequal zero. Thus, our hypotheses can be stated as follows:

H_0: $\rho = \rho_0$ ($\rho_0 \neq 0$)	H_1: $\rho \neq \rho_0$ ($\rho_0 \neq 0$)	non-directional
H_0: $\rho \leq$ or $\geq \rho_0$ ($\rho_0 \neq 0$)	H_1: $\rho >$ or $< \rho_0$ ($\rho_0 \neq 0$)	directional

ρ_0 can be transformed into a Fisher's Z-score. Knowing that the sampling distribution of Fisher's Z-scores is normally distributed with the expected value of E(Z) = μ_Z and known standard error σ_Z we calculate:

$$z_{emp} = \frac{Z - \mu_Z}{\sigma_Z}$$

$$Z = \frac{1}{2} \cdot \ln\left(\frac{1 + r}{1 - r}\right) = Z(r) \quad \text{Sample result Fisher's } Z \text{ transformed}$$

$$z = \frac{1}{2} \cdot \ln\left(\frac{1 + \rho_0}{1 - \rho_0}\right) = Z(\rho_0) \quad \text{Expected value Fisher's } Z \text{ transformed}$$

$$\sigma_Z = \sqrt{\frac{1}{n - 3}} \quad \text{Standard error of Fisher's } Z\text{-scores}$$

With this information, the significance is analog to a standard z-test introduced in Chapter 6, that is, analog to a test that compares a sample statistic with a population parameter. We simply compare the empirical z-value with a critical z-value at a given α-level. If the absolute empirical value exceeds the critical value, the result is significant and we reject the null hypothesis that $\rho = \rho_0$ ($\rho_0 \neq 0$).

Knowing the sampling distribution of correlations $\neq 0$ and also the standard error makes it easy to calculate a confidence interval for correlations $\neq 0$. Analog to all confidence intervals, this is defined as:

$$P(Z(r) - z_{(\alpha/2)} \cdot \sigma_Z \leq Z(\rho) \leq Z(r) + z_{(\alpha/2)} \cdot \sigma_Z) = 1 - \alpha$$

Let us consider an example. A researcher (c.f., Wheeless, Witt, Maresh, Bryand, & Schrodt, 2011) is interested in the relationship between instructor credibility and students' intent to persist in college. The researcher knows from prior research that these two variables correlate with $\rho_0 = 0.4$. In a sample of seven students (drawn randomly from a large lower-division communication course), she found a correlation of $r = 0.2$. Is this correlation significantly different from ρ_0? We want to test with $\alpha = 0.05$. Table 8.10 provides the computations.

Testing the Null Hypothesis $\rho_1 = \rho_2$ Using Fisher's *Z*-Scores

If we want to test if two correlations in two independent samples with sample sizes n_1 and n_2 are significantly different, we can calculate the following empirical z-value (knowing that Fisher Z-scores are normally distributed):

$$z = \frac{Z_1 - Z_2}{\sigma_{Z_1 - Z_2}} \qquad \sigma_{(Z_1 - Z_2)} = \sqrt{\frac{1}{n_1 - 3} + \frac{1}{n_2 - 3}}$$

Table 8.10. Example—Instructor credibility and students intent to persist in college

$H_0: \rho_0 = 0.4 \qquad H_1: \rho_1 \neq 0.4$

Step 1: Convert correlations into Fisher's Z-values using Table 5 in Appendix A

$$Z(r = 0.2) = 0.203; Z(\rho_0 = 0.4) = 0.424$$

Step 2: Compute standard error of $Z(r)$

$$\sigma_Z = \sqrt{\frac{1}{n-3}} = \sqrt{\frac{1}{7-3}} = 0.5$$

Step 3: Determine the standardized empirical test value

$$z_{emp} = \frac{Z(r) - Z(\rho_0)}{\hat{\sigma}_Z} = \frac{0.203 - 0.424}{0.5} = -0.442$$

Step 4: Compare with critical test value
For a two-sided test and for an α-level of 5% $\Rightarrow z_{crit} = -1.96$

Step 5. Summary and interpretation
The standardized empirical test value is not greater than the critical value. Thus, the correlation that the researcher found is not significantly different from what previous research has found.

The hypotheses and the test are analogous to the *t*-test for comparing two means in two independent samples (see Chapter 7). Again, let us consider a modification to the earlier example. This time we have correlation coefficients from two different ethnic populations, and we want to know whether the coefficients could represent the same population:

We know that there is a relationship between instructor credibility and students' intent to persist in college. In a sample of $n_1 = 50$ Caucasian students a researcher found a correlation of $r_1 = 0.6$. In an independent sample of $n_2 = 35$ Hispanic students, the researcher found a correlation $r_2 = 0.4$. Is the correlation among Caucasian students higher than among Hispanic students? Test with $\alpha = 0.05$. Table 8.11 provides the computations.

CORRELATION AND CAUSALITY

If we found a correlation between two variables *X* and *Y* there are several possible causal interpretations:

1. *X* has an (causal) effect on *Y*, that is, *X* is the cause and *Y* the effect.
2. *Y* has an (causal) effect on *X*, that is, *Y* is the cause and *X* the effect.

Table 8.11. Example—Instructor credibility and intent to persist in college

H0: $\rho_1 \leq \rho_2$ H1: $\rho_1 > \rho_2$ Directional Hypothesis

Step 1: Convert correlations into Fisher's Z-values using Table 5 in Appendix A

$$Z(r_1 = 0.6) = 0.693;\ Z(r_2 = 0.4) = 0.424$$

Step 2: Compute the standard error of the Z-score difference

$$\sigma_{Z_1-Z_2} = \sqrt{\frac{1}{n_1 - 3} + \frac{1}{n_2 - 3}} = \sqrt{\frac{1}{50 - 3} + \frac{1}{35 - 3}} = 0.23$$

Step 3: Determine the standardized empirical test value

$$z_{emp} = \frac{Z_1 - Z_2}{\sigma_{Z_1-Z_2}} = \frac{0.693 - 0.424}{0.23} = 1.17$$

Step 4: Compare with critical test value
For a one-sided test and for an α-level of 5% $\Rightarrow z_{crit} = 1.65$

Step 5: Summary and interpretations
The empirical test value is not greater than the critical value. Thus, the correlation among Caucasian students is not significantly higher than among Hispanic students.

3. Spurious correlation: *X* and *Y* are both affected by a third variable *Z* and have no direct connection. That is, *X* and *Y* are neither cause nor effect, controlling for *Z* results in no correlation between *X* and *Y*.
4. Chain relationship (complete mediation): *X* affects *Z* and *Z* affects *Y*. *X* is the cause for *Z*, but not the cause for *Y*. Controlling for *Z* results in no correlation between *X* and *Y*.
5. Interaction (moderation): The correlation between *X* and *Y* varies with levels of *Z*. Controlling for *Z* results in different correlations between *X* and *Y* according to levels of *Z*.
6. *X* affects *Y* directly and indirectly through *Z*.
7. *X* and *Y* mutually influence each other. That is, *X* and *Y* are cause and effect at the same time.

Figure 8.10 visualizes these possible causal interpretations.

Unfortunately, the Pearson correlation does not provide any information on which of these possible interpretations is correct. Many correlations in communication research are of type 3 (spurious); frequently, third variables are actually responsible for the correlation between two variables. For example, there is a positive correlation between education and the diversity of video games young adults own and play. A seemingly obvious conclusion could be that better educated people select a more diverse set of video games. However, available monetary resources (income) of people typically affect both variables: richer people tend to have a higher education and richer people have the resources to buy more (and a more diverse set) of video

Figure 8.10. Possible causal interpretations of a Pearson correlation

1.	$X \rightarrow Y$	X has an effect on Y
2.	$X \leftarrow Y$	Y has an effect on X
3.	Z $\swarrow \quad \searrow$ $X \qquad Y$	Spurious correlation
4.	$X \rightarrow Z \rightarrow Y$	Chain relation/mediation
5.	Z $\downarrow$ $X \rightarrow Y$	Interaction/moderation
6.	Z $\nearrow \quad \searrow$ $X \rightarrow Y$	Both direct and indirect effect
7.	$X \leftrightarrow Y$	X and Y are mutually affected

games. Thus, the correlation between education and video game diversity is spurious. We only see a correlation between the two variables because both variables are affected by a third variable (income).

The main takeaway that we'd like to emphasize is that a correlation is a necessary condition for a causal relationship, but not a sufficient condition. In other words, if we find a solid positive or negative Pearson correlation between two variables, those variables may stand in a causal relationship, but we cannot be certain. At the same time, if we can only find a Pearson correlation close to zero between two variables, it is very unlikely that the two variables are causally related. If it is difficult to demonstrate causality with simple correlation analysis, then you may ask how can researchers make a causal claim or find evidence for causality. Before we answer this question we need to define what causality actually means.

DEFINITION OF CAUSALITY

There are many definitions (or conceptualizations) of causality (c.f., Pearl, 2000). A basic, and frequently referenced, definition in communication research is the following: Two variables are causally related if three criteria are met:

1. There is a significant correlation between the two variables.
2. The variables stand in an appropriate time order. That is, a response follows a stimulus or the variable that is considered as the cause precedes the variable that is considered as the effect.
3. Alternative explanations for the correlation can be eliminated. That is, the correlation must not be spurious.

This definition of causality can be used to test causal relationships in specific research designs. Consider the following example. Von Salisch, Vogelgesang, Kristen, and Oppl (2011) were interested in studying whether the preference for playing violent video games among children leads to aggressive behavior (effect hypothesis) or whether already aggressive children (due to other reasons) prefer to play violent video games (selection hypothesis). The researcher designed a longitudinal study illustrated in Figure 8.11 to test which of the two causal hypotheses can be rejected.

Figure 8.11 demonstrates that the researchers asked the same sample of children twice, the first time in 2010 and the second time a year later. A study that collects data repeatedly from the same participants is called a panel study. Using a panel design enabled the researcher to analyze six different Pearson correlations: The cross-sectional correlations r_4 and r_3, the longitudinal correlations r_5 and r_6, and the cross-lagged correlations r_1 and r_2.

Of interest are particularly r_1 to r_4. Our first question is: What would we expect for r_1 compared to r_2 if the effect hypothesis were true? We would expect $r_1 > r_2$, because in this case, a model that considers preference for violent games at time t_1 and aggressive behavior at time t_2, or preference as the cause and aggressive behavior as the effect, fits the data better than a model that considers aggressive behavior as the cause and preference for violent games as the effect. With the same rationale, we would expect $r_1 < r_2$ if the selection hypothesis were true.

Our second question is: What would we expect for r_3 compared to r_4, if the effect hypothesis were true? We would expect $r_3 > r_4$, because one year additional exposure to violent video games should increase the correlation between playing violent video games and aggressive behavior if the effect hypothesis is true. In contrast, if the selection hypothesis were true, we would expect $r_3 \approx r_4$. Looking at the actual correlations in Figure 8.11 we can see that the results in this cross-lagged panel study suggest rejecting the effect hypothesis in favor of the selection hypothesis.

It should be noted that testing competing causal hypotheses with more elaborate research designs such as the cross-lagged panel design in Figure 8.11 is obsolete if causality is the result of logical reasoning. For example, it does not make logical sense that the mood of a person affects the weather conditions. It is, of course, the other way around: weather affects mood.

Figure 8.11. Cross-lagged panel study testing causal hypotheses

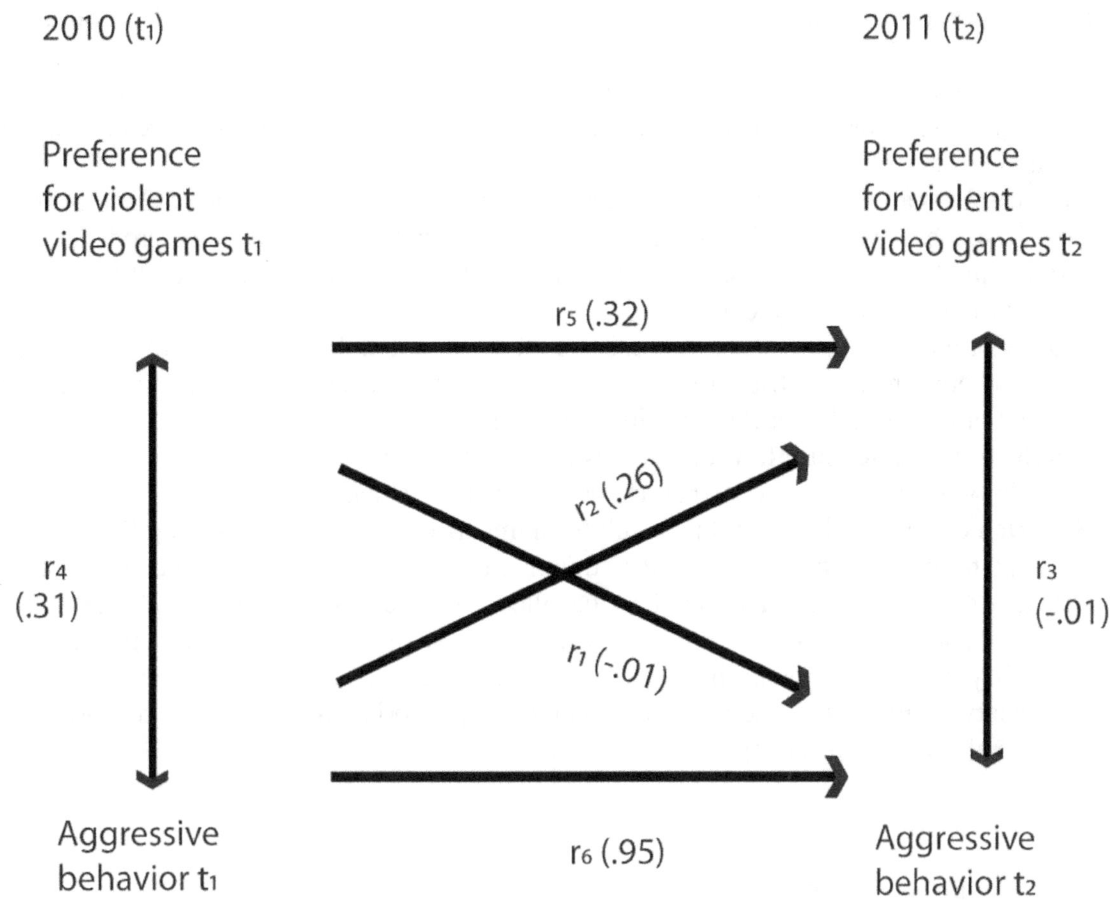

SUMMARY

The purpose of this chapter was to introduce specific tests for dependencies and relationships. We drew on both a descriptive (e.g., calculating regression coefficients, covariances, and correlations), as well as inferential perspectives (significance tests for regression and correlation coefficients). We have also discussed the challenges that researchers must consider about samples when employing these techniques, as well as not conflating correlation with causality. With a firm understanding of the concepts in this chapter, students are ready to move on to advanced statistical methods, including factorial analysis of variance, multiple and nonlinear regression, and multivariate statistical procedures. Although we do not discuss these advanced statistical methods in this book, we believe that with what students have learned from Chapters 6 through 8, they will be well-equipped to understand more challenging topics in statistics.

TERMS

- **Regression** is a procedure that is based on linear equations that describe the relationship between one dependent (criterion, response) variable and one or more independent (predictor, explanatory) variables.
- **Least squares line** is a line that best represents the relationship between two variables. The least square line corresponds to a linear equation and is the one that minimizes the sum of squared errors (residuals).
- **Covariance** is a measure that describes the strength and direction of a linear relationship between two metric variables. The covariance can be understood as the average product of corresponding deviations of values from their mean. The covariance depends on the variables' scales or units of measurements.
- **Pearson correlation** provides information on the strength and the direction of a linear relationship between two metric variables. In contrast to the covariance, correlations are independent of the variables' scales or units of measurement.
- **Coefficient of determination** informs us on how well explained our predicted/dependent variable (y) is by using the explanatory/independent variable(x).
- **Selection errors** are the errors that result from imperfect sampling techniques. Selection errors lead to problems of under- or overestimating regression and correlation coefficients.
- **Fisher's Z-transformation** is a method that allows researchers to compare and average correlations, and to test correlations that are expected to be unequal zero in the population.
- **Causality** is defined through three criteria of causality: 1) There is a significant correlation between two variables; 2) the variables stand in an appropriate time order; 3) the correlation between the variable is not spurious.

CHAPTER 8 PROBLEM SET

1. A researcher (c.f., Paek, 2009) is interested in the relationship between the intention to smoke and injunctive norms. Specifically, the researcher wants to predict the intention to smoke with data on perceived injunctive norms. Variables are measured on a seven-point scale. The data are as follows:
   a. Draw a scatterplot
   b. Assuming that this forms a linear function, calculate the slope and y-intercept.
   c. Compute the empirical t-value and compare with a critical t-value of $\alpha = 5\%$.
   d. Summarize and interpret the test outcome.
   e. Calculate the covariance.

Participants	Intention to smoke (1 = unlikely to 7 = very likely)	Perceived injunctive norms (1 = Not OK to 7 = perfectly fine)
1	1	2
2	3	4
3	2	3
4	1	3
5	4	4
6	2	5
7	1	1
8	5	6
9	6	7
10	7	5

2. What is the major disadvantage of covariance?
3. What is an advantage of correlation compared with covariance?
4. A researcher is interested in the relationship between perceived media influence (operationalized as youth's beliefs about the influence of smoking-based messages on peers) and intention to smoke. Variables are measured on a seven-point scale. You have the following data:

Participants	Intention to smoke (1 = unlikely to 7 = very likely)	Perceived Media influence (1 = Not at all to 7 = A lot)
1	1	3
2	3	4
3	2	5
4	1	2
5	4	4
6	2	3
7	1	1
8	5	5
9	6	7
10	7	6

   a. Calculate the correlation coefficient.
   b. Test the correlation coefficient for significance ($\alpha = 5\%$).
   c. Summarize and interpret the test outcome.
   d. Compute the coefficient of determination.
5. Explain the concept of a selection error. How might you check your data to determine whether a selection error has occurred?
6. A communication researcher is interested in the relationship between driver communication quality (e.g., driver politeness) and extensive policing (e.g., length of traffic stop). The researcher is applying for a grant and would like to summarize the correlation coefficients to include in her application. Prior to this application, the researcher has conducted two separate content analyses (one in a large metropolitan area and another in a medium sized city). She has the following two correlations between driver communication quality and extensive policing: $r_1 = -0.2$; $r_2 = -0.4$.
   a. What is the average Z-score?
   b. Convert the average Z-score intro an average Pearson correlation.
7. A researcher is interested in the use of fear appeals and attitude change. He wants to test whether two correlations from two independent samples in which the relationship between fear appeals and attitude change were studied are equal. The researcher has the following data: Sample 1, $r_1 = 0.3$, $n_1 = 50$. Sample 2, $r_2 = 0.6$, $n_2 = 55$.
   a. Conduct the appropriate significance test.
   b. Provide a summary of your findings and interpretations.
8. What are the criteria that allow you to determine causality? What are some considerations you would have in designing a study to determine causality?

REFERENCES

Cohen, J. (1988). *Statistical power analysis for the behavioral sciences* (2nd ed.). Hillsdale, NJ: Lawrence Erlbaum.

Dailey, R. M., Romo, L. K., & Thompson, C. M. (2011). Confirmation in couples' communication about weight management: An analysis of how both partners contribute to individuals' health behaviors and conversational outcomes. *Human Communication Research, 37*(4), 553–582.

Mendenhall, W., & Sincich, T. (2011). *A second course in statistics: Regression analysis* (7th Edition). Upper Saddle River, NJ: Prentice Hall.

Paek, H. (2009). Differential effects of different peers: Further evidence of the peer proximity thesis in perceived peer influence on college students' smoking. *Journal of Communication, 59*, 434–455.

Pearl, J. (2000). *Causality. Models, reasoning, and inference.* Cambridge: University Press.

Raines, S. A., & Keating, D. M. (2011). The social dimension of blogging about health: Health blogging, social support, and well-being. *Communication Monographs, 78*(4), 511–534.

Von Salisch, M., Vogelgesang, J., Kristen, A., & Oppl, C. (2011). Preference for violent electronic games and aggressive behavior among children. The beginning of the downward spiral? *Media Psychology,* 14(3), 233–258.

Weber, R., & Popova, L. (2012). Testing equivalence in communication research: Theory and applications. *Communication Methods and Measures,* 6(3), 190–213.

Wheeless, V. E., Witt, P. L., Maresh, M., Bryand, M. C., & Schrodt, P. (2011). Instructor credibility as a mediator of instructor communication and students' intent to persist in college. *Communication Education, 60*(3), 314–339.

CHAPTER 9

EPILOGUE

This final chapter offers something different. It is different from the rest of the book, and it is different from what is read in most books on statistics. For one thing, the reader may notice a change in "voice." This chapter was written by a guest author. Given the author switch, we first would like to introduce our guest author.

Timothy R. Levine (Tim) is a Professor in the Department of Communication at Michigan State University. Tim's research interests include deception, interpersonal communication, persuasion, and scale validation. He has published more than one hundred journal articles on topics related to communication. In 2011, he was honored with Michigan State University's prestigious Distinguished Faculty Award. He has published numerous journal articles on statistical practice and measurement validation in the field of communication. He is the founding co-editor of the journal *Communication Methods and Measures,* the communication discipline's flagship journal in matters of statistics, measurement, and research design. Without doubt, Tim is among the foremost experts in statistics in the field of communication and has the credentials of a true statistics expert in our field. We are very lucky that Tim—with an independent perspective—has agreed to contribute this critical chapter and we advise our readers to take Tim's recommendations to heart, be it as a communication researcher or a communication professional.

A CRITICAL PERSPECTIVE ON STATISTICS
TIMOTHY R. LEVINE, MICHIGAN STATE UNIVERSITY

This chapter offers a different perspective on quantitative methods for the social sciences in general and for communication research in particular. In the introduction to this chapter I am regarded as "among the foremost experts in statistics in the field of communication." On self-reflection and in contrast to this image of a statistical expert, I am struck by two interrelated things as I begin writing this chapter. First, if one reads my original research, and especially my more recent work, I think they will find that I use very simple and basic approaches to data analysis. My reader is more likely to encounter a frequency distribution than a structural equation model. In more than one hundred published articles, I have never published anything with a multivariate Analysis of Variance (MANOVA) or the like. So, I ask the reader to consider why this might be the case. Why does the guy who supposedly knows so much about statistics usually select the most elementary analyses in his own research? A second self-reflective point is that when thinking about my knowledge of statistics, I am struck more by what I don't know than what I know. I envision two sets. There is the set of what there is that can be currently known about statistics and there is the set of what I know about statistics. Contemplating how small the second set is in relation to the first is humbling. The proportion of what there is to know to what I know is surely tiny.

I know, I risk killing my credibility with these admissions. But, there is a very important point to be made here upfront: In statistics, like many things, a little knowledge can be dangerous. The problem is that it is often the very nature of ignorance to lack awareness of what is not known. There is so much to know, life is short, and humans have limited knowledge acquisition, integration, storage, and retrieval capacities. Besides, there are other useful things to learn about other than statistics. As a consequence, it can be the case when using or interpreting statistics that if we had more complete knowledge, we might do things differently or arrive at different conclusions. People publishing the results of statistical analyses risk being read by others more knowledgeable than themselves. People reading analyses done by others might, if only they knew a little more, catch some error, or reach some different conclusion. The book that you hold in your hands made an attempt to foster *understanding* of statistical analysis among communication researchers and professionals rather than merely advocating a "cookbook style" (and dangerous) application of statistical methods. I definitely support the book's mission and think this mission makes this book unique for communication scientists and communication professionals, but even if equipped with all the knowledge of this book, your understanding and knowledge of statistics remains incomplete. Where knowledge is incomplete, inadvertent and unnoticed error is of concern. In my view, we all are well served by greater awareness of our limitations. Not knowing everything is inevitable. Strive to reduce what is not known, but be aware that ignorance reduction is a lifelong pursuit that will never fully be accomplished. Awareness that there are things we don't know and that we don't know what we don't know has three desirable outcomes. It motivates continued learning, it advises caution where caution is surely warranted, and it reveals the statistically arrogant for what they are—self-delusional.

ON CRITICS AND CRITICISM

Because the title and function of this chapter involves "a critical perspective," it might be useful to discuss the critic, being critical, and doing criticism. Why I think a brief diversion into what might be called meta-criticism (i.e., a critical look at being critical) is necessary requires a little back-story.

It is my perception that the culture in the United States has undergone change over the past twenty years and that there is a parallel change in the culture of scholarship. I call the new order the "cult of esteem." What are paramount in the cult of esteem is feeling good about oneself and ensuring that others feel good about themselves, too. High self-esteem is viewed not only as the primary, and unquestionably desirable, goal state, but also as both insatiable and fragile. Thus, followers of the cult of esteem view the provision of continual praise as a virtue, obtaining praise regardless of merit as the sign of self-worth, and the finding of fault offensive.

In the cult of esteem, criticism is threatening. In scientific reasoning and methodology, however, criticism is absolutely essential. Thus, embracing a scientific approach means being out of step with the cult in matters of science, research, and scholarship. Unlike the ideas embraced by cults, in science, ideas must correspond with data and arguments must be logically coherent. Correspondence with data and coherence of arguments are the bottom lines in evaluating all scientific ideas, hypotheses, models, theories, and the like. When either of these is lacking, criticism is not only expected, it is required. Poor or underdeveloped ideas need to be culled and discarded. The goal is the creation and evaluation of defensible knowledge claims that pass the dual tests of coherence and correspondence. Critics test the strength of knowledge claims, and critics need to test knowledge claims. Scientific knowledge claims are not accepted at face value nor are they subjected to unconditional acceptance. Knowledge not subjected to critical appraisal is suspect.

Psychology has documented many biases in human cognitive processes, but perhaps none are as strong, pervasive, or well-documented as the confirmation bias. Humans tend to look for and more heavily weight evidence supporting our own views while ignoring, discounting, or under-appreciating disconfirming information. One recent theory even suggests that human reasoning capabilities evolved not to help us understand our world as it is, but to win arguments (Mercier & Sperber, 2011). Critics and criticism play an essential role in protecting against confirmation bias. It is usually easier for other people to see the weakness in our ideas, research, and analyses than to see those weaknesses on our own. Thus, we need good critics and criticism, not blanket praise.

A final point regarding the value of criticism is that it motivates both quality and progress. The idea that research will be subjected to criticism requires quality control to ensure that it survives the critics' best shots. However, no research project is ever perfect, so some criticism will always have merit. Overcoming valid criticism yields progress.

It is in the last sense that I intend this chapter. To the individual reader, I say now that you have hopefully learned what the book has to teach, be mindful that your knowledge is nevertheless inevitably incomplete. Put your new knowledge to good use, but do not be too smug in this knowledge and be open to learning more. To the larger communication research community, I say we can do better. In my view, there is plenty to criticize about statistical and methodological practices in the social sciences and in communication research. This does not mean, however, that current practices are worthless, fatally flawed, or valueless. Social science and communication research have come a long way since introspection was a primary form of data collection. Huge gains have been made in almost all areas of methods. This said, the status quo can be improved in numerous areas.

This chapter prioritizes four concerns that are presented as four rants. These include: (1) divorcing method and theory on one extreme and viewing the hypothetico-deductive (HD) approach as the only valid approach to social science on the other; (2) the prioritization of statistics over methods in social scientific education; (3) the over-reliance on null hypothesis significance testing and the under-use of descriptive statistics; and (4) the conflation of complexity and sophistication.

FOUR STATISTICAL RANTS

Rant Number One—Statistics, Substantive Focus, and Theory

My first criticism of statistics as taught and practiced in communication (and in the social sciences) is that it is sometimes considered as divorced from, or independent of, the substantive area in which the data exist. In contrast to made-up data that sometimes serve as examples in statistical texts and classes, real data have context. Real data exist in a larger ecology. Much like taking communication out of context, taking data out of context can be misleading. I refer to a rich understanding of the ecology in which data exists as "substantive feel." Statistical analyses that lack substantive feel might be called "dust bowl analyses."

As a point of clarification, when I advocate for substantive feel, what I mean is substantive feel for the data, not substantive feel for the statistics. Statistics can be applied in a wide variety of substantive areas, and understanding statistics is not tied to a particular application or another. For example, my first statistics professor used baseball examples to demonstrate various statistical principles and techniques. It did not matter that I lacked my professor's appreciation of baseball or that my research interests did not involve sports. The point was to understand statistics. Baseball was just one application interchangeable with others. One does not need to understand the history or rules of baseball in order to understand the statistical principles that can be demonstrated with baseball examples. However, the situation is different for the baseball sociologist who may use statistics in their research. No amount of complex statistics can compensate a baseball sociologist who does not have a deep understanding of the game and the society in which the game evolved.

Obviously, the label of dust bowl is meant to be derogatory. Substantive feel provides the data analyst and research consumer with an idea of which variables are important, a sense of why they are important, intuition about what is important to control, and about what findings mean. Findings always need to be understood as existing within a larger research context so substantive feel is as critical as statistical knowledge and dust bowl analysis is unfulfilling.

One should not confuse dust bowl data analysis with exploratory data analysis. Good exploratory data analysis is guided by informed curiosity and substantive feel. What makes exploratory analysis exploratory is that the goal is not to test hypotheses, but instead to see what's in the data. Exploratory analyses done with substantive feel are incredibly valuable and good practice. When a hypothesis is "confirmed" with data, we get a little more confidence in something that was already known. But, with exploratory analyses, there is the chance to come upon new insight. But gaining insight requires substantive feel, not a pre-existing hypothesis.

A practice related to dust bowl analysis that I find worrisome is the rise of the outside statistical expert. As statistics have proliferated in type and complexity, it is increasingly difficult for researchers to keep up with the latest statistical analyses and there is apparent need for statistical experts to analyze data for researchers. While it is the case that doing data analysis requires expertise, the outside statistical expert's interest and expertise is often in the statistics rather the substantive area in which the data exist. The result can be dust bowl analysis. A good *understanding* of basic principles in statistical analysis, as provided in this book, may prevent you from seeking outside experts or might help you to evaluate the true benefit of a statistical expert's advice. Only very few statistical procedures are indeed entirely new and most can be understood on a conceptual level by scholars and professionals equipped with a sound understanding of the basic principles.

There is an opposite extreme to dust bowl analysis that can also be problematic when overapplied. I'll call this other extreme "HD Extremism." The HD stands for the hypothetico-deductive approach to research. Whereas dust-bowl analyses are atheoretical, HD research involves highly formal theory testing. In HD research, existing formal theory is the starting point (see Chapter 6). From theory, a hypothesis or hypotheses are logically derived such that if the theory is true, then the derived hypotheses must also be true, but if the hypotheses are false, then the theory must be false. Thus, by testing the hypotheses, theory can be tested.

There is nothing wrong with HD science if the research goal is formal theory testing. Of course, formal testing theory is a necessary and worthwhile activity. Thus, the HD approach should be part of the skilled researcher's repertoire. The problems come when the HD approach is viewed as the *only* legitimate way to approach research. So, my objection is not to the HD approach per se, but instead to HD Extremism; that is, the extreme view that HD research is morally superior to exploratory analyses and that strict adherence to HD rules of conduct must be indoctrinated and enforced. Needless to say, hypothesis testing is a valuable research enterprise, but it is not the only valuable way to apply statistics.

While HD is a valid and useful research strategy, formal, well-developed theory that allows researchers to make a sound and concise argument is rarely the starting point in communication research. Instead, in actual practice, there is a research rationale (sometimes sound) and researchers fit some "theoretical frame" to their rationale. Hypotheses are not deduced from theory, and no theory is meaningfully tested, but the research is sold as HD. This malpractice ultimately reduces rather than increases knowledge and is done as a publication ritual rather than as a knowledge generation strategy. Because good exploratory research or replications have a lower chance to get published in journals, theory is added for the sake of appearance. This, of course, reinforces HD Extremism, even if authors themselves don't believe in HD as the only approach.

HD research has a major limitation that prevents it from being too widely applicable. The problem with HD is that it requires well developed theory as the starting point. As a consequence, it contains no mechanism for generating new ideas. That is, HD is good for evaluating already developed ideas, but not for coming up with new ideas. The best research programs need to be generative.

In my view, exploratory research with substantive feel is needed to generate new ideas, which once formally developed, can be tested with the HD approach. In between, an abductive approach makes sense. In abductive research, explanation follows from findings (c.f., Rozeboom, 1997; Weber, Sherry, & Mathiak, 2008). It involves coming up with the best explanation given existing findings, then designing new research to validate and replicate those findings. This allows research to lead to new and better theory, and not be restricted either to existing theory or a dust-bowl search for isolated findings. Even when existing theory exists and guides research, exploratory analysis is still advised because interesting findings outside the scope or expectations of the theory might prove valuable. Ultimately, science is about new discoveries, not just validating existing theory or formally testing hypotheses.

Rant Number Two—Focus on Data Analysis over Data Creation

Another of my concerns is that education in the quantitative social sciences and in the communication field emphasizes data analysis over data creation, especially on the graduate level. Students trained in the quantitative social sciences often equate methods with statistics and a large proportion of methods education involves statistics of one sort or another to the neglect of in-depth consideration of research design and measurement.

Here is the problem. The validity of the results from any statistical analysis depends critically on the quality of the data analyzed. Flawed analyses can be recalculated. Flawed data, however, are often beyond repair and no amount of fancy statistical analyses can save the day. Experienced statisticians have coined the phrase: "Garbage in, garbage out" for good reasons. A good "rule of thumb" is that it is better to have high quality data in the first place than to try to statistically correct or control problems. So, data creation seems fundamental to the research process. It is my belief that the weakest aspect of research findings in the social sciences stems from poor or potentially poor measurement. In my field of communication, personality scales of various sorts are widely used. In my experience, when given close scrutiny, few scales fair well (e.g., see Kotowski, Levine, Baker, & Bolt, 2009; Levine et al., 2003). Because it is unclear what, if anything, scores on these instruments tell us, statistical analyses of these scores yield results of uncertain meaning. Statistical knowledge can help reveal the validity problems, but it cannot fix the problem. That requires improvement in measurement creation and validation practices.

I fear that the problem of measurement is even bigger among communication professionals, for instance in public opinion research. It appears that public opinion researchers are much more skilled at sampling respondents than at question writing. Problems involve not only single-item measurement and a lack of pre-testing, but also a host of question effects (see an excellent summary by Schwarz, 1999).

A final related point is that in research design, it sometimes appears that people worry more about sampling respondents than about measurement or experimental control. For many critics, sampling is an easy target. Research is presumed less important or informative because student or convenience samples were used instead of obvious superior random or probability sampling techniques. There are several sub-issues here to discuss.

It is helpful to review distinctions between what Cook and Campbell (1979) call construct validity, internal validity, external validity, and statistical conclusions validity. At the risk of superficiality, I provide a brief overview of these terms, but I direct you to consult methods textbooks for a more detailed explanation. Construct validity has to do with the extent to which we can be sure that the scores on our variables reflect the intended constructs. That is, do the numbers mean what we think they mean? Construct validity is an issue of measurement quality. Internal validity refers to the quality of our inferences about the relationships between variables given the research design. Might some unknown third variable cause changes in the dependent variable rather than the independent variable (see also Chapter 8)? Are there alternative explanations for the findings? External validity involves generalizing the findings beyond the specifics of data analyzed (see Chapter 4). Would the findings generalize to other samples involving different research subjects, methods, contexts, and over time? Finally, statistical conclusions validity involves avoiding Type-I and Type-II errors (see Chapter 6) in statistical hypothesis testing.

The different aspects of validity are sometimes presented as opposing forces that require trade-offs. Sometimes, for example, it is presumed that boosting internal validity somehow requires reduction in external validity. Sometimes, too, prioritizing among types of validity is presented as a matter of debate. In line with the Campbell and Stanley (1963) classic, I view construct and internal validity as the sin qua non (without which, nothing) of research. If we don't know what the numbers mean, or if we cannot have confidence in what variables are in operation, then the quality of statistical analyses are moot, and there can be nothing meaningful to worry about generalization. Without good research design and measurement as a foundation, sampling and analysis are irrelevant. The findings cannot be valid and no amount of statistical wizardry will change that. As a journal reviewer and research consumer, I am often disheartened when I read research that collected data using suspect measurement scales and a hard-to-get or expensive sample of participants. What a waste.

Besides problematic measurement, I think the true value of experimental research is sometimes underappreciated. Experiments are studies meeting three criteria. In experiments, at least one independent variable is under experimental control (i.e., are deliberately manipulated by the researcher), there exists a comparison or control group, and subjects are randomly assigned to an experimental condition. Sometimes experimental findings are viewed as artificial and therefore dismissed as divorced from the real world. Reality, however, is not somehow switched off when a lab is entered. More importantly, however, experimental research has some huge advantages over non-experimental research in terms of internal validity. First, because key independent variables are under experimental control, time ordering is established, which is valuable in causal inference (see Chapter 8). Second, in experiments, extraneous variables can be held constant, allowing for greater confidence in control. In non-experimental research, extraneous variables can be statistically controlled, but such control is less effective because it (a) requires that a control variable was measured, and (b) because measurement is not perfect, the resulting statistical control is imperfect, too. In well-designed experiments, holding other things constant provides "perfect" control in that constants are statistically inert. Further, the "holding constant" approach can work for unanticipated contaminants. As long as things are constant, they are controlled. Finally, being able to create predictable variance in a dependent variable by purposively and systematically varying levels of the independent variable provides convincing evidence of understanding. It is one thing to observe things happening or to explain things happening. It is another to be able to make them happen or to predict things happening (see Chapters 6 and 7).

A final issue related to sampling is that all findings, regardless of sampling or statistical analyses, have limited generality. Random and probability sampling (see Chapter 4) is very useful in generalizing to some well-defined population, but no further. When we, for example, sample voters, if we sample well, we can generalize to some larger population of voters at the fixed point of time when the survey was done. Voting intentions, however, change over time. Therefore the findings from a poll generalize across people, but not over different elections or different time frames. Sophisticated sampling procedures allow for generalization across people, but not across contexts, specific question wordings, or over time. Sampling is always limited in such ways because generality from finite data is always limited. The important questions are (a) what do we want to generalize to, and (b) in what not-so-obvious ways is generalization limited?

In my view, the path to generalization is not so much an issue of method, sampling, or statistics. Because statistics and sampling-based generalizations are always limited by time and context, I find such generalization unfulfilling. Fortunately, I see a better option.

Although data is always finite, theory is not, at least within boundary conditions. So, if I know where and when a theory applies and when and where it does not, I know how far an idea can be generalized. This leads to an approach to research where one does not try to make generalizations, but test generalizations (c.f., Mook, 1983). This, too, is why I tend to worry first and foremost about internal validity and construct validity.

I will close this portion of the discussion with a disclaimer. It is not my position that sampling, statistical analysis, or generality are unimportant in social science and communication research. Instead, my argument is the more prudent position that good, informative analysis, requires good data, that design and measurement issues are sometimes underappreciated, and that good sampling enables generality only to the population sampled and not beyond.

Rant Number Three—Over-Reliance on $p < .05$ and Discrediting the Implausible Null Hypothesis

There are many approaches to statistical inference, and most if not all of them have their place. Unfortunately, one approach has gained dominance and is often viewed as both uncontroversial and as the way statistics should be practiced. I (and many other researchers) call this approach nil-null hypothesis significance testing (NNHST).

Here is how it is usually taught (see Chapter 6). There are two hypotheses, the null hypothesis, H_0, and the alternative or research hypothesis, H_1. H_0 is a nil-null, meaning that it reflects an expectation of no difference or no dependency. H_1 is the researcher's prediction and reflects a difference or dependency. Further, H_0 and H_1 are mutually exclusive and exhaustive. That is, if one is false, the other must be true. Once the hypotheses are specified, data are collected, and statistical tests are applied to the data. The statistics use probability functions to assign probabilities to the data presuming a distribution corresponding to the null hypothesis. If the probability of the resulting statistic is lower than a pre-determined maximum Type-I error, that is $p < \alpha$ (usually .05 or 5%), the null is rejected and it is inferred that the alternative is therefore supported.

Now, of course, the $p < .05$ criterion is arbitrary and just a convention, but that is not my objection. Instead, my first two criticisms are that in the social sciences, (a) there are usually differences or dependencies, that is, the nil-null is a priori implausible, and (b) although the statistical alternative and null hypotheses are logically exhaustive, this is not usually true for scientific hypotheses. The null is almost never literally true. So, debunking the null is uninformative. How many things in the real world are really correlated with exactly $r = .0000000000$? And, just because some correlation is not zero does not mean the researcher's thinking is correct. Is it not possible that the variables are correlated for some reason other than researcher's theory or logic? Statistical hypotheses may be written to be mutually exclusive and exhaustive, but theories and arguments that give rise to the tests are not usually mutually exclusive and exhaustive with all other theories or arguments. So, when a result is $p < .05$, I think "so what."

But, the novice researcher may say that in their experience, $p < .05$ is an achievement. After all, so many studies cannot even claim that. The set of all theses and dissertations yielding non-significant results is large indeed. Should no one be happy with $p < .05$?

But, the critic might say "ah, that is just a question of power" (see the useful discussion regarding the relevance of the power concept in Chapter 6). Achieving significance is more about your sample size than the quality of your ideas. If only there was a larger sample, then p could be less than .05 or even .001. The nil-null is almost always false, especially in non-experimental data, and $p < .05$ tells us more about the power of the test than the correctness of theory.

A corollary of this is that just because the Type-I error probability (p) is greater than .05 does not mean the null is viable. Non-significant findings might mean that the researcher's hypothesis is wrong, but it also might mean that statistical power was lacking (see "Non-Significant Results" in Chapter 6).

The usual solution is to do "power analyses," but these too are unsatisfying. One can calculate the power for a given significance test provided some sample size and some arbitrary effect size. But, what effect size should we use? In many research situations, it is not known what effect size to expect (but see Weber & Popova, in press). That is precisely why we are doing the research, because we don't know the relationship between the variables. If that relationship was known, then power could be calculated, but what then is the point of the research? Thus, power creates a "catch-22." When we most need to calculate statistical power, we lack the information that would make the calculations most informative.

I put it to the reader that the a priori implausibility of the nil-null, the fact that scientific research hypotheses (unlike statistical hypotheses) are not mutually exclusive and exhaustive, and the sensitivity of NNHST to sample size make NNHST unsuitable and undesirable as a research tool in many applications. And, these are not the only problems (see Levine, Weber, Hullett, Park, & Lindsey, 2008; Levine, Weber, Park, & Hullet, 2008). It is used because it is the convention, not because it is the best way to do research and analyze data.

Fortunately, there are alternatives and I will list a couple here. First, for many research applications, inferential statistics may not be needed at all. Maybe just some good statistical description of the data would make the point. But, if inference is desirable, two alternatives might be considered. These include confidence intervals (CIs) and Bayes factors (BFs).

With CIs, we usually report a 95% or 99% error range around our statistics (for more details, see Chapter 6). These can be done in such a way as to provide all the evidence of NNHST because if zero is outside the interval, the nil-null can be rejected (Tryon, 2001). In addition, CIs provide us with information about what findings (i.e., population parameter or effect sizes) are plausible. Bayes factors provide probabilities for both H_0 and H_1, that is, BFs directly provide the probability of the hypotheses rather than the probability of data assuming hypotheses. This has numerous advantages over NNHST, but also requires prior knowledge of effect sizes or of which effects a theory actually predicts (potentially leading to the same "catch-22″ problem as outlined previous). Bayes factors are not common in communication research yet, and they come with other disadvantages, but their popularity may grow in the future as knowledge in communication research and the quality of communication theories increases. For readers interested in Bayes factors, I recommend Dienes (2011). Wagenmakers, Wetzels, Borsboom, and van der Mass (2011) provide an example of an application in response to Bem's (2011) interesting use of NNHST.

Again, like my other criticisms, it is not my argument that NNHST is stupid or worthless. My concern is with its mis- and overuse, and with the misconception that it proves the single right way to practice statistics. I, myself, report NNHST in much of my own research. I try, however, to augment them with lots of descriptive statistics and confidence intervals and to not put too much weight on $p < .05$ *vs.* $p > .05$.

Rant Number Four—When Complexity is Not a Virtue

My fourth and final statistical practice gripe is in opposition to the idea that complex analyses are preferred to simpler analyses. Sometimes, it appears that more advanced analyses are chosen simply because they are more advanced. It seems that some people presume that people who do the most complex, or the latest, analyses must be smart, and the findings must be more valuable or enlightening. However, ultra-complex or the-latest-new-thing analyses make me skeptical.

If you follow published original communication research, you will know that the once omnipresent ANOVA (see Chapter 7) and correlation matrices (see Chapter 8) are increasingly being replaced with articles reporting multilevel modeling (MLM), Structural Equation Modeling (SEM), and meta-analysis. These techniques are gaining popularity for good reasons. Each solves problems and satisfies important statistical needs for communication researchers and professionals. However, these techniques create additional problems when misunderstood, misused, or misapplied.

The advent of MLM is an important advance because it allows researchers to analyze non-independent data. Traditional inferential statistics share a strong assumption of *independence of observation* (see Chapter 6). To introduce this idea, consider the probabilities associated with a string of coin flips. If we flip a coin, we have a 50–50 chance of getting a heads (see Chapter 3). If we flip the coin three times in a row, there is a .125 chance of getting three heads in a row ($5^3 = 0.5 \times 0.5 \times 0.5 = 0.125$), assuming, of course, that the outcome of the first flip did not affect the probability of subsequent flips. This is what is meant by independence of observations. If getting a heads on the first flip meant that each subsequent flip was now more or less likely than 0.5 to be a heads, that would change the expected probabilities, and understanding the probabilities would have to take the initial outcome and the nature of the contingency into account. Subsequent flips would be non-independent.

The degree of non-independence can be quantified with a statistic called the *intraclass correlation* (ICC). A positive ICC means that data points in the same class tend to be similar while a negative ICC means that data points in the same class tend to be dissimilar relative to data points in other classes. With the coin flip example, a negative ICC would mean that a heads is more likely to be followed by a tails. More relevant to communication, imagine that we are interested in the personalities of dating couples. If people tend to form relationships with partners with similar personalities, it would create a positive ICC, while opposites attracting would create a negative ICC. Only if the partner's personality traits were uncorrelated would the ICC be zero.

Non-independence of data becomes a potential concern when data points are nested in larger categories. Individuals might be nested within couples, families, or friendship circles. Students might be nested within classes, classes within majors, majors within colleges, and so forth. Fortunately, the existence of non-independence can be tested with the ICC, and if the ICC is meaningfully large, then the non-independence can be modeled with MLM. The two most popular MLM techniques in interpersonal research are David Kenny's social relations model, which is more ANOVA based, and hierarchical linear modeling (HLM), which is more regression based.

Currently, structural equation modeling (SEM) is more popular than MLM. SEM combines measurement modeling (confirmatory factor analysis, CFA) and path analysis (structural causal modeling), although it is advisable to do CFA and path analysis separately. Commonly used SEM software programs include AMOS, LISREL, EQS, and M-Plus.

CFA is generally superior to EFA (Exploratory Factor Analysis) when it is known which constructs specific items are supposed to measure, because CFA is better able to deal with correlated factors and better able to identify measurement confounds. Path analysis, on the other hand, is useful whenever one wishes to test a causal model with at least one mediated relationship. A mediated relationship is a series of causal relationships; e.g., x leads to y leads to z. Both CFA and path analysis require researchers to specify causal models in advance. The data are tested for how well they fit the model, and models are then accepted or rejected. The biggest problem with SEM, as it is used in communication research, is that it is more often used for model fitting than for model testing. When SEM is used for model fitting, replication with new data is advisable to avoid capitalization of chance. This replication is called *cross-validation,* and unfortunately, it is not common practice.

Meta-analysis is a set of procedures that allow researchers to cumulate findings across studies and to quantitatively summarize literatures. This is quite useful because the nature of significance testing with less-than-perfect statistical power makes most literatures appear to have mixed findings whether or not consistent patterns exist. To the extent to which the findings from a literature converge on a single conclusion, the findings are said to be homogeneous. When findings are homogeneous, they can be averaged as a meaningful summary of the findings. When findings are heterogeneous, however, researchers need to try and find out why. Sometimes there exist some subsets of the findings that tend to cluster around one outcome while other subsets cluster around another point. This suggests an important moderator.

While a well-done meta-analysis can be extremely informative, a few words of caution are advisable. First, average effects can only be safely interpreted when the findings of individual studies are homogeneous. Second, the results of meta-analysis depend on the quality of the original research. If there is some artifact or bias that runs through an entire literature, meta-analysis will not catch it. Third, meta-analysis is susceptible to publication bias, that is, significant results have a higher chance to be published in journals, and thus, the existence of publication bias tends to inflate the average effects reported in meta-analysis (Levine, Asada, & Carpenter, 2009).

Back to my concerns regarding complexity, just because someone reports a complex analysis does not mean they understand what they are doing. One of the curses of modern statistical software is that button-pushing and pull-down menus have made complex analyses evermore available to those who lack *understanding* of statistics. There are those who believe that knowing how to run software and read a printout is sufficient, and that deeper understanding of the basic principles and algorithms behind statistical analyses is unnecessary. This ties into my earlier argument about a little knowledge being dangerous. Knowing how to use software is not enough.

Second, sometimes a complex analysis is an indication for a dust-bowl analysis by an outside statistical expert. When a person doing an analysis knows a great deal about statistics, but little about the topic being researched, there can be a temptation to show off. Often, one metaphorically does not need a bazooka to hunt small rodents when a mouse trap might be both more efficient and safer.

Third, even if the analysis method is deeply understood by a data analyst with rich substantive feel, still the more "moving parts," the more room there is for something to break. The more transformations done on data, the further we get from the data. I prefer more immediacy with the data and the more "dummy proof," the better.

Fourth, while not always the case, in some cases, the more complex the analysis, the more difficult the interpretation. Complex analyses sometimes produce complex results that lack coherence. The result can be a jumbled mess or, in Abelson's (1995) words, a "blob." Other times, there is nothing wrong with the results per se, but the added complexity makes the results more difficult for the research consumer to understand. Of course, sometimes more advanced analyses help clarify complex data. My point is not that complexity is bad, but simply that complexity can be undesirable.

A good example of over-complexity in communication research practice is the current popularity of structural equation models (SEM). SEM can be very useful in testing a priori models involving mediated statistical relationships (see previous). But, SEM is very susceptible to the complexity bug. Here is what I see in the journals that really bothers me: I see a very complex model with arrows going everywhere. When the results are presented, the SEM is not used so much for rigorous model testing, but instead for model fitting. What frequently happens is that the a-priori model either provides a poor fit to the data, or some other (modified) model provides a better fit. Some paths are dropped, some paths are added, and some error terms are sometimes allowed to co-vary suggesting some unknown confounding factor. The end result is a model that fits the current data very well. But, what should we make of the original model? Should it be abandoned? And, because the model was fit to some existing data, it is unclear (and doubtful) that it will replicate.

My solution is, whenever possible, to follow Einstein's principle: Keep analyses simple, just not too simple. Keeping it appropriately simple means that there is less to go wrong. I am less likely to be extending beyond my knowledge, and I am less likely to miss something important. Keeping analyses simple means that my results will be assessable to a greater number of people, and that my results will be more transparent. The way I keep analyses simple is by reporting many descriptive statistics, that is, means, dispersion, and the nature or shape of distributions (see Chapter 2). If statistical inference is required, I prefer to provide confidence intervals over NNHST. I keep my designs as simple as possible to keep the number of statistical tests manageable and straightforward. As I know from the first author of this book, experienced statisticians usually take these three suggestions to heart and always prefer simple (but not too simple) statistical procedures. I recommend that you do the same. If you are in a position in which you are primarily a consumer of statistical analyses (e.g., as a communication professional) or you are dependent on the advice of a self-acclaimed statistics expert, then be aware of the observation that mostly amateur or wannabe statisticians tend to disguise their shortcomings by applying overly complex statistical analyses.

SUMMARY

This chapter provides a critical perspective on statistical analyses. Statistical analyses are incredibly useful, and *understanding* statistics provides new and powerful tools for understanding our world. Tools, however, can be misapplied with dangerous results. It is hoped that this chapter will warn the reader of some potential pitfalls.

TERMS

- **Research hypothesis** is the statement about characteristics of variables in the population of interest. Research hypotheses can be about differences or dependencies. The type of hypothesis determines the statistical test used.
- **Null hypothesis** is the counterpart to the research hypothesis, and claims that the statement in the research hypothesis is not true, that there is no difference or dependency between variables.
- **Type-I error** refers to accepting the research hypothesis when the null hypothesis is true.
- **Type-II error** refers to accepting the null hypothesis when the research hypothesis is true.
- **Exploratory research** is research guided by curiosity rather than pre-existing formal hypotheses. Exploratory research is used to generate rather than test hypotheses.
- **Experimental design** involves research in which there is at least on independent variable that is manipulated by the research, there exists a comparison or control group, and there is random assignment to experimental condition.
- **Hypothetico-deductive method** is formal approach to theory testing where hypotheses are logically derived from existing theory.
- **Abductive research** is a research strategy that starts with an interesting finding. The most plausible explanation that accounts for the results is generated, and tested with subsequent research.
- **Effect size** is defined as the least difference between two population parameters that can be considered practically relevant.
- **Causal relationship** is one in which the cause precedes the effect, the effect has statistical dependency on the cause, and in which the statistical dependency is not a function of some third variable.

- **Statistical independence of observation** is an assumption that is important in many types of statistical analysis. It means that each data point is not dependent on each other.
- **Multi-level modeling** involves a number of statistical analyses for testing nested data that violate assumptions of statistical independence.
- **Structural Equation Modeling** is a set of statistical analyses designed to test measurement and causal models.
- **Meta-analysis** is a secondary research analysis strategy that involves cumulating results across multiple previous studies.

REFERENCES

Abelson, R. P. (1995). *Statistics as principled argument.* Hillsdale, NJ: LEA.

Bem, D. J. (2011). Feeling the future: Experimental evidence for anomalous retroactive influences on cognition and affect. *Journal of Personality and Social Psychology, 100,* 407–425.

Campbell, D. T., & Stanley, J. C. (1963). *Experimental and quasi-experimental designs for research.* Boston: Houghton Mifflin.

Cook, T. D., & Campbell, D. T. (1979). *Quasi-experimentation.* Boston: Houghton.

Kotowski, M. R., Levine, T. R., Baker, C., & Bolt, J. (2009). A multi-trait multi-method validity assessment of the verbal aggressiveness and argumentativeness scales. *Communication Monographs, 76,* 443–462.

Dienes, Z. (2011). Bayesian versus orthodox statistics: Which side are you on? *Perspectives on Psychological Science,* 6(3), 274–290.

Levine, T. R., Asada, K. J., & Carpenter, C. (2009). Sample size and effect size are negatively correlated in meta-analysis: Evidence and implications of a publication bias against non-significant findings. *Communication Monographs, 76*(3), 286–302.

Levine, T. R., Bresnahan, M., Park, H. S., Lapinski, M. K., Wittenbaum, G., Shearman, S., et al. (2003). Self report measures of self-construals lack validity. *Human Communication Research, 29,* 210–252.

Levine, T. R., Weber, R., Park, H. S., & Hullett, C. R. (2008). A communication researcher's guide to null hypothesis significance testing and alternatives. *Human Communication Research, 34,* 188–209.

Levine, T. R., Weber, R., Hullett, C. R., Park, H. S., & Lindsey, L. (2008). A critical assessment of null hypothesis significance testing in quantitative communication research. *Human Communication Research, 34,* 171–187.

Mercier, H. & Sperber, D. (2011). Why do humans reason? Arguments for an argumentative theory. *Behavioral and Brian Sciences,* 34, 57–74.

Mook, D. G. (1983). In defense of external invalidity. *American Psychologist, 38,* 379–387.

Rozeboom, W. W. (1997). Good science is abductive, not hypothetico-decuctive. In L. L. Harlow, S. A. Mulaik, & J. H. Steiger (Eds.), *What if there were no significance tests?* (pp. 335–392). Mahwah: Lawrence Erlbaum.

Schwarz, N. (1999). Self-reports: How the questions shape the answers. *American Psychologist, 54,* 93–105.

Tryon, W. W. (2001). Evaluating statistical difference, equivalence, and indeterminacy using inferential confidence intervals: An integrated alternative method of conducting null hypothesis statistical tests. *Psychological Methods,* 6, 371–386.

Wagenmakers, E. J., Wetzels, R., Borsboom, D., & van der Mass, H. (2011). Why psychologists must change the way they analyze their data: The case of Psi: Comments on Bem (2011). *Journal of Personality and Social Psychology, 100,* 426–432.

Weber, R., & Popova, L. (in press). Testing equivalence in communication research: Theory and applications. *Communication Methods and Measures.*

Weber, R., Sherry, J., & Mathiak, K. (2008). The neurophysiological perspective in mass communication research. Theoretical rationale, methods, and applications. In M. J. Beatty, J. C. McCroskey & K. Floyd (Eds.), *Biological dimensions of communication: Perspectives, methods, and research* (pp. 41–71). Cresskill, NJ: Hampton Press.

APPENDIX A
TABLES

Table A.1. Standard normal distribution: Right tail probabilities

The table contains right-tail probabilities (areas under the curve) *from* a z-score to plus infinity. Reading example (see the probability in the shaded cell): A z-score of 0.35 (i.e., first decimal 0.3; second decimal 0.05 = 0.35) cuts exactly 0.3632 or 36.32% from the right tail of a standard normal distribution. One can also say, the probability that a z-score of a standardized normal distribution falls in an area under the curve *from* $z = 0.35$ to plus infinity is 0.3632 or 36.32%.

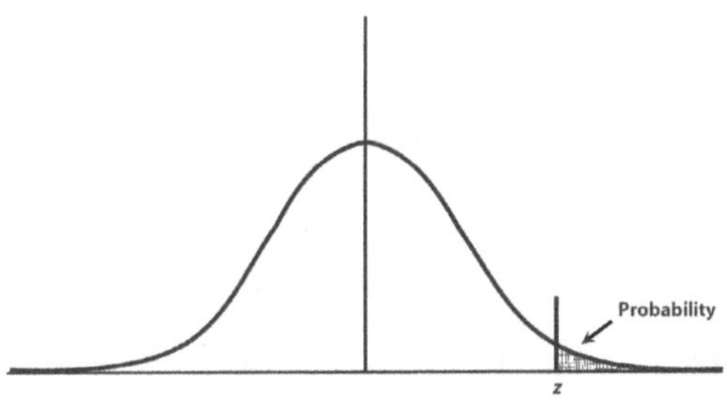

	Second Decimal Place of z									
z	**0**	**0.01**	**0.02**	**0.03**	**0.04**	**0.05**	**0.06**	**0.07**	**0.08**	**0.09**
0	0.5000	0.4960	0.4920	0.4880	0.4840	0.4801	0.4761	0.4721	0.4681	0.4641
0.1	0.4602	0.4562	0.4522	0.4483	0.4443	0.4404	0.4364	0.4325	0.4286	0.4247
0.2	0.4207	0.4168	0.4129	0.4090	0.4052	0.4013	0.3974	0.3936	0.3897	0.3859
0.3	0.3821	0.3783	0.3745	0.3707	0.3669	**0.3632**	0.3594	0.3557	0.3520	0.3483
0.4	0.3446	0.3409	0.3372	0.3336	0.3300	0.3264	0.3228	0.3192	0.3156	0.3121
0.5	0.3085	0.3050	0.3015	0.2981	0.2946	0.2912	0.2877	0.2843	0.2810	0.2776
0.6	0.2743	0.2709	0.2676	0.2643	0.2611	0.2578	0.2546	0.2514	0.2483	0.2451
0.7	0.2420	0.2389	0.2358	0.2327	0.2296	0.2266	0.2236	0.2206	0.2177	0.2148

Standard normal distribution: Right tail probabilities—*Continued*

	Second Decimal Place of z									
z	**0**	**0.01**	**0.02**	**0.03**	**0.04**	**0.05**	**0.06**	**0.07**	**0.08**	**0.09**
0.8	0.2119	0.2090	0.2061	0.2033	0.2005	0.1977	0.1949	0.1922	0.1894	0.1867
0.9	0.1841	0.1814	0.1788	0.1762	0.1736	0.1711	0.1685	0.1660	0.1635	0.1611
1	0.1587	0.1562	0.1539	0.1515	0.1492	0.1469	0.1446	0.1423	0.1401	0.1379
1.1	0.1357	0.1335	0.1314	0.1292	0.1271	0.1251	0.1230	0.1210	0.1190	0.1170
1.2	0.1151	0.1131	0.1112	0.1093	0.1075	0.1056	0.1038	0.102	0.1003	0.0985
1.3	0.0968	0.0951	0.0934	0.0918	0.0901	0.0885	0.0869	0.0853	0.0838	0.0823
1.4	0.0808	0.0793	0.0778	0.0764	0.0749	0.0735	0.0721	0.0708	0.0694	0.0681
1.5	0.0668	0.0655	0.0643	0.063	0.0618	0.0606	0.0594	0.0582	0.0571	0.0559
1.6	0.0548	0.0537	0.0526	0.0516	0.0505	0.0495	0.0485	0.0475	0.0465	0.0455
1.7	0.0446	0.0436	0.0427	0.0418	0.0409	0.0401	0.0392	0.0384	0.0375	0.0367
1.8	0.0359	0.0351	0.0344	0.0336	0.0329	0.0322	0.0314	0.0307	0.0301	0.0294
1.9	0.0287	0.0281	0.0274	0.0268	0.0262	0.0256	0.025	0.0244	0.0239	0.0233
2	0.0228	0.0222	0.0217	0.0212	0.0207	0.0202	0.0197	0.0192	0.0188	0.0183
2.1	0.0179	0.0174	0.017	0.0166	0.0162	0.0158	0.0154	0.015	0.0146	0.0143
2.2	0.0139	0.0136	0.0132	0.0129	0.0125	0.0122	0.0119	0.0116	0.0113	0.011
2.3	0.0107	0.0104	0.0102	0.0099	0.0096	0.0094	0.0091	0.0089	0.0087	0.0084
2.4	0.0082	0.008	0.0078	0.0075	0.0073	0.0071	0.0069	0.0068	0.0066	0.0064
2.5	0.0062	0.006	0.0059	0.0057	0.0055	0.0054	0.0052	0.0051	0.0049	0.0048
2.6	0.0047	0.0045	0.0044	0.0043	0.0041	0.004	0.0039	0.0038	0.0037	0.0036
2.7	0.0035	0.0034	0.0033	0.0032	0.0031	0.003	0.0029	0.0028	0.0027	0.0026
2.8	0.0026	0.0025	0.0024	0.0023	0.0023	0.0022	0.0021	0.0021	0.002	0.0019
2.9	0.0019	0.0018	0.0018	0.0017	0.0016	0.0016	0.0015	0.0015	0.0014	0.0014
3	0.00135									

Table values generated using R. R Development Core Team (2011). R: A language and environment for statistical computing. R Foundation for Statistical Computing, Vienna, Austria.

Table A.2. *t* distribution: Critical values

The tables contains critical *t*-values for right-tail probabilities (areas under the curve) ranging from 10% (or 0.1) to 0.1% (or 0.001). Due to the symmetry of confidence intervals, confidence levels correspond to 1 minus 2 times the right tail probability. Reading example (see the critical value in the shaded cell): In a *t*-distribution with 7 degrees of freedom, a *t*-value of 2.365 cuts exactly 0.025 or 2.5% from the right tail of the distribution. One can also say, the probability that a *t*-value of a *t*-distribution with 7 degrees of freedom falls in an area under the curve *from* $t = 2.365$ to plus infinity is 0.025 or 2.5%.

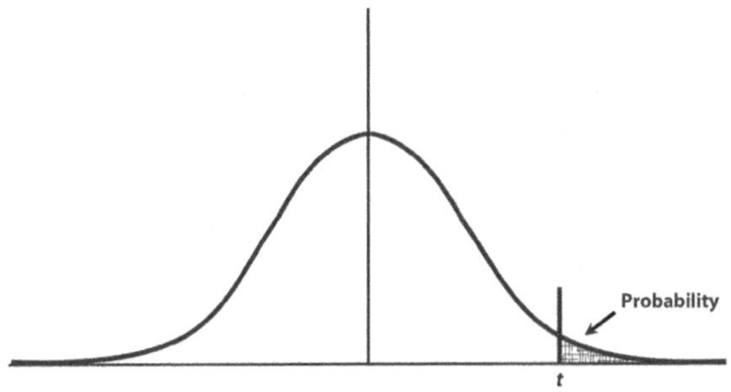

	Confidence Level					
	80%	**90%**	**95%**	**98%**	**99%**	**99.8%**
			Right-Tail Probabilities			
df	**10%**	**5%**	**2.5%**	**1%**	**0.5%**	**0.1%**
1	3.078	6.314	12.706	31.821	63.657	318.309
2	1.886	2.920	4.303	6.965	9.925	22.327
3	1.638	2.353	3.182	4.541	5.841	10.215
4	1.533	2.132	2.776	3.747	4.604	7.173
5	1.476	2.015	2.571	3.365	4.032	5.893
6	1.440	1.943	2.447	3.143	3.707	5.208
7	1.415	1.895	**2.365**	2.998	3.499	4.785
8	1.397	1.860	2.306	2.896	3.355	4.501
9	1.383	1.833	2.262	2.821	3.25	4.297
10	1.372	1.812	2.228	2.764	3.169	4.144
11	1.363	1.796	2.201	2.718	3.106	4.025
12	1.356	1.782	2.179	2.681	3.055	3.930
13	1.350	1.771	2.160	2.650	3.012	3.852

t distribution: Critical values—*Continued*						
	Confidence Level					
	80%	90%	95%	98%	99%	99.8%
	Right-Tail Probabilities					
df	10%	5%	2.5%	1%	0.5%	0.1%
14	1.345	1.761	2.145	2.624	2.977	3.787
15	1.341	1.753	2.131	2.602	2.947	3.733
16	1.337	1.746	2.120	2.583	2.921	3.686
17	1.333	1.740	2.110	2.567	2.898	3.646
18	1.330	1.734	2.101	2.552	2.878	3.610
19	1.328	1.729	2.093	2.539	2.861	3.579
20	1.325	1.725	2.086	2.528	2.845	3.552
21	1.323	1.721	2.08	2.518	2.831	3.527
22	1.321	1.717	2.074	2.508	2.819	3.505
23	1.319	1.714	2.069	2.500	2.807	3.485
24	1.318	1.711	2.064	2.492	2.797	3.467
25	1.316	1.708	2.060	2.485	2.787	3.450
26	1.315	1.706	2.056	2.479	2.779	3.435
27	1.314	1.703	2.052	2.473	2.771	3.421
28	1.313	1.701	2.048	2.467	2.763	3.408
29	1.311	1.699	2.045	2.462	2.756	3.396
30	1.310	1.697	2.042	2.457	2.750	3.385
40	1.303	1.684	2.021	2.423	2.704	3.307
50	1.299	1.676	2.009	2.403	2.678	3.261
60	1.296	1.671	2.000	2.390	2.660	3.232
70	1.294	1.667	1.994	2.381	2.648	3.211
80	1.292	1.664	1.990	2.374	2.639	3.195
90	1.291	1.662	1.987	2.368	2.632	3.183
100	1.290	1.660	1.984	2.364	2.626	3.174
∞	1.282	1.645	1.960	2.326	2.576	3.090

Table values generated using R. R Development Core Team (2011). R: A language and environment for statistical computing. R Foundation for Statistical Computing, Vienna, Austria.

Table A.3. *F*-distribution: Critical values

The tables contains critical F-values for right-tail probabilities (areas under the curve) ranging from 5% (or 0.05) to 0.1% (or 0.01). Reading example (see the critical value in the shaded cell): In an F-distribution with 6 numerator degrees of freedom and 5 denominator degrees of freedom, an F-value of 4.95 cuts exactly 0.05 or 5% from the right tail of the distribution. One can also say, the probability that an F-value of an F-distribution with 6 numerator degrees of freedom (e.g., in ANOVA, the degrees of freedom of the treatment variance) and 5 denominator degrees of freedom (e.g., in ANOVA, the degrees of freedom of the error variance; see Chapter 7) falls in an area under the curve *from* $F = 4.95$ to plus infinity is 0.05 or 5%.

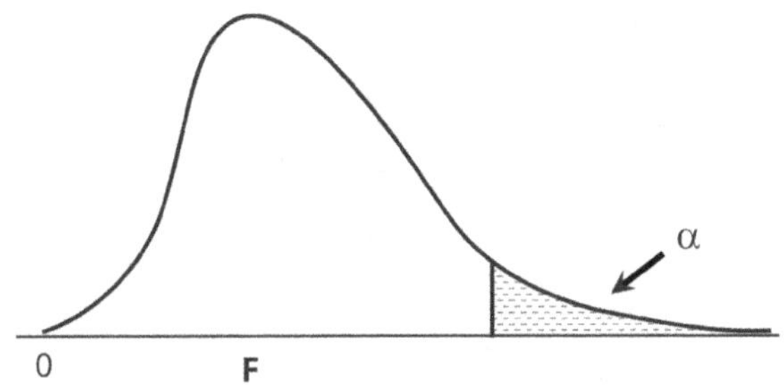

	$\alpha = 5\%$									
	***df*1 (numerator)**									
***df*2 denominator**	**1**	**2**	**3**	**4**	**5**	**6**	**8**	**12**	**24**	**∞**
1	161.448	199.5	215.707	224.583	230.162	233.986	238.883	243.906	249.052	254.314
2	18.513	19	19.164	19.247	19.296	19.33	19.371	19.413	19.454	19.496
3	10.128	9.552	9.277	9.117	9.013	8.941	8.845	8.745	8.639	8.526
4	7.709	6.944	6.591	6.388	6.256	6.163	6.041	5.912	5.774	5.628
5	6.608	5.786	5.409	5.192	5.05	**4.95**	4.818	4.678	4.527	4.365
6	5.987	5.143	4.757	4.534	4.387	4.284	4.147	4	3.841	3.669
7	5.591	4.737	4.347	4.12	3.972	3.866	3.726	3.575	3.41	3.23
8	5.318	4.459	4.066	3.838	3.687	3.581	3.438	3.284	3.115	2.928
9	5.117	4.256	3.863	3.633	3.482	3.374	3.23	3.073	2.9	2.707
10	4.965	4.103	3.708	3.478	3.326	3.217	3.072	2.913	2.737	2.538
11	4.844	3.982	3.587	3.357	3.204	3.095	2.948	2.788	2.609	2.404
12	4.747	3.885	3.49	3.259	3.106	2.996	2.849	2.687	2.505	2.296

F-distribution: Critical values—*Continued*										
	$\alpha = 5\%$									
	*df*1 (numerator)									
*df*2 denominator	1	2	3	4	5	6	8	12	24	∞
13	4.667	3.806	3.411	3.179	3.025	2.915	2.767	2.604	2.42	2.206
14	4.6	3.739	3.344	3.112	2.958	2.848	2.699	2.534	2.349	2.131
15	4.543	3.682	3.287	3.056	2.901	2.79	2.641	2.475	2.288	2.066
16	4.494	3.634	3.239	3.007	2.852	2.741	2.591	2.425	2.235	2.01
17	4.451	3.592	3.197	2.965	2.81	2.699	2.548	2.381	2.19	1.96
18	4.414	3.555	3.16	2.928	2.773	2.661	2.51	2.342	2.15	1.917
19	4.381	3.522	3.127	2.895	2.74	2.628	2.477	2.308	2.114	1.878
20	4.351	3.493	3.098	2.866	2.711	2.599	2.447	2.278	2.082	1.843
21	4.325	3.467	3.072	2.84	2.685	2.573	2.42	2.25	2.054	1.812
22	4.301	3.443	3.049	2.817	2.661	2.549	2.397	2.226	2.028	1.783
23	4.279	3.422	3.028	2.796	2.64	2.528	2.375	2.204	2.005	1.757
24	4.26	3.403	3.009	2.776	2.621	2.508	2.355	2.183	1.984	1.733
25	4.242	3.385	2.991	2.759	2.603	2.49	2.337	2.165	1.964	1.711
26	4.225	3.369	2.975	2.743	2.587	2.474	2.321	2.148	1.946	1.691
27	4.21	3.354	2.96	2.728	2.572	2.459	2.305	2.132	1.93	1.672
28	4.196	3.34	2.947	2.714	2.558	2.445	2.291	2.118	1.915	1.654
29	4.183	3.328	2.934	2.701	2.545	2.432	2.278	2.104	1.901	1.638
30	4.171	3.316	2.922	2.69	2.534	2.421	2.266	2.092	1.887	1.622
40	4.085	3.232	2.839	2.606	2.449	2.336	2.18	2.003	1.793	1.509
60	4.001	3.15	2.758	2.525	2.368	2.254	2.097	1.917	1.7	1.389
120	3.92	3.072	2.68	2.447	2.29	2.175	2.016	1.834	1.608	1.254
∞	3.841	2.996	2.605	2.372	2.214	2.099	1.938	1.752	1.517	1

F-distribution: Critical values—*Continued*										
	$\alpha = 1\%$									
	*df*1 (numerator)									
*df*2 denominator	**1**	**2**	**3**	**4**	**5**	**6**	**8**	**12**	**24**	∞
1	4052.18	4999.5	5403.35	5624.58	5763.65	5858.99	5981.07	6106.32	6234.63	6365.86
2	98.503	99	99.166	99.249	99.299	99.333	99.374	99.416	99.458	99.499
3	34.116	30.817	29.457	28.71	28.237	27.911	27.489	27.052	26.598	26.125
4	21.198	18	16.694	15.977	15.522	15.207	14.799	14.374	13.929	13.463
5	16.258	13.274	12.06	11.392	10.967	10.672	10.289	9.888	9.466	9.02
6	13.745	10.925	9.78	9.148	8.746	8.466	8.102	7.718	7.313	6.88
7	12.246	9.547	8.451	7.847	7.46	7.191	6.84	6.469	6.074	5.65
8	11.259	8.649	7.591	7.006	6.632	6.371	6.029	5.667	5.279	4.859
9	10.561	8.022	6.992	6.422	6.057	5.802	5.467	5.111	4.729	4.311
10	10.044	7.559	6.552	5.994	5.636	5.386	5.057	4.706	4.327	3.909
11	9.646	7.206	6.217	5.668	5.316	5.069	4.744	4.397	4.021	3.602
12	9.33	6.927	5.953	5.412	5.064	4.821	4.499	4.155	3.78	3.361
13	9.074	6.701	5.739	5.205	4.862	4.62	4.302	3.96	3.587	3.165
14	8.862	6.515	5.564	5.035	4.695	4.456	4.14	3.8	3.427	3.004
15	8.683	6.359	5.417	4.893	4.556	4.318	4.004	3.666	3.294	2.868
16	8.531	6.226	5.292	4.773	4.437	4.202	3.89	3.553	3.181	2.753
17	8.4	6.112	5.185	4.669	4.336	4.102	3.791	3.455	3.084	2.653
18	8.285	6.013	5.092	4.579	4.248	4.015	3.705	3.371	2.999	2.566
19	8.185	5.926	5.01	4.5	4.171	3.939	3.631	3.297	2.925	2.489
20	8.096	5.849	4.938	4.431	4.103	3.871	3.564	3.231	2.859	2.421
21	8.017	5.78	4.874	4.369	4.042	3.812	3.506	3.173	2.801	2.36
22	7.945	5.719	4.817	4.313	3.988	3.758	3.453	3.121	2.749	2.305
23	7.881	5.664	4.765	4.264	3.939	3.71	3.406	3.074	2.702	2.256
24	7.823	5.614	4.718	4.218	3.895	3.667	3.363	3.032	2.659	2.211

F*-distribution: Critical values—*Continued										
	α = 1%									
	***df*1 (numerator)**									
***df*2 denominator**	**1**	**2**	**3**	**4**	**5**	**6**	**8**	**12**	**24**	**∞**
25	7.77	5.568	4.675	4.177	3.855	3.627	3.324	2.993	2.62	2.169
26	7.721	5.526	4.637	4.14	3.818	3.591	3.288	2.958	2.585	2.131
27	7.677	5.488	4.601	4.106	3.785	3.558	3.256	2.926	2.552	2.097
28	7.636	5.453	4.568	4.074	3.754	3.528	3.226	2.896	2.522	2.064
29	7.598	5.42	4.538	4.045	3.725	3.499	3.198	2.868	2.495	2.034
30	7.562	5.39	4.51	4.018	3.699	3.473	3.173	2.843	2.469	2.006
40	7.314	5.179	4.313	3.828	3.514	3.291	2.993	2.665	2.288	1.805
60	7.077	4.977	4.126	3.649	3.339	3.119	2.823	2.496	2.115	1.601
120	6.851	4.787	3.949	3.48	3.174	2.956	2.663	2.336	1.95	1.381
∞	6.635	4.605	3.782	3.319	3.017	2.802	2.511	2.185	1.791	1

F*-distribution: Critical values—*Continued										
	α = 0.1%									
	***df*1 (numerator)**									
***df*2 denominator**	**1**	**2**	**3**	**4**	**5**	**6**	**8**	**12**	**24**	**∞**
1	405284.1	499999.5	540379.2	562499.6	576404.6	585937.1	598144.2	610667.8	623497.5	636619.4
2	998.5	999	999.167	999.25	999.3	999.333	999.375	999.417	999.458	999.5
3	167.029	148.5	141.108	137.1	134.58	132.847	130.619	128.316	125.935	123.469
4	74.137	61.246	56.177	53.436	51.712	50.525	48.996	47.412	45.766	44.051
5	47.181	37.122	33.202	31.085	29.752	28.834	27.649	26.418	25.133	23.785
6	35.507	27	23.703	21.924	20.803	20.03	19.03	17.989	16.897	15.745
7	29.245	21.689	18.772	17.198	16.206	15.521	14.634	13.707	12.732	11.696
8	25.415	18.494	15.829	14.392	13.485	12.858	12.046	11.194	10.295	9.334
9	22.857	16.387	13.902	12.56	11.714	11.128	10.368	9.57	8.724	7.813
10	21.04	14.905	12.553	11.283	10.481	9.926	9.204	8.445	7.638	6.762

F-distribution: Critical values—*Continued*										
	$\alpha = 0.1\%$									
	*df*1 (numerator)									
*df*2 denominator	1	2	3	4	5	6	8	12	24	∞
11	19.687	13.812	11.561	10.346	9.578	9.047	8.355	7.626	6.847	5.998
12	18.643	12.974	10.804	9.633	8.892	8.379	7.71	7.005	6.249	5.42
13	17.815	12.313	10.209	9.073	8.354	7.856	7.206	6.519	5.781	4.967
14	17.143	11.779	9.729	8.622	7.922	7.436	6.802	6.13	5.407	4.604
15	16.587	11.339	9.335	8.253	7.567	7.092	6.471	5.812	5.101	4.307
16	16.12	10.971	9.006	7.944	7.272	6.805	6.195	5.547	4.846	4.059
17	15.722	10.658	8.727	7.683	7.022	6.562	5.962	5.324	4.631	3.85
18	15.379	10.39	8.487	7.459	6.808	6.355	5.763	5.132	4.447	3.67
19	15.081	10.157	8.28	7.265	6.622	6.175	5.59	4.967	4.288	3.514
20	14.819	9.953	8.098	7.096	6.461	6.019	5.44	4.823	4.149	3.378
21	14.587	9.772	7.938	6.947	6.318	5.881	5.308	4.696	4.027	3.257
22	14.38	9.612	7.796	6.814	6.191	5.758	5.19	4.583	3.919	3.151
23	14.195	9.469	7.669	6.696	6.078	5.649	5.085	4.483	3.822	3.055
24	14.028	9.339	7.554	6.589	5.977	5.55	4.991	4.393	3.735	2.969
25	13.877	9.223	7.451	6.493	5.885	5.462	4.906	4.312	3.657	2.89
26	13.739	9.116	7.357	6.406	5.802	5.381	4.829	4.238	3.586	2.819
27	13.613	9.019	7.272	6.326	5.726	5.308	4.759	4.171	3.521	2.754
28	13.498	8.931	7.193	6.253	5.656	5.241	4.695	4.109	3.462	2.695
29	13.391	8.849	7.121	6.186	5.593	5.179	4.636	4.053	3.407	2.64
30	13.293	8.773	7.054	6.125	5.534	5.122	4.581	4.001	3.357	2.589
40	12.609	8.251	6.595	5.698	5.128	4.731	4.207	3.642	3.011	2.233
60	11.973	7.768	6.171	5.307	4.757	4.372	3.865	3.315	2.694	1.89
120	11.38	7.321	5.781	4.947	4.416	4.044	3.552	3.016	2.402	1.543
∞	10.828	6.908	5.422	4.617	4.103	3.743	3.266	2.742	2.132	1

Table values generated using R. R Development Core Team (2011). R: A language and environment for statistical computing. R Foundation for Statistical Computing, Vienna, Austria.

Table A.4. Chi-square distribution: Critical values

The tables contains critical χ^2-values for right-tail probabilities (areas under the curve) ranging from 25% (or 0.25) to 0.1% (or 0.001). Reading example (see the critical value in the shaded cell): In a χ^2-distribution with 1 degree of freedom, a χ^2-value of 3.841 cuts exactly 0.05 or 5% from the right tail of the distribution. One can also say, the probability that a χ^2-value of a χ^2-distribution with 1 degree of freedom falls in an area under the curve *from* $\chi^2 = 3.841$ to plus infinity is 0.05 or 5%. Note that the table is constructed for two-sided tests. For example, for a two-sided test with an α-level of x% we find the critical value in the column of x%. For a one-sided test with an α-level of x% we find the critical value in the column of $2 \cdot x$%.

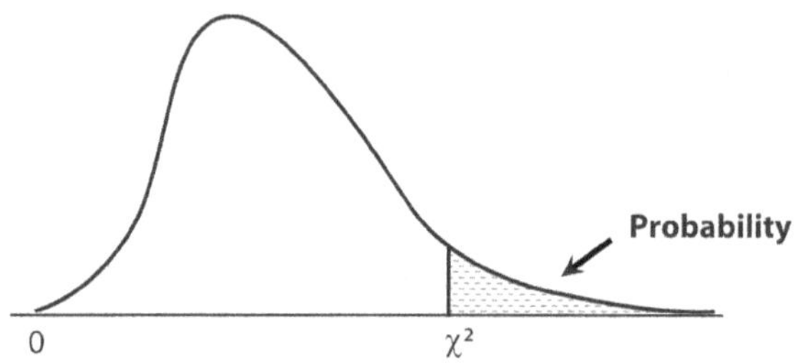

df	Right Tail Probability						
	25%	**10%**	**5%**	**2.5%**	**1%**	**0.5%**	**0.1%**
1	1.323	2.706	**3.841**	5.024	6.635	7.879	10.828
2	2.773	4.605	5.991	7.378	9.21	10.597	13.816
3	4.108	6.251	7.815	9.348	11.345	12.838	16.266
4	5.385	7.779	9.488	11.143	13.277	14.86	18.467
5	6.626	9.236	11.07	12.833	15.086	16.75	20.515
6	7.841	10.645	12.592	14.449	16.812	18.548	22.458
7	9.037	12.017	14.067	16.013	18.475	20.278	24.322
8	10.219	13.362	15.507	17.535	20.09	21.955	26.124
9	11.389	14.684	16.919	19.023	21.666	23.589	27.877
10	12.549	15.987	18.307	20.483	23.209	25.188	29.588
11	13.701	17.275	19.675	21.92	24.725	26.757	31.264
12	14.845	18.549	21.026	23.337	26.217	28.3	32.909
13	15.984	19.812	22.362	24.736	27.688	29.819	34.528

Chi-square distribution: Critical values—*Continued*							
	Right Tail Probability						
df	**25%**	**10%**	**5%**	**2.5%**	**1%**	**0.5%**	**0.1%**
14	17.117	21.064	23.685	26.119	29.141	31.319	36.123
15	18.245	22.307	24.996	27.488	30.578	32.801	37.697
16	19.369	23.542	26.296	28.845	32	34.267	39.252
17	20.489	24.769	27.587	30.191	33.409	35.718	40.79
18	21.605	25.989	28.869	31.526	34.805	37.156	42.312
19	22.718	27.204	30.144	32.852	36.191	38.582	43.82
20	23.828	28.412	31.41	34.17	37.566	39.997	45.315
21	24.935	29.615	32.671	35.479	38.932	41.401	46.797
22	26.039	30.813	33.924	36.781	40.289	42.796	48.268
23	27.141	32.007	35.172	38.076	41.638	44.181	49.728
24	28.241	33.196	36.415	39.364	42.98	45.559	51.179
25	29.339	34.382	37.652	40.646	44.314	46.928	52.62
30	34.8	40.256	43.773	46.979	50.892	53.672	59.703
40	45.616	51.805	55.758	59.342	63.691	66.766	73.402
50	56.334	63.167	67.505	71.42	76.154	79.49	86.661
60	66.981	74.397	79.082	83.298	88.379	91.952	99.607
70	77.577	85.527	90.531	95.023	100.425	104.215	112.317
80	88.13	96.578	101.879	106.629	112.329	116.321	124.839
90	98.65	107.565	113.145	118.136	124.116	128.299	137.208
100	109.141	118.498	124.342	129.561	135.807	140.169	149.449

Table values generated using R. R Development Core Team (2011). R: A language and environment for statistical computing. R Foundation for Statistical Computing, Vienna, Austria.

Table A.5. Fisher's Z-values

The table lists Pearson Correlation Coefficients (r) and their corresponding Fisher's Z value. The transformation of r into Z and Z into r is based on the following formulas:

$$Z = \frac{1}{2} \cdot \ln \frac{1+r}{1-r} \Leftrightarrow r = \frac{e^{2Z}-1}{e^{2Z}+1}$$

Note: The table lists only positive correlations. For $r < 0$ we select the corresponding Z-value for $r > 0$ and multiply by -1.That is, we put a minus sign in front of the Z-value.

r	Z	r	Z	r	Z	r	Z	r	Z
0	0	0.2	0.203	0.4	0.424	0.6	0.693	0.8	1.099
0.005	0.005	0.205	0.208	0.405	0.43	0.605	0.701	0.805	1.113
0.01	0.01	0.21	0.213	0.41	0.436	0.61	0.709	0.81	1.127
0.015	0.015	0.215	0.218	0.415	0.442	0.615	0.717	0.815	1.142
0.02	0.02	0.22	0.224	0.42	0.448	0.62	0.725	0.82	1.157
0.025	0.025	0.225	0.229	0.425	0.454	0.625	0.733	0.825	1.172
0.03	0.03	0.23	0.234	0.43	0.46	0.63	0.741	0.83	1.188
0.035	0.035	0.235	0.239	0.435	0.466	0.635	0.75	0.835	1.204
0.04	0.04	0.24	0.245	0.44	0.472	0.64	0.758	0.84	1.221
0.045	0.045	0.245	0.25	0.445	0.478	0.645	0.767	0.845	1.238
0.05	0.05	0.25	0.255	0.45	0.485	0.65	0.775	0.85	1.256
0.055	0.055	0.255	0.261	0.455	0.491	0.655	0.784	0.855	1.274
0.06	0.06	0.26	0.266	0.46	0.497	0.66	0.793	0.86	1.293
0.065	0.065	0.265	0.271	0.465	0.504	0.665	0.802	0.865	1.313
0.07	0.07	0.27	0.277	0.47	0.51	0.67	0.811	0.87	1.333
0.075	0.075	0.275	0.282	0.475	0.517	0.675	0.82	0.875	1.354
0.08	0.08	0.28	0.288	0.48	0.523	0.68	0.829	0.88	1.376
0.085	0.085	0.285	0.293	0.485	0.53	0.685	0.838	0.885	1.398
0.09	0.09	0.29	0.299	0.49	0.536	0.69	0.848	0.89	1.422
0.095	0.095	0.295	0.304	0.495	0.543	0.695	0.858	0.895	1.447
0.1	0.1	0.3	0.31	0.5	0.549	0.7	0.867	0.9	1.472

Fisher's Z-values—*Continued*									
r	Z	*r*	Z	*r*	Z	*r*	Z	*r*	Z
0.105	0.105	0.305	0.315	0.505	0.556	0.705	0.877	0.905	1.499
0.11	0.11	0.31	0.321	0.51	0.563	0.71	0.887	0.91	1.528
0.115	0.116	0.315	0.326	0.515	0.57	0.715	0.897	0.915	1.557
0.12	0.121	0.32	0.332	0.52	0.576	0.72	0.908	0.92	1.589
0.125	0.126	0.325	0.337	0.525	0.583	0.725	0.918	0.925	1.623
0.13	0.131	0.33	0.343	0.53	0.59	0.73	0.929	0.93	1.658
0.135	0.136	0.335	0.348	0.535	0.597	0.735	0.94	0.935	1.697
0.14	0.141	0.34	0.354	0.54	0.604	0.74	0.95	0.94	1.738
0.145	0.146	0.345	0.36	0.545	0.611	0.745	0.962	0.945	1.783
0.15	0.151	0.35	0.365	0.55	0.618	0.75	0.973	0.95	1.832
0.155	0.156	0.355	0.371	0.555	0.626	0.755	0.984	0.955	1.886
0.16	0.161	0.36	0.377	0.56	0.633	0.76	0.996	0.96	1.946
0.165	0.167	0.365	0.383	0.565	0.64	0.765	1.008	0.965	2.014
0.17	0.172	0.37	0.388	0.57	0.648	0.77	1.02	0.97	2.092
0.175	0.177	0.375	0.394	0.575	0.655	0.775	1.033	0.975	2.185
0.18	0.182	0.38	0.4	0.58	0.662	0.78	1.045	0.98	2.298
0.185	0.187	0.385	0.406	0.585	0.67	0.785	1.058	0.985	2.443
0.19	0.192	0.39	0.412	0.59	0.678	0.79	1.071	0.99	2.647
0.195	0.198	0.395	0.418	0.595	0.685	0.795	1.085	0.995	2.994

Table values generated using "psychometric" R package. Fletcher, T. D. (2010). Psychometric: Applied psychometric theory. R package version 2.2. http://CRAN.R-project.org/package=psychometric.

APPENDIX B

PROBLEM SET ANSWERS

CHAPTER 1

1. A population is the theoretically specified aggregation of study elements, and a sample is a special subset of a population observed in order to make inferences about the population itself.
2. A parameter is a numerical summary of a given variable in a population. A statistic is a numerical summary of sample data. A statistic is usually an estimator of a parameter.
3. The different levels of measurement include nominal, ordinal, interval, and ratio. The level of measurement is important because it determines which statistical tests can be applied.
4. a-d require student to find a study.
5. a-c require student to find news agency report.

CHAPTER 2

1. a.

Attitudes toward Twitter

Answers/Values	Frequency	Proportion	Percentage
A fad that will go away	420	.406	40.58
Heard of it, but don't know what it is	375	.362	36.23
An important (useful) tool	170	.164	16.43
Never heard of it (don't know)	70	.068	6.76
Total	1,035	1.00	100

1. b.

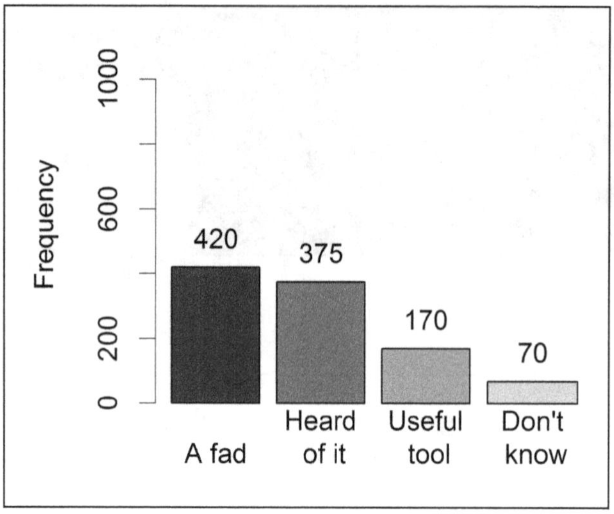

2. a.

Interval	Frequency	Relative Frequency	Percentage	Cumulative Frequency
0–9	5	0.098	9.8	9.8
10–19	9	0.1764	17.64	27.45
20–29	5	0.098	9.8	37.25
30–39	4	0.0784	7.84	45.1
40–49	6	0.1176	11.76	56.86
50–59	8	0.1568	15.68	72.55
60–69	3	0.0588	5.88	78.43
70–79	6	0.1176	11.76	90.2
80–89	5	0.098	9.8	100
Total	**51**	**1.0**	**100**	
Note: Due to rounding, values may not add up to 1 or 100%				

2. b.

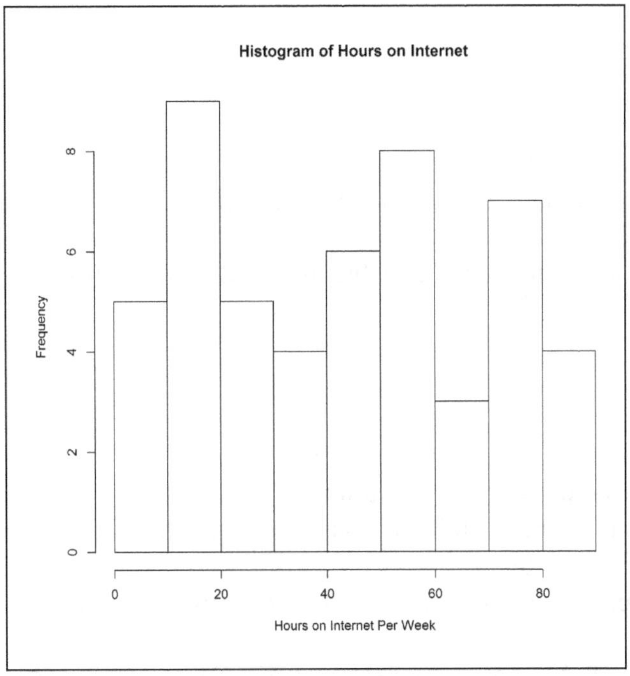

3. A bar chart visualizes categorical data by drawing a bar over each category. The height of the bar represents the proportion or percentage of that category. The histogram visualizes data at the interval/ratio level. A bar is drawn over each interval of numbers, with the height representing the number of observations in that interval. The bars are not separated in order to show that the histogram represents the distribution of an interval/ratio variable.

4. a. $n = 30$ observations; stem-and-leaf figure

```
1|16
2|24477
3|00113446799
4|59
5|3
6|189
7|5
8|024
9|08
```

4. b. 1Q = 30; Md = 36.5; 3Q = 68

5. The primary difference is that the stem-and-leaf plot provides information on both the shape of the distribution and the actual values of the variable under consideration—information which is lost in a histogram.

6. $N = 24$ observations
  a. Mn = 27.33
  b. Md = 22.50
  c. Mo = 9
  d. Mode < Median < Mean; The distribution is skewed to the right.

7. If the mean is greater than the median, and the mode is smaller than the median, the distribution is skewed to the right.

8. If the median is larger than the mean but smaller than the mode, the distribution is skewed to the left.

9. $n = 9$ observations
  a. Range = 8
  b. Variance = 5.33
  c. Standard deviation = 2.31
  d. Sum of squares = 48 (based on $n - 1$ observations)

10. **a.** Unlike standard deviation, the range fails to consider the center of a distribution and how values deviate from the center.
  b. An advantage of IQR over standard deviation is its insensitivity to outliers.

11. John (0.5714) > Chris (0.33)

12. **a.** Campaign 1: Mean = 15.38, Median = 16, Mode = 17; Campaign 2: Mean = 11.69, Median = 11, Mode = 11.
  b. Campaign 1 has a left-skewed distribution; Campaign 2 has a near symmetric distribution.
  c. Campaign 1: Range = 10, variance = 8.30, standard deviation = 2.88; Campaign 2: Range = 7, variance = 3.94, standard deviation = 1.98
  d. See following figure.

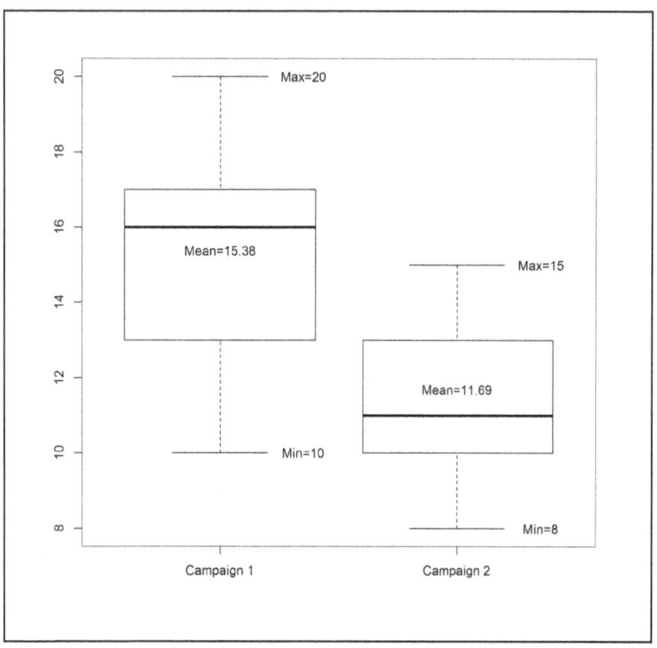

e. Campaign 1 has a larger range of values, a distribution that is skewed to the left, and more variation about the mean, than campaign 2. Campaign 1, however, seems to be superior to Campaign 2 in that the measures of central tendency (mean, median, and mode) are all larger than for Campaign 2. We would therefore recommend Campaign 1, as this seemed to resonate better with the audience.

CHAPTER 3

1. There are 52 cards in a deck; there are 13 spades (S), so $P(S) = 13/52$; there are 12 face (F) cards (King, Jack, Queen), so $P(F) = 12/52$; there are 3 face cards that are also spades. Using the rule of addition:

$P(S \cup F) = P(S) + P(F) - P(S \cap F)$
$P(S \cup F) = 13/52 + 12/52 - 3/52 = 22/52 = 11/26 = 0.423 = 42.3\%$

2. There are 20 large (L) balls and 40 (S) small balls in an urn

a. Without replacement:

$P(L1) = 20/60$; $P(L2 \mid L1) = 19/59$; $P(L3 \mid L2) = 18/58$.
Based on the rule of multiplication:

$P(L1 \cap L2 \cap L3) = P(L1) * P(L2 \mid L1) * P(L3 \mid L2)$
$P(L1 \cap L2 \cap L3) = 20/60 * 19/59 * 18/58 = 0.033$

b. With replacement:

$P(L1 \cap L2 \cap L3) = P(L1) * P(L2) * P(L3)$
$P(L1 \cap L2 \cap L3) = 20/60 * 20/60 * 20/60 = 0.037$

3. See table below

Toss 1	Toss 2	Toss 3	Toss 4	# Heads	# Tails	P_{result}
H	H	H	H	4	0	0.0625
H	H	H	T	3	1	0.0625
H	H	T	H	3	1	0.0625
H	T	H	H	3	1	0.0625
H	H	T	T	2	2	0.0625
H	T	H	T	2	2	0.0625
H	T	T	H	2	2	0.0625
H	T	T	T	1	3	0.0625
T	T	T	T	0	4	0.0625
T	T	T	H	1	3	0.0625
T	T	H	T	1	3	0.0625
T	H	T	T	1	3	0.0625
T	T	H	H	2	2	0.0625
T	H	T	H	2	2	0.0625
T	H	H	T	2	2	0.0625
T	H	H	H	3	1	0.0625

Probability distribution of flipping 4 coins	
Outcome	**Probability**
4 Heads, no Tails	1/16
3 Heads, 1 Tail	4/16, or 1/4
2 Heads, 2 Tails	6/16, or 3/8
1 Head, 3 Tails	4/16, or 1/4
4 Tails, no Head	1/16
Total	1

4. A continuous variable can take on any value within the limits of the variable and is characterized by an infinite number of possible values. The probability of a single event does not exist or is 0; thus for continuous variables we have to calculate the probability that values of a variable fall within a specific interval.
5. Mean = 1.75; sd = .45
   a. Let 2 be $z1$, and 1 be $z2$. We recommend that you draw a normal distribution visualizing $z1$ and $z2$ and the areas under the curves (AUCs). The probability that an individual's score falls between $1(z2 = -1.66)$ and $2(z1 = 0.55)$ is: AUC(z-scores $\leq z1$) − AUC(z-scores $\leq z2$) = [1 − AUC(z-scores $> z1$)] − AUC(z-scores $> |z2|$) = $(1 - 0.2912) - 0.0485 = 0.7088 - 0.0485 = 0.6603$ or 66.03%.
   b. The probability that an individual's score falls between $2(z = 0.55)$ and $3(z = 2.77)$ is: AUC(z-scores > 0.55) − AUC(z-scores > 2.77) = $0.2912 - 0.0028 = 0.2884$ or 28.84%.

CHAPTER 4

1. We sample in order to save resources (e.g., time, money), and sampling, when done well will provide a more accurate estimate of a population's parameter.
2. In nonprobability sampling we do not know the probability that a member of a target population was or will be selected as a study element, whereas in probability sampling we do have such information.
3. There are several ways in which a researcher may bias the selection of study elements. A researcher might be prejudiced toward specific people and prefer not to select or talk to them for the study, or a researcher may prefer to select people to whom he has easy access, or a researcher may choose to include study elements that support his hypothesis.
4. If variables that are used to define the strata are relevant for representativeness then those variables are important predictors for the dependent variables of interest. If strata are important predictors for the dependent variables, then study participants within a stratum are more homogeneous with respect to the dependent variable.
5. See table below

	Male (.49)			Female (.51)			
	18–34 (.41)	35–49 (.24)	50+ (.35)	18–34 (.41)	35–49 (.24)	50+ (.35)	
High school (.32)	64.29	37.63	54.88	66.91	39.17	57.12	320
Some college (.37)	74.33	43.51	63.46	77.37	45.29	66.05	370
BA or higher (.31)	62.28	36.46	53.17	64.82	37.94	55.34	310
	200.9	117.6	171.5	209.1	122.4	178.5	1,000

CHAPTER 5

1. The population distribution refers to the distribution of values in the population of interest; the sample distribution refers to values observed in a sample. The sampling distribution, in contrast, refers to possible values a statistic can take; they are not for individual observations but for possible values of a statistic (e.g., the mean).
2. Sample sizes of 30 and higher provide a reasonable approximation of a normal distribution.
3. The standard error is the standard deviation of a statistic's sampling distribution.
4. Often times the variance (or standard deviation) of a variable in a population is unknown.
5. Mean = 180; Standard deviation = 60
   a. The standard error is 42.43 for $n = 2$; 26.83 for $n = 5$; 10.95 for $n = 30$. As the sample size increases, the standard error decreases.
   b. Standard error is 6 for $n = 100$; $\bar{y}$: $z = (190 - 180)/6 = 1.67$. Using the standard normal distribution table in Table 1 in Appendix A, we find that the corresponding right-tail probability above 1.67 is 0.0475, or 4.75%.
6. Median = 60; Standard deviation = 10; $n = 100$; $\sigma_{Md} = 1.25 * \left(\frac{10}{\sqrt{100}}\right) = 1.25$
7. Accuracy judgment: 52%; $n = 500$; $\sigma_{\%} = \sqrt{\frac{52\% \cdot 48\%}{500}} = 2.23\%$

CHAPTER 6

1. Mean = 66; Standard deviation = 15; $n = 100$
   a. 95 % CI: 66 ± (1.96 * 1.5) = [63.06, 68.94]
   b. 99 % CI: 66 ± (2.58 * 1.5) = [62.13, 69.87]
2. Accuracy = 56.1%; $n = 130$
   a. Standard error = 4.352%
   b. 95 % CI: 56.1 ± (1.96 * 4.352) = [47.57, 64.63]
   c. 99 % CI: 56.1 ± (2.58 * 4.352) = [44.87, 67.33]
3. Directionality refers to the direction (above or below) of the alternative hypothesis with respect to the null hypothesis. Specificity refers to the size of an effect. Directional, specific hypotheses provide the highest amount of information, and therefore contain the highest informational value in empirical research.
4. When you reject the null hypothesis based on sample data and it is true in the population you commit a Type-I error.
5. When you do not reject the null hypothesis but the research hypothesis is true in the population you commit a Type-II error.

6. An effect size is the least difference between two population parameters that can be considered practically relevant. This is a content (substantial) decision and not a statistical (methodological) one; researchers must decide what effect size is relevant for practice. Theory or prior explorative research should inform about reasonable effect sizes.
7. The researcher's decision to modify his hypothesis in light of his results is unscientific. He would need to conduct a new study with his hypothesis that television viewing leads to more optimistic world views.
8. a. The media researcher uses a directional research hypothesis; that is, he indicates the direction of the effect. In words, the alternative hypothesis is: The more hours of television one watches, the more optimistic is their worldview. There is a positive dependency between worldview and television watching. $H_1 : \rho > 0$
   b. The media researcher uses a directional null hypothesis (which is simply negation of the research hypothesis). The null hypothesis assumes that there is no or even a negative relationship between television watching and optimistic worldview. $H_0 : \rho \leq 0$
9. a. A one-tailed test is the test needed for a directional hypothesis, that is, a hypothesis where the researcher has some information about the direction of the effect.
   b. A two-tailed test is the test needed for a non-directional hypothesis, that is, one where the researcher posits some difference, but does not know the direction of the effect. The alpha levels are split between the two tails of the sampling distribution.
10. Reducing the Type-I error in effect increases the chances for a Type-II error. Likewise, reducing the chances of a Type-II error increases the chances for a Type-I error.
11. A specific hypothesis is necessary in order to assess the power of a statistical test.
12. Primarily the sample size.
13. α-level of 5%; $n = 100$; standard deviation = 8; $\varepsilon = 0.375$
    a. The power of the statistical test would be 98.21%
    b. The optimal sample size is 44.0896
14. z is greater than or equal to 3. According to Appendix A, this corresponds to 0.00135, or 0.1%

CHAPTER 7

1. a. $df = 8$, not significant; fail to reject H_0
   b. $df = 20$, not significant; fail to reject H_0
   c. $df = 28$, significant at $\alpha = 5\%$; reject H_0
   d. $df = 30$, significant at $\alpha = 5\%$; reject H_0
2. a. Assuming a non-directional hypothesis: H_1: The means of two groups are different from each other, or H_1: $\mu_1 - \mu_2 \neq 0$ ($\mu_1 \neq \mu_2$) and H_0: $\mu_1 - \mu_2 = 0$ ($\mu_1 = \mu_2$).
   b. $t_{emp} = -2.43$; $t_{crit} = 2.101$ or -2.101
   c. The empirical t-value falls within the null hypothesis rejection region. Therefore, we reject the null hypothesis in favor of the research hypothesis of a difference of means.
3. a. The mean of differences across all pairs is greater than 0. H_1: $\mu_d > 0$ and H_0: $\mu_d \leq 0$.
   b. $t_{emp} = 0.921$; t_{crit} ($df = 14$, $\alpha = 5\%$) $= 1.761$

c. Because the empirical *t*-value does not cross the critical *t*-value, we do not reject the null hypothesis.

4. a. See table below

	Attitude		
Gender	Positive	Negative	**Totals**
Male	Cell a: 48	Cell b: 12	60
Female	Cell c: 37	Cell d: 3	40
	85	15	$N = 100$

b. H_1: The frequency distribution of "attitude" and "gender" in the population are related. That is, the two categorical variables are dependent. H_0: The frequency distribution of "attitude" and "gender" are independent in the population.

c. Step 1: expected frequencies: a: 51; b: 9; c: 34; d: 6.
Step 2: $\chi^2_{emp} = 2.94$.
Step 3: χ^2_{crit} $(df = 1, \alpha = 1\%) = 6.63$.
Step 4: The empirical chi-square value is less than the critical chi-square value. Thus, the result is non-significant.
Step 5: This example violates the assumption of expected frequencies being greater than 10. The test result should be interpreted with caution.

5 a. H_a: The campaign changes student preferences in favor of using condoms (directional).

b. Step 1: expected frequencies = 30.
Step 2: $\chi^2_{emp} = 26.67$.
Step 3: $\chi^2_{crit}(df = 1, 2 * \alpha = 10\%) = 2.71$.
Step 4: Empirical chi-square value is larger than the critical chi-square value. Thus, the result is significant.
Step 5: The expected frequencies for cells b and c are greater than 10. The statistical test result can be considered as valid.

6. a. True
b. True
c. False
d. False

7. a. True
b. False
c. False
d. True
e. False

CHAPTER 8

1. a. See figure below

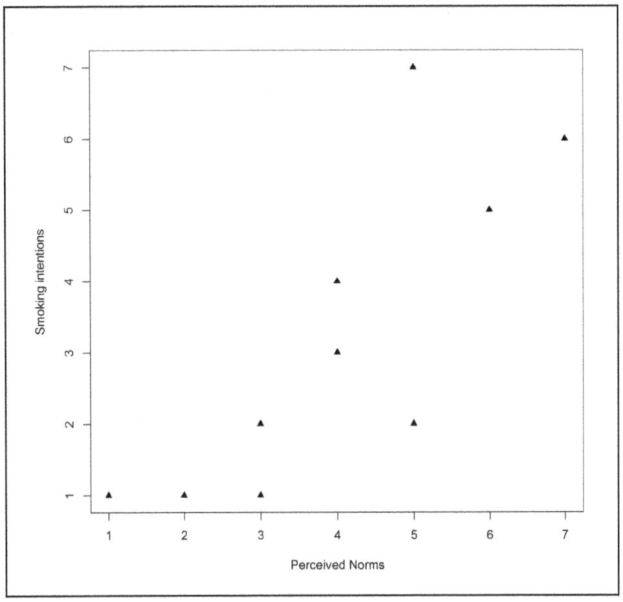

b. $\hat{y}_i = 0.966x_i - 0.667$

c. $t_{emp} = 3.79$; t_{crit} $(df = 8, \alpha = 5\ \%) = 1.86$

d. The empirical value exceeds the critical value, and so we reject the null hypothesis in favor of the research hypothesis.

e. Covariance = 2.9

2. The major disadvantage of covariance is that it depends on the scales or units of the variables; that is, it is dependent upon the variances of the variables.
3. Correlations are insensitive to scales or units of measurement because they are *z*-standardized.
4. a. $r = 0.857$

b. $t_{emp} = 4.707$; t_{crit} $(df = 8, \alpha = 5\%) = 1.86$

c. The empirical value exceeds the critical value, and so we reject the null hypothesis in favor of the research hypothesis.

d. $R^2 = 0.734$

5. Selection errors occur due to some biases in sampling methods. Selection errors include: 1) sample doesn't cover the entire range; 2) the sample covers extreme groups; 3) outliers bias the regression line.
6. a. The average *Z*-score is: 0.3135

b. The average *r* value is: 0.305

7. **a.** $\sigma_2 = 0.2012$; $Z_1 = 0.310$; $Z_2 = 0.693$; $t_{emp} = 1.903$; $t_{crit} = 1.65$ ($\alpha = 5\%$; one-sided)
   b. The empirical t-value is larger than the critical t-value. We conclude that the correlation is significantly greater in one sample than the other.
8. Causality (in a simple way) is usually defined through three criteria: 1) there is a significant correlation between variables or variables co-vary; the variables stand in appropriate time order; alternative explanations for associations among the variables can be eliminated.

APPENDIX C

COURSE SCHEDULES—RECOMMENDATIONS

QUARTER-SYSTEM (10 WEEKS SCHEDULE)

Week	Topics	Chapter(s)
01	Introduction & Basics of Measurement	Intro & 1
02	Descriptive Statistics	2
03	Probability & Probability Distributions	3
04	Sampling Methods & Sampling Distributions	4 & 5
05	Estimation (Predicting Parameters) & Confidence Intervals	6
	Midterm Examination	
06	Testing Parameters, Effect Size, & Test Power	6
07	Specific Significance Tests (*z*-test, *t*-test)	7
08	Specific Significance Tests (Chi-square Tests)	7
09	Linear Regression and Correlation	8
10	Correlation, Causality, and a Critical Perspective on Statistics	8 & 9
	Final Examination	

SEMESTER-SYSTEM (14 WEEKS SCHEDULE)

Week	Topics	Chapter(s)
01	Introduction & Basics of Measurement	Intro & 1
02	Descriptive Statistics	2
03	Probability & Probability Distributions	3
04	Sampling Methods	4
05	Sampling Distributions	5
06	Estimation (Predicting Parameters) & Confidence Intervals	6
07	Testing Parameters	6
	Midterm Examination	
08	Testing Parameters, Effect Size, & Test Power	6
09	Specific Significance Tests (*z*-test, *t*-test)	7
10	Specific Significance Tests (Chi-square tests)	7
11	Linear Regression and Correlation	8
12	Correlation & Causality	8
13	Introduction to Advanced Topics (ANOVA, MLM, SEM)	7&9
14	Critical Perspective on Statistics	9
	Final Examination	

USING R AND R-COMMANDER FOR STATISTICAL ANALYSES

APPENDIX D

This section provides you with hands-on training using a statistical computing program, "R" (R Core Development Team, 2011). There are several statistics programs available, but we focus on R because it doesn't require you to purchase an expensive license and R runs on all available computer platforms. You can work in R using its command line (and many of the powerful functions in R are done through this); however, we use one of the user-friendly graphical user interfaces in R, called R-Commander or Rcmdr (Fox, 2011). As of this writing, there are different graphical user interfaces (GUIs) with point-and-click functionality that are available to use, and that run from within or alongside R (e.g., pmg, rattle, Rcmdr, RExcel, RKWard). Given space constraints, we cannot provide overviews for each one. We have selected Rcmdr because it works for both Windows and Mac operating systems, and is fairly simple to use. We direct students who are curious about these other GUIs to do further research. If you think one may be a better fit for your needs, we encourage you to try these other GUIs out. For an overview of GUI projects, you are encouraged to visit: **http://www.sciviews.org/_rgui/** and under "Table of Contents," click on "RGUIs," and then "GUIs: overview of GUI projects." In addition, we introduce you to some supplemental packages to assist with analyses (Kim, 2011; Nimon, 2011).

This training will proceed as follows. First, we will guide you through how to install R and install and load Rcmdr. We will show you how to do this for both Windows and Apple operating systems. The process differs in some important ways, but we will clearly denote which instructions are for which system. You need only read installation instructions for your operating system, unless you plan to install R and Rcmdr on different computers with Windows and Mac operating systems. Subsequently, we will guide you through 12 problems, from entering data in Rcmdr to analyzing and interpreting results of a multiple regression. Windows- and Mac-based versions of Rcmdr differ in one key aspect—how data can be entered, and we will address this early on in these instructions. Specifically, problems include:

- Problem 1: Entering and retrieving data
- Problem 2: Descriptive statistics
- Problem 3: Finding standard scores (i.e., z-scores)
- Problem 4: One sample t-test
- Problem 5: Two independent samples t-test
- Problem 6: Two dependent samples t-test
- Problem 7: $k \times l$ chi-square test

- Problem 8: Pearson correlation
- Problem 9: Partial correlation
- Problem 10: One-way Analysis of Variance (ANOVA)

Advanced Examples

- Problem 11: Factorial ANOVA
- Problem 12: Multiple regression

Each of the problems will describe the research aim/problem, overview assumptions of the test, provide data as it should be entered into Rcmdr, and offer a summary of the results.

Before we proceed, a few notes are in order. First, we do not offer a CD or a web site with the data in this appendix because we believe that you should become familiar with entering data, as this is something you will do time and again as a communication researcher or a communication professional involved in research. Each problem has fewer than 100 data points, making data entry not overly burdensome. In addition, we provide a conclusion statement so that you may check your work. Second, all examples are derived from actual research studies in communication. These are not the actual data from those studies; the data were generated for these tutorials. We should note that the effects may be much stronger here than in the studies on which these problems are modeled.

INSTALLING R AND RCMDR FOR WINDOWS

System Requirements

R for Windows runs on Windows XP or later.

STEP-BY-STEP

1. Go to http://cran.r-project.org
2. Choose your operating system (Windows).
3. Click on base and then choose the most recent version of R.
4. Once the exe file has downloaded, open it and follow the installation prompt. Be sure to create a desktop shortcut so you can easily access R.

Running R with Rcmdr for Windows

STEP-BY-STEP

1. Once you have successfully installed R, load the R Console. You will see a window like the one in Figure D1.

Figure D1. RGui console Current version as of (09/27/2012)

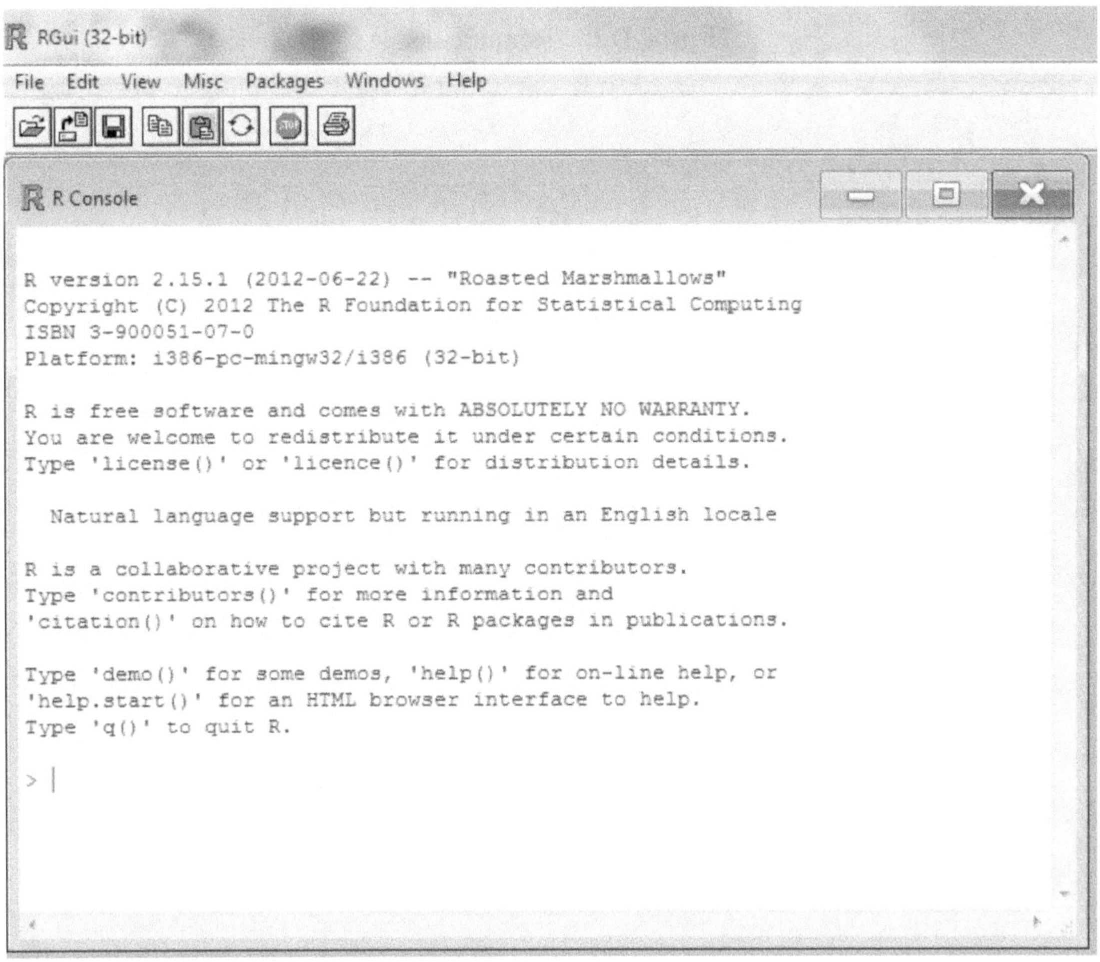

2. Once you have R open, you will need to install Rcmdr.
3. Go to **Packages,** and **Install packages.**
4. Select a CRAN mirror. It is typically best (in terms of Internet connection and download speed) to select a mirror site nearest to you geographically.
5. Select the **Rcmdr** package from the menu.

6. Once Rcmdr has been installed, you must now load the package. Note, once a package is installed you do not need to install it again next time you start R.
7. Go to the **Packages** menu, **Load package, Rcmdr.** Note: Every time you start R, you'll need to load (not install) any packages you're going to use; the first time you load **Rcmdr,** some installations require the installation of additional packages. R will tell you, just click on "OK" when R tells you that additional packages are required.

Figure D2 shows the screen you'll get once you've successfully loaded Rcmdr:

Figure D2. Rcmdr console

Loading R and Rcmdr after Successful Installation

To load R and Rcmdr for subsequent uses:

1. Open R by double clicking on the desktop icon you created during installation.
2. Go to the **Packages** menu, **Load package, Rcmdr.**

Exiting from R and Rcmdr

1. Go to **File, Exit, Exit from Commander and R.**
2. Click **OK** to the "Exit?" prompt.

Uninstalling R and Its Packages for Windows

These instructions are for completely removing R and its packages from your machine.

TO UNINSTALL R:

The labels for each step may vary depending on which version of Windows you are running, but the steps are essentially the same.

1. Click on the **Windows Start button,** and then **Control Panel.**
2. Click on **Uninstall a Program (or Add/Remove Programs).**
3. Scroll down and find **R for Windows,** tick the check box and click on **Uninstall.**
4. Click **Yes** to the prompt to remove R and all of its components. Click **OK** at the end of the uninstallation process.

TO REMOVE R PACKAGES:

1. Find your R library (you determined the location for these folders when you installed R). The folder will be called "R" unless you named it something different. Often the default installation location is "My Documents" or "Program Files." If this is not the case and you don't remember, you can search for "R" in the search bar located on the Windows start menu.
2. Once you've navigated to the parent folder for "R," select "R" and delete the entire folder. With R uninstalled there is no need to keep R libraries or folders. Note: If you have saved any datasets in your R library, this step will also delete your datasets.

INSTALLING R AND RCMDR FOR MAC

System Requirements

R and Rcmdr for Mac run on Mac OSX version 10.5 (Leopard) or higher. In order to run Rcmdr, you will need to install the Tcl/tk package for your Mac and make sure that X11 is installed on your machine. Most Macs have X11 already installed. If you experience problems after the installation of R and the Tcl/tk package, it is likely that the X11 package is missing on your machine. For this unusual situation, we provide instructions on how to install X11 for Mac under the troubleshooting heading.

STEP-BY-STEP—INSTALLATION OF R

1. Go to **http://cran.r-project.org**
2. Choose your operating system (Mac OS X).
3. Scroll down; under the heading "Files," click to download the pkg file that has "(latest version)" next to it.
4. Once the file has downloaded, open it and follow the installation prompt.
5. Once R has been installed, do NOT open it yet. First install the Tcl/tk package below.

STEP-BY-STEP—INSTALLATION OF THE TCL/TK PACKAGE

1. Go to **http://cran.r-project.org**
2. Choose your operating system (Mac OS X).
3. Scroll down; under the heading "Sub-directories," select **'tools.'**
4. Under the heading "Files," download the dmg file containing the Tcl/tk package. Once downloaded, open the file and follow the installation prompt.

Running R with Rcmdr for Mac

STEP-BY-STEP

1. Go to your "Applications" folder. If you would like a desktop shortcut, copy and paste the "R" icon to your desktop. Otherwise, you will always need to open R from your Applications folder.
2. Open R by clicking on the "R" icon in your Applications folder, or by the shortcut you installed on your desktop. Figure D3 shows the Mac R Console once you have successfully opened R.

Figure D3. Mac R Console Current version as of (09/27/2012)

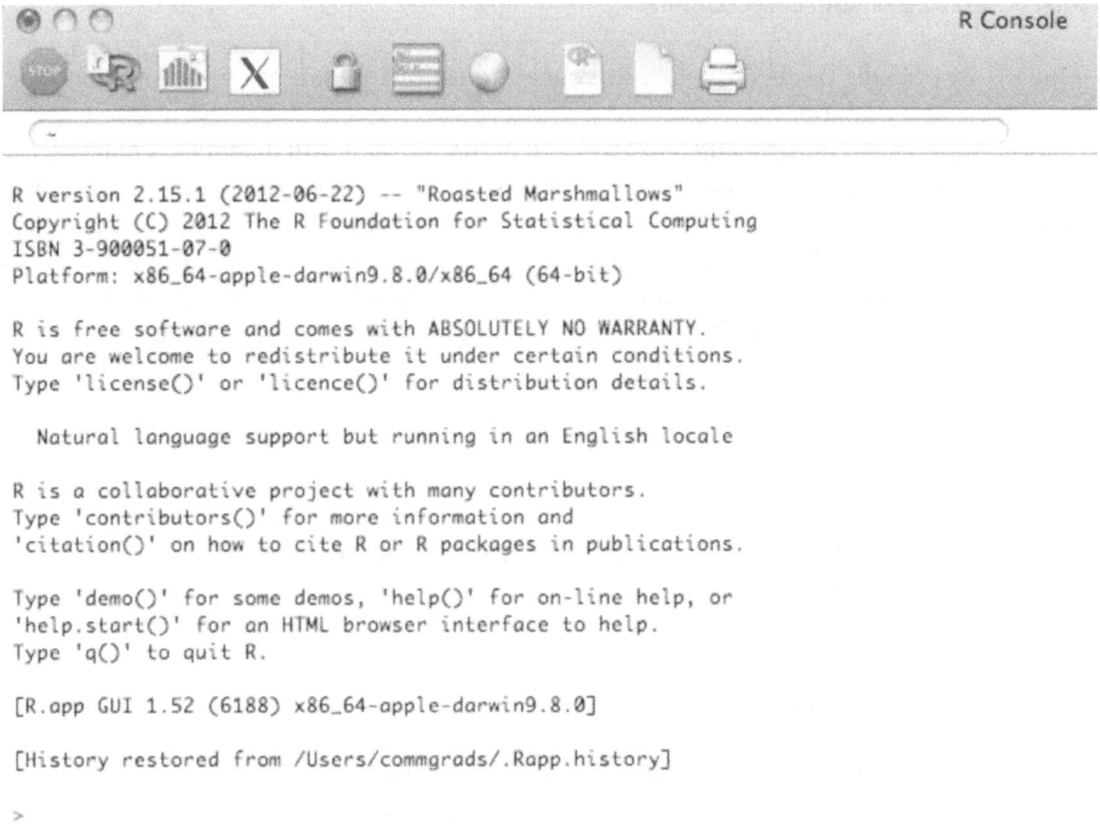

3. In the R Console (see Figure D3), select **Packages & Data,** and **Package Installer.**
4. Select CRAN (binaries) from the dropdown menu, click on **Get List,** and then select a CRAN mirror. It is typically best (in terms of Internet connection and download speed) to select a mirror site nearest to you geographically.
5. Scroll down to the **Rcmdr** package. It is very important to tick the checkbox to **Install dependencies,** select an install location (the default is "At System Level"), and then hit **Install Selected.** Note, once a package is installed you do not need to install it again next time you start R. Close the window by clicking on the red "x" in the upper left-hand corner.
6. Now you are ready to load Rcmdr and run it for the first time. Go to **Packages & Data,** and then to **Package manager.** This will display a window with all of the packages that are available for you to use on your Mac. Scroll down until you see Rcmdr. The left column ("status") will let you know whether it is loaded or not loaded. If Rcmdr is "not loaded" tick the check box to change its status to "loaded." You will know that Rcmdr has successfully loaded if the Rcmdr user interface (see Figure D2) opens up. Once Rcmdr has loaded, close the Package manager by clicking on the red "x" in the upper left-hand corner.

Loading R and Rcmdr after Successful Installation

To load R and Rcmdr for subsequent uses:

1. Open R by clicking on the "R" icon in your Applications folder, or by the shortcut you installed on your desktop.
2. Go to **Packages & Data,** and then to **Package manager.** This will display a window with all of the packages that are available for you to use on your Mac. Scroll down until you see Rcmdr. The left column ("status") will let you know whether it is loaded or not loaded. If Rcmdr is "not loaded" tick the check box to change its status to "loaded." You will know that Rcmdr has successfully loaded if the Rcmdr user interface (see Figure D2) opens up. Once Rcmdr has loaded, close the Package manager by clicking on the red "x" in the upper left-hand corner.

Exiting from R and Rcmdr

1. On the Rcmdr menu, click on **File, Exit, Exit from Commander and R.**
2. Click "OK" to the "Exit?" prompt.

Troubleshooting Installation for Mac

If Rcmdr fails to load, there are two potential issues: 1) you might not have loaded all of the packages that Rcmdr is dependent on, or 2) you may need to check whether X11 is installed on your Mac.

REINSTALLING RCMDR

If you get an error message that indicates that one or more packages are missing (e.g., "Error: package 'car' required by 'Rcmdr' could not be found"), follow the instructions above (Running R with Rcmdr) again and make sure that you tick the checkbox **"Install dependencies."** You do not need to remove Rcmdr—the reinstallation will overwrite the previous file.

INSTALLING X11

1. Run R and click on the "X" atop the R Console. This will check to ensure X11 is installed on your computer. If X11 is installed, you should see an "xterm" window come up. If not, then follow the next instructions.
2. Depending on which version of Mac OSX you have, you need to install the appropriate version of X11.
   a. X11 should always be included on OSX 10.7. If it is missing under earlier versions (10.5 and 10.6), it can be installed from the Mac installation disc with the instructions below:
      i. Insert the OS X installation disc
      ii. Double click on **Optional Installs**

   iii. Double click on **Optional Installs.mpkg,** and click **Continue** and accept the licensing agreement.
   iv. Click on the triangle next to **Applications** to expand the list of applications.
   v. Tick the checkbox "X11," and then click **Continue** and **Install.** Close the dialogue box when the installation is complete.
   b. For OSX 10.8 or later, you can download X11 (XQuartz), from: **http://xquartz.macosforge.org/landing/**
      i. Download the dmg file for XQuartz.
      ii. Open the file by double clicking it; then double click on **XQuartz.pkg** to run the installer, and click through all of the defaults.
3. Close R, and relaunch. Follow step 5 under "Running R with Rcmdr" to load the Rcmdr package.

Uninstalling R and Rcmdr for Mac

The typical installation of R will install the GUI (R.app) and the R framework. Thus, uninstalling R and its packages consists of two parts for Mac. Installations typically install the GUI in the /Applications folder and the R framework in the/Library/Frameworks folder. How to access these files depends on whether you installed R at the user (accessible to only one user) or the system-level (accessible to all users on your computer). These instructions assume that you installed at the system level.

TO REMOVE THE GUI:

1. Locate the /Applications folder.
2. Highlight the R.app folder and drag it into the Trash Bin.

TO REMOVE THE R FRAMEWORK:

1. Locate the /Library/Frameworks folder.
2. Highlight the R framework folder and drag it into the Trash Bin. (Your operating system may require that you enter an administrative password to complete this step.)

Alternative Installation Processes for Mac

If you prefer the Windows version of R to the Mac version, there are two other possible ways to install R on a Mac that we will mention. But a word of caution: these are not for the "faint at heart," and we will not provide detailed instructions here, as there are more detailed sources available on the web.

VIRTUAL MACHINE

Macs that have OSX 10.5 or higher are equipped with Boot Camp, which allows you to run Windows programs provided that you have a Windows installation disc. Visit **http://support.apple.com/kb/HT1461** for instructions on running Boot Camp. Once you have Windows running, you can install R according to the instructions we have provided in the section "Downloading and installing R and Rcmdr for Windows."

WINE

Alternatively, you can download a program called WINE, which allows you to run Windows applications on other operating systems without using a virtual machine. The student is directed to **http://wiki.winehq.org/MacOSX/** for instructions on how to install WINE on Mac, and run programs within Mac. Once you have WINE running, you can install R according to the instructions we have provided according to the section "Downloading and installing R and Rcmdr for Windows."

INTRODUCTION: PROGRAM INSTRUCTIONS FOR RCMDR

Before entering data and running statistical tests, you should familiarize yourself with Rcmdr. For data entry, Rcmdr uses a spreadsheet arrangement, with columns that represent the names of variable, and rows that refer to cases or subject numbers. The intersection of a column and row is a cell. Rcmdr for Windows and Mac look quite similar, although in the Mac version some menu options are only available once data are loaded.

Menu bar

The Menu Bar goes from File on the far left to Help on the far right. Most of the analyses will be run from this bar.

Tool bar

The Tool Bar goes from Active data set to Active model.

Script Window

As we mentioned, R is a command-line program. The Script Window shows the code required to run any task in R, from loading data, to running any number of statistical tests.

Output Window

The Output Window shows: 1) the result from any script (in dark blue) and 2) the command that was used to generate the output (in red).

Message Window

The message window provides: 1) error messages (in red), 2) warning messages (in green), and 3) other information (in blue).

Data Editor

In the Windows version of Rcmdr, the Data Editor can be accessed either by creating a new data set or loading/importing an existing data set, and then clicking on Edit Data set. You can add single observations using the Data Editor. In the Mac version of Rcmdr, the Data Editor can only be launched once a data set is imported.

TO REMOVE A ROW

1. Click on **Data, Active data set, Remove row(s) from active data set.**
2. Under **Indices or quoted names of rows to remove,** enter the row number or numbers. For example, if you want to remove the 5th row, enter 5. If you want to remove a range of rows, enter the range. For example, 8 through 10 would be 8:10. Enter a name for the new active data set. Keep it the same if you want to override the old data set.

TO REMOVE A VARIABLE (COLUMN)

1. Click on **Data, Manage variables in active data set, Delete variables from data set.**

TO RENAME VARIABLES (COLUMN)

1. Click on **Data, Manage variables in active data set, Rename variables.**

   -OR-

2. Click on **Edit Data set.**
3. By clicking on the column name in the Data Editor, you can change the variable name.

TO CHANGE THE TYPE OF VARIABLE

1. By clicking on the column name in the data editor, you can change the type of variable (e.g., "numeric" and "character"). For variables measured at the interval and ratio level, you should select "numeric." For variables measured at the ordinal or nominal level, you should select "character." It is good practice to CAPITALIZE column names, and use lowercase for factor levels. Figure D4 shows the R Data Editor.

Figure D4. R Data Editor

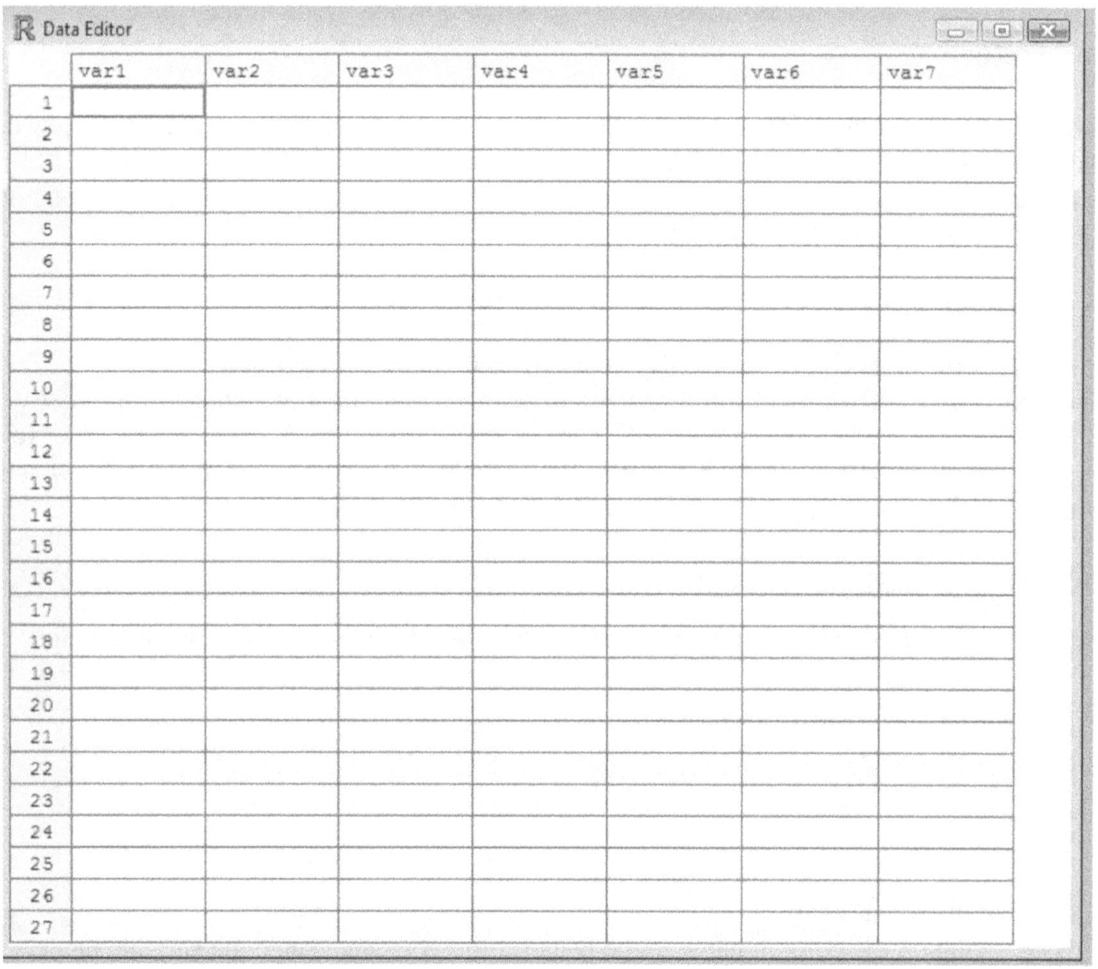

PROBLEM 1: ENTERING AND RETRIEVING DATA

Entering data is a key concern, and we will provide more detail here. The Windows and Mac versions differ in an important way. In Windows, you can enter data directly into the Data Editor in R. In Mac, you need to import data before you can edit it in the Data Editor. The reason for this difference is that the Mac version of R freezes up when there is a blank data set, and so the developer of Rcmdr did not provide the option to create a new data set in Mac. The step-by-step instructions that follow are included for importing data from a Mac and work for Windows users as well.

R and Rcmdr support several different data file types. We are having you enter the data in a plain text editor (.txt). You could also import data from other statistical programs (e.g., SPSS, SAS, Minitab), and spreadsheet and database programs (e.g., Microsoft Excel, Microsoft Access). The options are available under the same menu (**Data, Import data**) in both Windows and Mac versions. Because most machines come with some plain text editor, and in the interest of ease of presentation and space constraints, this is our method of choice. We encourage you to try out other methods for importing your data as well.

As you prepare your data sets with a plain text editor, there are a few things to keep in mind:

1. The first row of your text data set should always contain your variable names (in CAPITAL LETTERS).
2. Use the "Tab" key to separate entries within a row.
3. Use the Enter/Return key to move to create or insert a new row.
4. Use the arrow keys to move the cursor up, down, left or right to move to cells that already contain entered data.

Example: Dangerous conversations

A researcher is interested in exploring the frequency (in days per week) with which individuals who identify as either Republican or Democrat engage in discussions with those who identify as the opposite party. See Eveland & Hively (2009).

Step-by-step

WINDOWS

1. Click **Data, New data set.**
2. Enter PROB1 as the name for the new data set.
3. Enter the data from Figure D5 into the Data Editor.

MAC

1. In your text editor of choice (TextEdit, Microsoft Word, or OpenOffice Writer) enter the data as pictured in Figure D5, and save as PROB1.txt.
2. Click on **Data, Import Data, from text file, clipboard, or URL…** Once the data set is imported, check the Message Window for any errors.
3. Enter PROB1 as name of the data set. Now click on **Data, Active data set, Save active data set** as PROB1.

Figure D5. Problem 1 data

Data Editor

	DISCUSS	var2	var3	var4
1	4			
2	2			
3	5			
4	4			
5	2			
6	5			
7	2			
8	3			
9	2			
10	2			
11				
12				
13				
14				
15				

BOTH WINDOWS/MAC

4. After data are entered, check to see if there are any errors. If you discover an error, put the cursor over that score and enter the correct value.
5. Click **Data, Active data set, Save active set** as PROB1.
6. Close Rcmdr by going to **File, Exit, Exit from Commander and R.**
7. Load R and Rcmdr again.
8. Click **Data, Load data set,** PROB1.rda.

PROBLEM 2: DESCRIPTIVE STATISTICS

Example: Aggression and video games

A researcher is interested in the relationship between violent video games and hostile thoughts. After participants played a violent video game, the researcher had them fill in word fragments representing neutral and aggressive words. The participants' scores are aggressive words completed by participants divided by the total of fragments that could yield aggressive words. For example, a score of 0/1 would indicate that a participant chose to fill no/all word fragments with aggressive words. Compute descriptive statistics (mean, standard deviation, quartiles), a frequency distribution, and a histogram (see Chapter 2). See Eastin, Appiah, and Cicchirillo (2009).

STEP-BY-STEP (BASIC STATISTICS)

Windows

1. Click **Data, New data set.**
2. Enter PROB2 as the name for the new data set.
3. Enter the data shown in Figure D6 into the Data Editor.

Mac

1. In your text editor of choice (TextEdit, Microsoft Word, or OpenOffice Writer) enter the data as pictured in Figure D6, and save as PROB2.txt.
2. Click on **Data, Import data, from text file, clipboard, or URL** ... Once the data set is imported, check the Message Window for any errors.
3. Enter PROB2 as name of the data set and check **Variable names in file.**

Figure D6. Problem 2 data

Data Editor

	AGGRESS	var2	var3	var4
1	0.6			
2	0.3			
3	0.6			
4	0.4			
5	0.6			
6	0.1			
7	0.6			
8	0.7			
9	0.6			
10	0.6			
11	0.5			
12	0.3			
13	0.3			
14	0.5			
15	0.3			
16	0.4			
17	0.5			
18	0.3			
19	0.1			
20	0.4			

BOTH WINDOWS/MAC

4. Click **Data, Active data set, Save active set as** PROB2.
5. Click **Statistics, Summaries, Numerical Summaries.**
6. Select AGGRESS, and check **Mean, Standard Deviation,** and **Quantiles.** By default, quantiles will have the values 0 (minimum), 0.25, 0.50 (median), 0.75, and 1 (maximum) which stands for the 25th, the 50th, and the 75th percentiles or the first, second, and third quartiles (see Chapter 2). By changing these values you can compute any percentile you are interested in.
7. Click **OK.**

Output (basic statistics)

Figure D7 shows the selected output from R, which includes the mean, standard deviation, minimum value, quartiles, and maximum value.

Figure D7. Output for problem 2

```
 mean          sd  0% 25%  50% 75% 100%  n
0.435 0.1725200 0.1 0.3 0.45 0.6  0.7 20
```

Step-by-step (histogram)

1. Click on **Graphs, Histogram.**
2. Leave number of bins on **<auto>,** and select **Frequency counts** for axis scaling.
3. Click **OK.**

Output (histogram)

Figure D8 is a histogram of the aggression data. The frequency counts are on the *y*-axis, and the bins are along the *x*-axis.

Figure D8. Histogram for problem 2

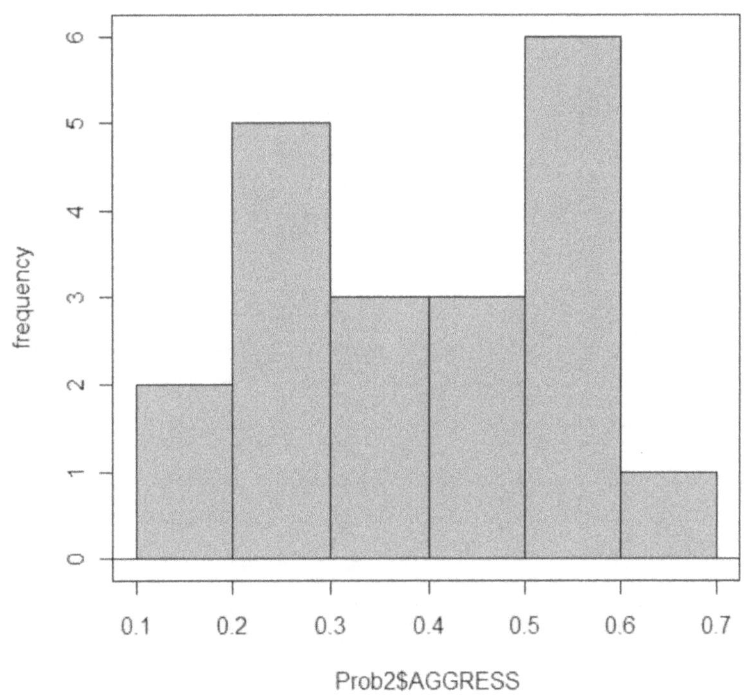

PROBLEM 3: FINDING STANDARD SCORES

Example: Advertisement believability

A researcher is interested in how individuals assess the believability of advertisements. A higher overall score indicates that ads are more believable. Transform the scores into *z*-scores. See Chang (2006).

STEP-BY-STEP

WINDOWS

1. Click **Data, New data set.**
2. Enter PROB3 as the name for the new data set.
3. Enter the data shown below in Figure D9 into the Data Editor.

MAC

1. In your text editor of choice (TextEdit, Microsoft Word, or OpenOffice Writer) enter the data as pictured in Figure D9, and save as PROB3.txt.
2. Click on **Data, Import data, from text file, clipboard, or URL.** Once the data set is imported, check the Message Window for any errors.
3. Enter PROB3 as name of the data set and check **Variable names in file.**

Figure D9. Problem 3 data

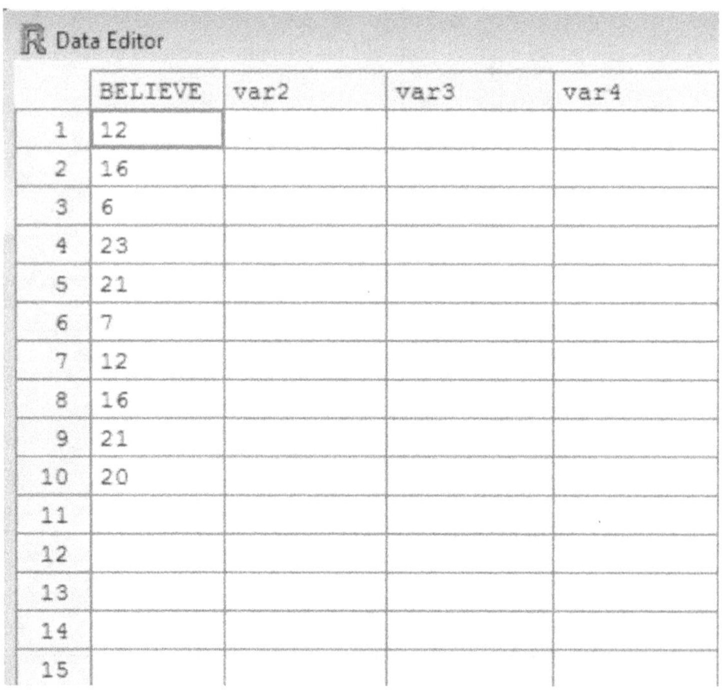

	BELIEVE	var2	var3	var4
1	12			
2	16			
3	6			
4	23			
5	21			
6	7			
7	12			
8	16			
9	21			
10	20			
11				
12				
13				
14				
15				

BOTH WINDOWS/MAC

4. Click **Data, Active data set, Save active set.**
5. Click **Data, Manage variables in active data set, Standardize variables.**
6. Select BELIEVE.
7. Click **View edit set.**

Output

After clicking on "Edit data set" you will see a new variable in the active data set called Z.BELIEVE (see Figure D10). The new variable contains the *z*-transformed scores (*z*-scores) of the original variable BELIEVE.

Figure D10. Problem 3 with ad believability scores *z*-transformed

Data Editor

	BELIEVE	Z.BELIEVE	var3	var4
1	12	-0.5663172		
2	16	0.09993833		
3	6	-1.565700		
4	23	1.265885		
5	21	0.9327577		
6	7	-1.399137		
7	12	-0.5663172		
8	16	0.09993833		
9	21	0.9327577		
10	20	0.7661939		
11				
12				
13				
14				
15				

PROBLEM 4: ONE SAMPLE T-TEST

Background

The one sample *t*-test evaluates whether a sample comes from a population where the population mean μ equals a specified value. The test is utilized with interval level data and is employed when the researcher does not know the population standard deviation.

Assumptions

The one sample *t*-test assumes that: 1) the sample has been randomly selected from the population, and 2) the sample size is sufficiently large ($n \geq 30$). In case of a small sample ($n < 30$), the test assumes that the variable in the population is normally distributed. If either of these assumptions is violated the reliability of the test statistic is compromised.

Example: Expressive action

A researcher was interested in the extent to which individuals would take expressive action under stress. Prior research has established a population mean of μ_0=24 under normal circumstances; i.e., without stress. A higher score indicates a greater self-reported likelihood that an individual would take expressive action. Observing $n = 10$ participants under stress, the researcher is interested in testing the hypothesis that individuals operating under stress do not represent the general population with $\mu_0 = 24$. Thus, the hypotheses are as follows: H1: $\mu \neq \mu_0$; H0: $\mu = \mu_0$. See Boyle, et al. (2006).

STEP-BY-STEP

WINDOWS

1. Click **Data, New data set.**
2. Enter PROB4 as the name for the new data set.
3. Enter the data shown below in Figure D11 in the Data Editor.

MAC

1. In your text editor of choice (TextEdit, Microsoft Word, or OpenOffice Writer) enter the data as pictured in Figure D11, and save as PROB4.txt.
2. Click on **Data, Import data, from text file, clipboard, or URL.** Once the data set is imported, check the Message Window for any errors.
3. Enter PROB4 as name of the data set and check **Variable names in file.**

BOTH WINDOWS/MAC

4. Click **Data, Active data set, Save active set.**

Figure D11. Problem 4 data

Data Editor

	RESPONSE	var2	var3	var4
1	25			
2	10			
3	12			
4	42			
5	18			
6	29			
7	10			
8	40			
9	21			
10	25			

5. Click **Statistics, Means, Single-sample *t*-test.**
6. Select RESPONSE as the variable.
7. Enter the null hypothesis (H_0) parameter μ, which is 24, and a confidence level of **0.95,** which corresponds to a two-tail significance level of $\alpha = 0.05$ or 5%.
8. Click **OK.**

Output

Figure D12 shows the output for the one sample *t*-test, which includes the data on which the analysis is based, *t*-value, degrees of freedom, statement of the alternative hypothesis (H_1), the 95% confidence interval, and the estimated population mean. Note that a *p*-value smaller than α (here 0.05 or 5%) rejects H_0, and a *p*-value greater than α does not reject H_0 (see Chapter 6).

Figure D12. Single-sample t-test output

```
        One Sample t-test

data:  PROB4$RESPONSE
t = -0.2208, df = 9, p-value = 0.8302
alternative hypothesis: true mean is not equal to 24
95 percent confidence interval:
 15.00334 31.39666
sample estimates:
mean of x
     23.2
```

Output Interpretation

A one sample t-test was computed comparing a sample mean (i.e., a population mean estimate) with a known population mean of $\mu_0 = 24$. Because no significant difference was found ($t(9) = -0.2208, p > 0.05$), the null hypothesis H_0 was not rejected. Therefore, this sample mean is likely a realization of the known population mean. The researcher *cannot* say that individuals operating under stress belong to a different population with a different mean parameter μ. Note: Not rejecting the null hypothesis does not mean accepting the null hypothesis; i.e., $\mu_0 = 24$ (see Chapter 6). Still, due to the lack of strong evidence against the null hypotheses, the researcher would probably recommend to (preliminarily) maintain the assumption that individuals operating under stress are no different regarding expressive action than individuals of the general population.

PROBLEM 5: TWO INDEPENDENT SAMPLES T-TEST

Background

The two independent samples t-test evaluates whether two independent samples represent two populations with different means.

Assumptions

The two independent sample t-test is used with interval/ratio data for the dependent variable and nominal data for the independent or group variable. The test assumes the following: 1) each sample has been randomly selected from the population it represents; 2) the sample sizes are sufficiently large ($n_1 + n_2 \geq 50$) (in case of small samples ($n_1 + n_2 > 50$), the test assumes that the variable in the population is normally distributed); and 3) the variance of the variable in the population that sample 1 represents is equal to the variance that sample 2 represents (homogeneity of variances). If the above assumptions are violated, the reliability of the test may be compromised.

Example: Self-presentation and F-t-F v. Email

A researcher is interested in determining whether male and female students differ in their preferences for self-presentational goal accomplishment. A higher score indicates a greater preference for face-to-face communication. Thus, the hypotheses are as follows: H_1: $\mu 1 - \mu 2 \neq 0$; H_0: $\mu 1 - \mu 2 = 0$. See Taylor, Jowi, Schreirer, & Bertelsen (2011).

STEP-BY-STEP (TESTING THE HOMOGENEITY OF VARIANCES: LEVENE'S TEST)

WINDOWS

1. Click **Data, New data set.**
2. Enter PROB5 as the name for the new data set.
3. Enter the data shown in Figure D13 into the Data Editor.

Figure D13. Problem 5 data

Data Editor

	IMPRESSION	GENDER	var3	var4
1	2.8	F		
2	2.8	F		
3	2.6	F		
4	2.8	F		
5	2.6	F		
6	2.4	F		
7	3	F		
8	3.4	F		
9	3	F		
10	2.2	F		
11	3.6	M		
12	2.8	M		
13	3	M		
14	3.6	M		
15	3	M		
16	3.4	M		
17	3	M		
18	3	M		
19	3.6	M		
20	3	M		

MAC

1. In your text editor of choice (TextEdit, Microsoft Word, or OpenOffice Writer) enter the data as pictured in Figure D13, and save as PROB5.txt.
2. Click on **Data, Import data, from text file, clipboard, or URL.** Once the data set is imported, check the Message Window for any errors.
3. Enter PROB5 as name of the data set and check **Variable names in file.**

BOTH WINDOWS/MAC

4. Click **Data, Active data set, Save active set.**
5. Click **Statistics, Variances, Levene's test.**
6. Choose GENDER as the Groups variable and IMPRESSION as the Response variable; select **Mean** as the Center.

Output (Levene's test)

Levene's test is a test of homogeneity of variance, which is one of the assumptions of the *t*-test. Levene's test produces an *F* ratio, which if not significant ensures that the homogeneity requirement has been met. Figure D14 shows the output for the Levene's Test for Homogeneity of Variance, which includes degrees of freedom, an *F* value, and a *p*-value.

Figure D14. Levene's test for homogeneity of variances

```
Levene's Test for Homogeneity of Variance (center = mean)
      Df F value Pr(>F)
group  1  0.1825 0.6743
      18
```

In the output, the Levene *F* ratio is only 0.1825 and the significance level is .6743, which is clearly well above the .05 level. Thus, we can say that the requirement of homogeneity of variance has been satisfied.

Note: We have learned in Chapter 6 that non-significant results cannot be interpreted as evidence for the null hypothesis. While this is correct, there are cases in which it is difficult to apply a correct so-called "equivalence test." The Levene Test would qualify as such a case. Therefore, one indirectly concludes the equivalence of population variances (the null hypothesis assumption) if the statistical result is *obviously* or *clearly* non-significant. We recommend that *p*-values above 0.5 should be interpreted as "obviously" non-significant and *p*-values above 0.8 as "clearly" non-significant. If a result is *slightly* non-significant; i.e., $0.05 > p\text{-value} < 0.5$, we caution to assume homogeneity of variances in the population. However, as we mention in Chapter 6, a violation of this assumption is only important if the sample sizes were small. For large samples the independent samples *t*-test decides reliably even if this assumption is violated.

7. Click **Statistics, Means, Independent samples *t*-test.**
8. Select GENDER as the Groups variable, and IMPRESSION as the Response variable.
9. Click **Assume equal variances.**
10. Set the alternative hypothesis as **two-sided.**
11. Set the confidence interval as **0.95.**
12. Click **OK.**

Output (*t*-test)

Figure D15 shows the output from the test, which includes the data on which the analysis is based, *t*-value, degrees of freedom, statement of the alternative hypothesis, width of 95% confidence interval, and the estimated mean.

Figure D15. Two independent sample t-test

```
        Two Sample t-test

data:  IMPRESSION by GENDER
t = -3.0251, df = 18, p-value = 0.007277
alternative hypothesis: true difference in means is not equal to 0
95 percent confidence interval:
 -0.7455780 -0.1344220
sample estimates:
mean in group F mean in group M
           2.76            3.20
```

Output Interpretation

The two independent sample *t*-test was found to be significant ($t(18) = -3.025, p < 0.01$); that is, there is good empirical evidence for the rejection of the null hypothesis. Because the estimated population mean for women is 2.76 and for men 3.20, the researcher concludes that women tend to prefer email for self-presentational goals, whereas men tend to prefer face-to-face for self-presentational goals.

PROBLEM 6: TWO DEPENDENT SAMPLES T-TEST-RATIO

Background

The *t*-test for two dependent samples evaluates whether two dependent samples represent two populations with different mean values. Usually repeated measurements (a measurement before and after an intervention) or so-called matched samples are considered dependent samples (see Chapter 7).

Assumptions

The *t*-test for two dependent samples is used with interval/ratio data for the dependent variable and nominal data for the independent or group variable. The test assumes the following: 1) each sample has been randomly selected from the population it represents; 2) the sample sizes are sufficiently large ($n \geq 30$); n is the number of measurement pairs; the difference between a first and second measurement, for example, defines a measurement pair), in case of small samples ($n < 30$), the test assumes that the measurement differences are normally distributed in the population; and 3) the variance of the dependent variable in the population represented by the first measurement equals the variance of the dependent variable in the population represented by the second measurement.

Example: Media influence and promiscuous behavior

A researcher is interested in whether viewing a media depiction of promiscuous sexual behavior leads to increased promiscuous sexual behavior among viewers. In an experiment 10 viewers are exposed to a movie showing promiscuous sexual behavior. Two measurements on the intention of engaging in promiscuous sexual behavior are taken, one before exposure (PRE) and one after exposure (POST). Scores indicate a participant's intention of having a one night stand. Thus, the hypotheses are as follows: H_1: μ(pre-post) < 0;. H_0: μ(pre-post) ≥ 0. See Nabi and Clark (2008).

STEP-BY-STEP

WINDOWS

1. Click **Data, New data set.**
2. Enter PROB6 as the name for the new data set.
3. Enter the data in Figure D16 into the Data Editor.

Figure D16. Problem 6 data set

Data Editor

	PRE	POST	var3	var4
1	2	3		
2	2	3		
3	2	3		
4	3	3		
5	3	4		
6	3	4		
7	3	4		
8	3	6		
9	3	6		
10	4	6		
11				
12				
13				
14				
15				

MAC

1. In your text editor of choice (TextEdit, Microsoft Word, or OpenOffice Writer) enter the data as pictured in Figure D16, and save as PROB6.txt.
2. Click on **Data, Import data, from text file, clipboard, or URL.** Once the data set is imported, check the Message Window for any errors.
3. Enter PROB6 as name of the data set and check **Variable names in file.**

BOTH WINDOWS/MAC

4. Click **Data, Active data set, Save active set.**
5. Click **Statistics, Means,** and **Paired-T test.**
6. Select Pre as First Variable and Post as Second Variable.
7. Select alternative hypothesis as **Difference < 0** (Note: this is because it is hypothesized that pre scores will be less than post scores, so pre minus post will produce a negative mean value).
8. Set the confidence level as **0.95.**
9. Click **OK.**

Output Interpretation

Figure D17 provides the output for the two dependent samples *t*-test, which includes the data on which the analysis is based, *t*-value, degrees of freedom, statement of the alternative hypothesis, width of a 95% confidence interval, and the estimated mean of the measurement differences.

Figure D17. Output for the two dependent samples t-test

```
        Paired t-test

data:  PROB6$PRE and PROB6$POST
t = -4.5826, df = 9, p-value = 0.0006615
alternative hypothesis: true difference in means is less than 0
95 percent confidence interval:
       -Inf -0.8399747
sample estimates:
mean of the differences
                   -1.4
```

The *t*-statistic of the two dependent samples *t*-test was significant ($t = -4.58$, $df = 9$, $p < 0.001$), thus the analysis found empirical support for the research hypothesis. The scores were significantly higher in the post-test (the mean of the measurement differences is negative), showing participants self-reported increased intention to engage in promiscuous sexual behavior after exposure to the movie.

PROBLEM 7: K × L CHI-SQUARE TEST

Background

A *k* by *l* chi-square test analyzes the cell frequencies that result from a cross-tabulation of two categorical (nominal level) variables with one variable having "*k*" categories/levels and the other variable "*l*" categories/levels. The test is concerned with the central question of whether the observed frequency differences across cells occur by chance (i.e., correspond to expected frequencies according a null hypothesis assumption) or not. Simplified, the *k* by *l* chi-square test tests whether there is a relationship between two categorical variables in the population.

Assumptions

The *k* by *l* Chi-square test assumes the following: 1) categorical/nominal data used for mutually exclusive categories in the analysis; 2) the data represent a random sample of independent observations.

Example: Indirect Aggression and Justification for Actions

A researcher is interested in whether there is a relationship between indirect aggression and justification (justified or unjustified) of such aggression in children's cartoons. According to the researcher, indirect aggression includes such behaviors as social exclusion (e.g., gossip, ignoring, dirty looks), guilt induction (e.g., undue pressure, blame unjustly, judge work unjustly), and physical indirect aggression (e.g., plotting, kidnapping, vandalism). Thus the hypotheses of this test are: H_1: There *is* a relationship between indirect aggression and justification in the population of children's cartoons; H_0: There *is no* relationship between indirect aggression and justification in the population of children's cartoons. See Coyne and Whitehead (2008).

STEP-BY-STEP (RAW DATA, GENERATING THE CROSS-TABLE)

Windows

1. Click **Data, New data set.**
2. Enter PROB7 as the name for the new data set.
3. Enter the data shown below in Figure D18 in the Data Editor.

Mac

1. In your text editor of choice (TextEdit, Microsoft Word, or OpenOffice Writer) enter the data as pictured in Figure D18, and save as PROB7.txt.
2. Click on **Data, Import data, from text file, clipboard, or URL.** Once the data set is imported, check the Message Window for any errors.
3. Enter PROB7 as name of the data set and check **Variable names in file.**

Both Windows/Mac

4. Click **Data, Active data set, Save active set**

Figure D18. Problem 7 data

Data Editor

	AGGRESS	JUSTIFY	var3	var4
1	SOCIAL	YES		
2	SOCIAL	YES		
3	SOCIAL	YES		
4	SOCIAL	YES		
5	SOCIAL	YES		
6	SOCIAL	YES		
7	SOCIAL	YES		
8	SOCIAL	YES		
9	SOCIAL	YES		
10	SOCIAL	YES		
11	SOCIAL	YES		
12	SOCIAL	YES		
13	SOCIAL	YES		
14	SOCIAL	YES		
15	SOCIAL	YES		
16	SOCIAL	NO		
17	SOCIAL	NO		
18	SOCIAL	NO		
19	SOCIAL	NO		
20	GUILT	YES		
21	GUILT	YES		
22	GUILT	NO		
23	GUILT	NO		
24	GUILT	NO		
25	GUILT	NO		
26	GUILT	NO		
27	GUILT	NO		
28	GUILT	NO		
29	GUILT	NO		
30	PHYSICAL	YES		
31	PHYSICAL	YES		
32	PHYSICAL	YES		
33	PHYSICAL	NO		
34	PHYSICAL	NO		
35	PHYSICAL	NO		

36	PHYSICAL	NO		
37	PHYSICAL	NO		
38	PHYSICAL	NO		
39	PHYSICAL	NO		
40	PHYSICAL	NO		

5. Select **Statistics, Contingency Tables, Two-Way Table.**
6. Select **AGRESS** as the Row variable and **JUSTIFY** as the Column variable, and **Row percentages** for Compute Percentages.
7. Check **Chi-square test of independence, Components of chi-square statistic,** and **Print expected frequencies.**
8. Click **OK.**

Output

Figure D19 shows the cross table and row percentages, and Figure D20 shows the chi-square test values, expected counts, and residuals (i.e., contributions to chi-square).

Figure D19. Cross table of aggregation type by justification

```
> .Table <- xtabs(~AGGRESS+JUSTIFY, data=PROB7)

> .Table
          JUSTIFY
AGGRESS     NO YES
  GUILT      8   2
  PHYSICAL   8   3
  SOCIAL     4  15

> rowPercents(.Table) # Row Percentages
          JUSTIFY
AGGRESS       NO  YES Total Count
  GUILT     80.0 20.0   100    10
  PHYSICAL  72.7 27.3   100    11
  SOCIAL    21.1 78.9   100    19
```

Figure D20. Chi-square output

```
        Pearson's Chi-squared test

data:  .Table
X-squared = 12.2411, df = 2, p-value = 0.002197


> .Test$expected # Expected Counts
          JUSTIFY
AGGRESS     NO YES
  GUILT    5.0 5.0
  PHYSICAL 5.5 5.5
  SOCIAL   9.5 9.5

> round(.Test$residuals^2, 2) # Chi-square Components
          JUSTIFY
AGGRESS      NO  YES
  GUILT    1.80 1.80
  PHYSICAL 1.14 1.14
  SOCIAL   3.18 3.18
```

Output Interpretation

A *k* by *l* chi-square test (here a 3 × 2) was computed analyzing the relationship between the "indirect aggression" and "justification" of actions. A significant chi-square test statistic ($\chi^2(1) = 12.24$, $p < 0.01$) indicates that the researcher found empirical support for the research hypothesis stating that there *is* a relationship between indirect aggression and justification in the population of children's cartoons. In order to interpret this relationship it is helpful to compare the observed frequencies/counts in Figure D19 with the expected frequencies/counts from Figure D20. The comparison shows that acts of social exclusion appear justified more than expected and unjustified less than expected. Physical indirect aggression and guilt induction are not justified more than expected and justified less than expected. Overall, then, the researcher concludes that acts of social exclusion tend to be justified more often, and indirect physical aggression and guilt induction appear unjustified more often in the population of children's cartoons.

A look at the row percentages may also help to interpret the findings. Row percentages add up to 100% within a row of the cross-table. That is, 21.1% of cartoons portraying social exclusion frame this portrayal as unjustified and 78.9% as justified while 72.7% of cartoons portraying physical aggression frame this depiction as unjustified and 27.3% as justified. The row percentages of guilt induction are with 80%/20% more similar to physical aggression than to social exclusion.

PROBLEM 8: PEARSON CORRELATION R

Background

The Pearson correlation represents the strength of relationship between two variables that are measured on an interval- or ratio level. The use of the Pearson correlation coefficient (also called product-moment correlation) assumes a linear relationship between two variables. The Pearson correlation cannot be applied (or may lead to meaningless results) if the relationship between the two variables is non-linear.

Assumption

Strictly speaking, a Pearson correlation analysis requires the same assumption as a multiple regression analysis (see Problem 12). In practice however, researchers work with a simplified set of assumptions: 1) The two variables stem from a bivariate normally distributed population; and 2) there is a linear relationship between the two variables.

Example: Relationship uncertainty

A researcher is interested in the relationship between "relationship uncertainty" and "verbal disfluency" when leaving a message for a date request. Higher scores indicate greater relationship uncertainty and higher verbal disfluency. The researcher hypothesizes that greater relationship uncertainty increases verbal disfluencies. Thus, the hypotheses are as follows: H1: $\rho > 0$; H_0: $\rho \leq 0$. See Knobloch (2006).

STEP-BY-STEP (CORRELATION COEFFICIENT)

WINDOWS

1. Click **Data, New data set.**
2. Enter PROB8 as the name for the new data set.
3. Enter the data shown in Figure D21 into the Data Editor.

Figure D21. Problem 8 data

Data Editor

	UNCERTAINTY	DISFLUENCY	var3	var4
1	1	0		
2	1.44	1		
3	3.94	1		
4	2.12	2		
5	3.62	3		
6	3.81	4		
7	3.94	4		
8	1.5	5		
9	4.62	5		
10	5.88	6		
11				
12				
13				
14				
15				

MAC

1. In your text editor of choice (TextEdit, Microsoft Word, or OpenOffice Writer) enter the data as pictured in Figure D21, and save as PROB8.txt.
2. Click on **Data, Import data, from text file, clipboard, or URL.** Once the data set is imported, check the Message Window for any errors.
3. Enter PROB8 as name of the data set and check **Variable names in file.**

BOTH WINDOWS/MAC

4. Click **Data, Active data set, Save active set.**
5. Click **Statistics, Summaries, Correlation test.**
6. Select DISFLUENCY and UNCERTAINTY.
7. For Type of Correlation, select **Pearson product-moment.**
8. For Alternative Hypothesis, select **two-sided.**
9. Click **OK.**

Output (correlation coefficient)

Figure D22 shows the output of the test. The output includes the *t*-value, degrees of freedom, *p*-value, and statement of the alternative or research hypothesis.

Figure D22. Problem 9 Pearson product-moment correlation

```
        Pearson's product-moment correlation

data:  PROB8$DISFLUENCY and PROB8$UNCERTAINTY
t = 2.3527, df = 8, p-value = 0.02324
alternative hypothesis: true correlation is greater than 0
95 percent confidence interval:
 0.1347950 1.0000000
sample estimates:
      cor
0.6394937
```

Output interpretation (correlation coefficient)

A Pearson correlation coefficient between individuals relationship uncertainty and verbal disfluency scores was computed and found to be 0.64 and significant with $p < 0.05$. It can be concluded that individuals' relationship uncertainty scores and their verbal disfluency scores when leaving a message for their partner stand in a strong linear relationship and thus the research hypothesis found strong empirical evidence. The researcher would interpret that indeed, the higher the uncertainty, the greater the number of verbal disfluencies produced by the speaker. At this point, a scatterplot may help to visualize the results.

STEP-BY-STEP (SCATTERPLOT)

1. Click **Graphs** and select **Scatterplot.**
2. Select UNCERTAINTY as the *x*-variable, and DISFLUENCY as the *y*-variable.
3. Uncheck **Marginal boxplots.**
4. Uncheck **Smooth line.**
5. Click **OK.**

Output (scatterplot)

Figure D23. Scatterplot of relationship uncertainty and verbal disfluency

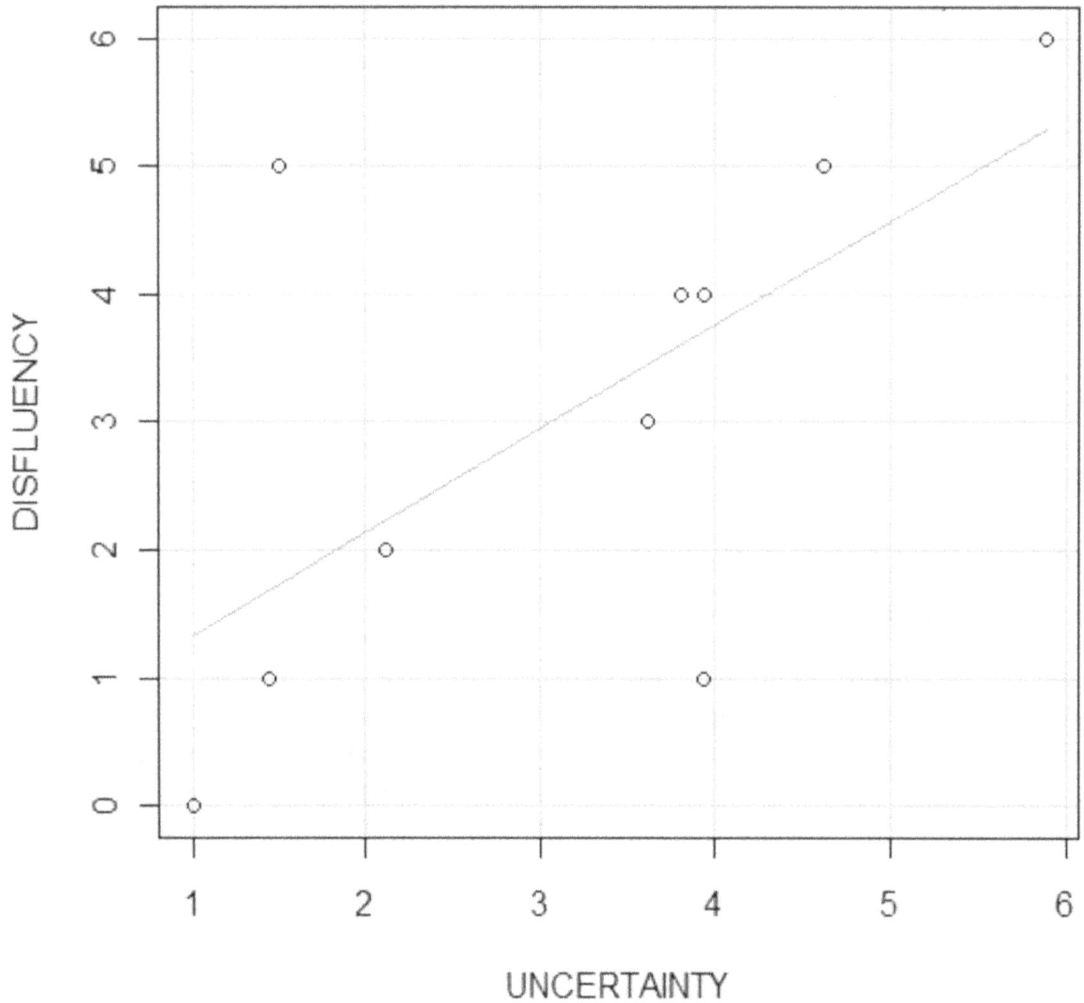

The scatterplot in Figure D23 visualizes that the relationship between disfluency and uncertainty can be well, albeit not perfectly, represented by a linear function or a line. No outliers that may lead to a misrepresentation of this relationship (see Chapter 8) can be observed.

PROBLEM 9: PARTIAL CORRELATION

Background

A Pearson correlation between two variables can be spurious if both variables are affected by a third variable in the same way. A partial correlation analysis can detect a spurious relationship between variables. A partial correlation coefficient indicates the (true) correlation between variables by adjusting each variable for the impact of the third variable (see Chapter 8).

Assumptions

A partial correlation analysis is based on the same assumptions as a Pearson correlation analysis (see problem 8): 1) All variables are bivariate normally distributed in the population; and 2) the relationship between the variables is of linear nature.

Example: Driver and officer communication quality

A researcher is interested in the relationship between driver communication quality and traffic stop length (in minutes). Lower scores for driver and officer communication indicate that they exhibited less accommodative behavior. The researcher hypothesizes that driver communication quality influences the length of traffic stop. Specifically, less accommodative drivers experience longer traffic stops. Both driver communication and traffic stop correlate with officer communication quality in the same way: less accommodative communication by the officer leads to less accommodative communication by the driver. Moreover, traffic stops of less accommodative officers usually take longer. Hence, a correlation between driver communication quality and the length of a traffic stop may be spurious to the extent that both variables are affected by officer communication quality. A partial correlation analysis tests the correlation between driver communication quality and the length of a traffic stop with the influence of officer communication quality being statistically controlled. The hypotheses of a partial correlation analysis are: H_1: $\rho_{(\text{driver,length} \cdot \text{officer})} < 0$; H_0: $\rho_{(\text{driver,length} \cdot \text{officer})} \geq 0$. See Dixon, Schell, Giles, & Drogos (2008).

STEP-BY-STEP

WINDOWS

1. Click **Packages, Install packages, CRAN Mirror (nearest you), ppcor.**
2. Click **Packages, Load packages,** and **ppcor.**
3. Click **Data, New data set.**
4. Enter PROB9 as the name for the new data set.
5. Enter the data shown in Figure D24 into the Data Editor.

MAC

1. In the main R Console, select **Packages & Data,** and **Package Installer.**
2. In the dropdown menu select CRAN Binaries. Scroll down to the **ppcor** package. It is very important to tick the checkbox box to **Install dependencies,** and then **Install Selected.** Go to **Packages & Data,** and **Package Manager.** Scroll down till you see ppcor. The left column ("status") will let you know whether it is loaded or not loaded. If ppcor is "not loaded" tick the check box to change its status to "loaded." You may need to hit the refresh button at the top right-hand corner to update the status. Once ppcor has loaded you can minimize the Package manager.
3. In your text editor of choice (TextEdit, Microsoft Word, or OpenOffice Writer) enter the data as pictured in Figure D24, and save as PROB9.txt.
4. Click on **Data, Import data, from text file, clipboard, or URL.** Once the data set is imported, check the Message Window for any errors.
5. Enter PROB9 as name of the data set and check **Variable names in file.**

Figure D24. Problem 9 data set

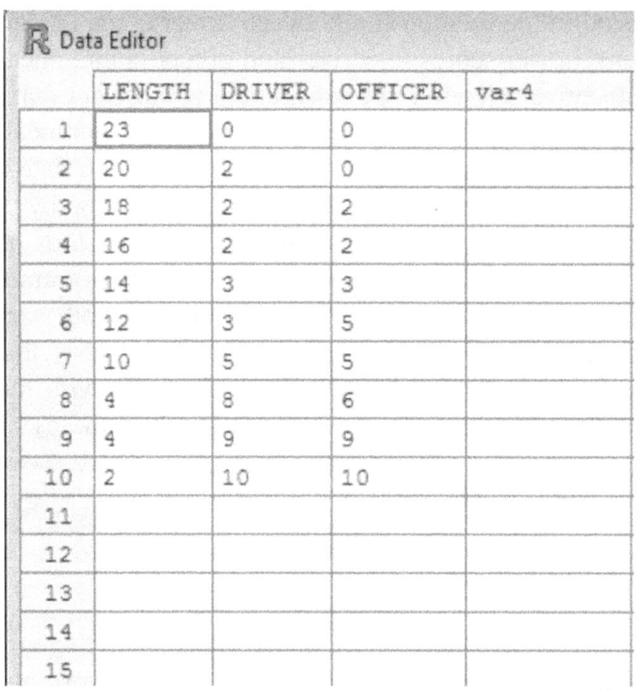

Data Editor

	LENGTH	DRIVER	OFFICER	var4
1	23	0	0	
2	20	2	0	
3	18	2	2	
4	16	2	2	
5	14	3	3	
6	12	3	5	
7	10	5	5	
8	4	8	6	
9	4	9	9	
10	2	10	10	
11				
12				
13				
14				
15				

BOTH WINDOWS/MAC

6. Click **Data, Active data set, Save active set.**
7. In the Script Window, enter the code below and click **Submit:**
   attach(PROB9)
   pcor.test(DRIVER,LENGTH, OFFICER, method = c("pearson"))

Output Interpretation

Figure D25 provides the output of the partial correlation analysis. The table displays the correlation between driver communication quality and length of a traffic stop, controlling for officer communication quality. As expected the partial correlation coefficient (−0.67) is the direction hypothesized and is significant ($p < 0.05$).

Figure D25. Output of driver by length controlling for officer communication quality

```
> library(ppcor, pos=4)

> attach (PROB9)

> pcor.test(DRIVER,LENGTH, OFFICER, method = c("pearson"))
    estimate      p.value statistic  n gp  Method
1 -0.6701435 0.01690411 -2.388787 10   1 pearson
```

PROBLEM 10: ONE-WAY ANALYSIS OF VARIANCE

Background

A one-way analysis of variance (ANOVA) can be understood as a generalization of the two independent samples *t*-test (see Problem 5). Thus, a one-way ANOVA evaluates whether more than two, that is, *p* independent samples represent *p* populations of which at least two are characterized by a different mean parameter.

Assumptions

The one-way analysis of variance is used with interval/ratio level data for the dependent variable and nominal level data for the independent or group variable. A one-way ANOVA assumes the following: 1) each sample has been randomly selected from the population it represents; 2) the population distributions from which samples derive are normally distributed; 3) the dependent variable's variances in the underlying populations are equal or homogeneous. If any of the assumptions are violated, the reliability of the test statistic may be compromised.

Example: Communication utility of online shopping

A researcher is interested in whether three different groups of shoppers have different perceptions about communication utility of online shopping. A higher score indicates a higher perceived communication utility of online shopping among shoppers. Thus, the hypotheses are as follows: H_1: $\mu_i \neq \mu_i{}'$; H_0: $\mu_1 = \mu_2 = \mu_3$. See Li, Kuo, and Russell (1999).

STEP-BY-STEP

Windows

1. Click **Data, New data set.**
2. Enter PROB10 as the name for the new data set.
3. Enter the data from Figure D26 below into the Data Editor.

Mac

1. In your text editor of choice (TextEdit, Microsoft Word, or OpenOffice Writer) enter the data as pictured in Figure D26, and save as PROB10.txt.
2. Click on **Data, Import data, from text file, clipboard, or URL.** Once the data set is imported, check the Message Window for any errors.
3. Enter PROB10 as name of the data set and check **Variable names in file.**

Figure D26. Problem 10 data

Data Editor

	COMMUTIL	SHOPPER	var3	var4
1	3.33	BUYER		
2	3.33	BUYER		
3	3.33	BUYER		
4	3.67	BUYER		
5	4	BUYER		
6	3.67	OCCASIONAL		
7	4	OCCASIONAL		
8	4	OCCASIONAL		
9	4	OCCASIONAL		
10	4.33	OCCASIONAL		
11	4.67	FREQUENT		
12	4.67	FREQUENT		
13	5	FREQUENT		
14	5	FREQUENT		
15	5	FREQUENT		

4. Click **Data, Active data set, Save active set.**
5. Click **Statistics, Means, One-way ANOVA.**
6. Enter **PROB10.MODEL** as the name for the model.
7. Select **SHOPPER** as **Groups variable** and **COMMUTIL** as **Response variable.**
8. Check **Pairwise comparison of means.**
9. Click **OK.**

Output

Figure D27 shows the output of the test, which includes four tables. The first table provides degrees of freedom, sum of squared errors, and mean squared errors for SHOPPER and Residuals, an F value, and a p-value for SHOPPER. The second table provides the mean, standard deviation, and number of cases in each group. The first table only tells us that there is a significant mean difference, but it doesn't tell us between which groups. This is the information provided by the third and fourth tables. The third table offers information about the estimated mean differences between the three groups of online shoppers, and the fourth table gives us a confidence interval for those mean differences.

Figure D27. One-way ANOVA output

```
> summary(PROB10.MODEL)
            Df Sum Sq Mean Sq F value  Pr(>F)
SHOPPER      2 4.5956 2.29779  38.893 5.7e-06 ***
Residuals   12 0.7090 0.05908
---
Signif. codes:  0 '***' 0.001 '**' 0.01 '*' 0.05 '.' 0.1 ' ' 1

> numSummary(PROB10$COMMUTIL , groups=PROB10$SHOPPER, statistics=c("mean",
+   "sd"))
            mean        sd n
buyer      3.532 0.3001999 5
frequent   4.868 0.1807484 5
occasional 4.000 0.2333452 5

         Simultaneous Tests for General Linear Hypotheses

Multiple Comparisons of Means: Tukey Contrasts


Fit: aov(formula = COMMUTIL ~ SHOPPER, data = PROB10)

Linear Hypotheses:
                           Estimate Std. Error t value Pr(>|t|)
frequent - buyer == 0        1.3360     0.1537   8.691   <0.001 ***
occasional - buyer == 0      0.4680     0.1537   3.044   0.0253 *
occasional - frequent == 0  -0.8680     0.1537  -5.646   <0.001 ***
---
Signif. codes:  0 '***' 0.001 '**' 0.01 '*' 0.05 '.' 0.1 ' ' 1
(Adjusted p values reported -- single-step method)

         Simultaneous Confidence Intervals

Multiple Comparisons of Means: Tukey Contrasts


Fit: aov(formula = COMMUTIL ~ SHOPPER, data = PROB10)

Quantile = 2.6679
95% family-wise confidence level


Linear Hypotheses:
                           Estimate lwr      upr
frequent - buyer == 0       1.33600  0.92587  1.74613
occasional - buyer == 0     0.46800  0.05787  0.87813
occasional - frequent == 0 -0.86800 -1.27813 -0.45787
```

Output Interpretation

A one-way ANOVA was computed comparing the communication utility scores for three different groups of shoppers. A significant F ratio was found between groups ($F = 38.89, p < 0.001$), thus the research hypothesis (H_1: $\mu_i \neq \mu_i'$) was supported by the data. A post-hoc group mean comparison analysis was conducted using Tukey mean contrasts. Significant differences in the means of communication utilities were found between the groups "frequent online shopper" and "buyer" (label "frequent-buyer == 0″; $t = 8.69, p < 0.001$), "occasional shopper" and "buyer" (label "occasional-buyer == 0″; $t = 3.044$, $p < 0.0257$), and "occasional shopper" and "frequent shopper" (label "occasional-frequent == 0″; $t =$?5.646, $p < 0.001$)." The labels of these mean comparisons contain the null hypothesis assumption. For example, "occasional-buyer == 0″ stands for "μ(occasional shopper) − μ(buyer) = 0″; i.e., no mean difference between these groups in the population. Under "estimate," table three outputs the estimated mean difference between the groups. Table four displays the 95% confidence interval limits (see Chapter 6) of those mean estimates.

PROBLEM 11: FACTORIAL ANOVA

Background

Factorial analysis of variance (factorial ANOVA) can be understood as a generalization of a one-way ANOVA (see Problem 10). The procedure allows the analysis of more than one independent variable (factor) as well as an analysis of interactions among the independent variables (factors). For each combination of the factor groups an independent sample of data is used to estimate a corresponding population mean-parameter. Thus, a factorial ANOVA evaluates whether independent samples across combinations of factor groups represent the corresponding populations of which at least two are characterized by different mean parameters.

Assumptions

The factorial ANOVA is used with interval/ratio level data for the dependent variable and nominal level data for the independent variables (factors). A factorial ANOVA assumes the following: 1) each sample has been randomly selected from the population it represents; 2) the population distributions from which samples derive are normally distributed; 3) the dependent variable's variances in the underlying populations across factor group combinations are equal or homogeneous. If any of the assumptions are violated, the reliability of the test statistic may be compromised.

Example: Humor and product evaluations

A researcher is interested in studying the effects of gender and two different levels of humor in advertisements on product evaluations. Higher scores on product evaluations indicate greater negative sentiment toward the advertisement. Thus, the hypotheses are for gender, H1: $\mu_M \neq \mu_F$, H_0: $\mu_M = \mu_F$; for humor, H_1: $\mu_{Low} \neq \mu_{High}$, H_0: $\mu_{Low} = \mu_{High}$; and for the interaction of gender and humor, H_1: $\mu_{ij} \neq \mu_i + \mu_j - \mu$, H_0: $\mu_{ij} = \mu_i + \mu_j - \mu$ (μ_{ij} stands for the population mean for the combination of group "*i*" of factor "gender" and group "*j*" of factor "humor"; μ stands for the grand mean (i.e., mean of all data across all factor combinations in the population). See Perry et. al. (1997).

STEP-BY-STEP (BASIC ANALYSIS)

Windows

1. Click **Data, New data set.**
2. Enter PROB11 as the name for the new data set.
3. Enter the data shown in Figure D28 into the Data Editor.

Mac

1. In your text editor of choice (TextEdit, Microsoft Word, or OpenOffice Writer) enter the data as pictured in Figure D28, and save as PROB11.txt.
2. Click on **Data, Import data, from text file, clipboard, or URL.** Once the data set is imported, check the Message Window for any errors.
3. Enter PROB11 as name of the data set and check **Variable names in file.**

Both Windows/Mac

4. Click **Data, Active data set, Save active set.**
5. Click **Statistics, Means, Multi-way ANOVA.**
6. Enter **PROB11.MODEL** as the name for the model.
7. Select **GENDER** and **HUMOR** as Factors and **EVAL** as Response Variable.
8. Click **OK.**

Figure D28. Problem 11 data

Data Editor

	EVAL	GENDER	HUMOR	var4
1	3	M	LOW	
2	3.5	M	LOW	
3	4	M	LOW	
4	4	M	LOW	
5	4	M	LOW	
6	1	M	HIGH	
7	1	M	HIGH	
8	1.5	M	HIGH	
9	2	M	HIGH	
10	2	M	HIGH	
11	1.5	F	LOW	
12	2	F	LOW	
13	2.5	F	LOW	
14	2.5	F	LOW	
15	2.5	F	LOW	
16	2	F	HIGH	
17	2.5	F	HIGH	
18	2.5	F	HIGH	
19	3	F	HIGH	
20	3	F	HIGH	

Output (basic analysis)

Figure D29 shows the output of the test, which includes four tables. The first table provides an ANOVA table for the main effects (GENDER, HUMOR) and their interaction term (GENDER:HUMOR). The subsequent tables provide the mean, standard deviation, and number of cases, respectively, for each factor combination of GENDER and HUMOR.

Figure D29. Factorial ANOVA

```
> Anova(PROB11.MODEL)
Anova Table (Type II tests)

Response: EVAL
              Sum Sq Df F value    Pr(>F)
GENDER          0.20  1  0.9697 0.3394136
HUMOR           4.05  1 19.6364 0.0004191 ***
GENDER:HUMOR    8.45  1 40.9697  8.78e-06 ***
Residuals       3.30 16
---
Signif. codes:  0 '***' 0.001 '**' 0.01 '*' 0.05 '.' 0.1 ' ' 1

> tapply(PROB11$EVAL, list(GENDER=PROB11$GENDER, HUMOR=PROB11$HUMOR), me
+    na.rm=TRUE) # means
      HUMOR
GENDER HIGH LOW
     F  2.6 2.2
     M  1.5 3.7

> tapply(PROB11$EVAL, list(GENDER=PROB11$GENDER, HUMOR=PROB11$HUMOR),
+    na.rm=TRUE) # std. deviations
      HUMOR
GENDER    HIGH       LOW
     F 0.41833 0.4472136
     M 0.50000 0.4472136

> tapply(PROB11$EVAL, list(GENDER=PROB11$GENDER, HUMOR=PROB11$HUMOR),
+    function(x) sum(!is.na(x))) # counts
      HUMOR
GENDER HIGH LOW
     F    5   5
     M    5   5
```

Output Interpretation (basic analysis)

A factorial ANOVA was conducted in an attempt to test differences in product evaluations. A significant main effect was found for humor ($F(1,16) = 19.64$, $p < 0.000$). The effect for gender was not significant ($F(1,16) = 0.97$, $p < 0.34$). There was a significant interaction between gender and humor ($F(1,16) = 40.97$, $p < 0.000$), indicating that specific combinations of gender groups with humor level affect product evaluations. Looking at the factor combination means (table 3) it seems that using humor was especially successful for male viewers as their mean product evaluation score was the lowest (1.5). At this point, it is recommended to graph the effects for a closer examination of the main effects and the interaction.

STEP-BY-STEP (EFFECT EXAMINATION)

1. Click **Models, Graphs, Effect plots.** This will show an effect plot for the current model, which should be PROB11.MODEL.

Output (effect examination)

Figure D30 shows an interaction plot. Note that evaluation (the dependent variable) is along the *y*-axis and that higher values correspond to a more negative evaluation.

Figure D30. Factorial ANOVA – Interaction charts

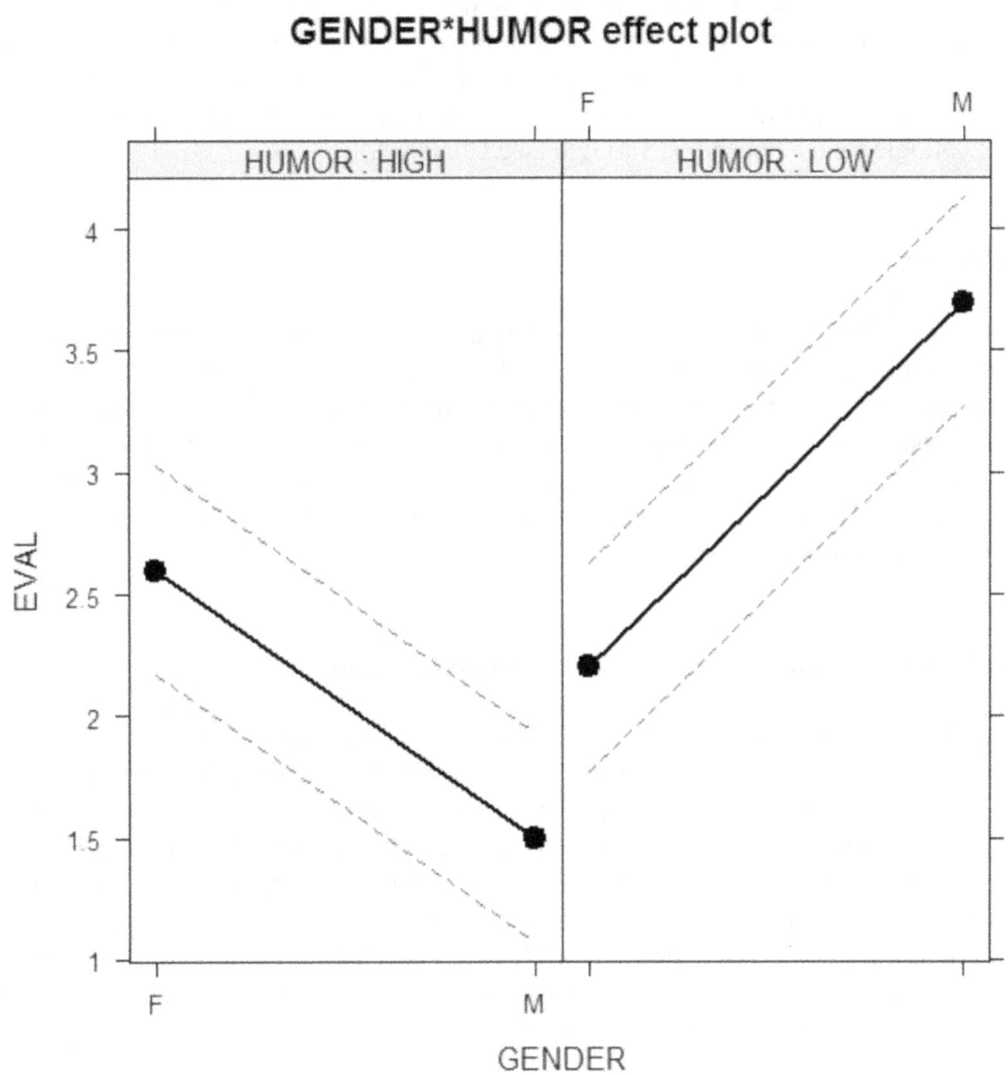

Output Interpretation (effect examination)

Figure D30 visualizes the interaction of the factors gender and humor. One can see (left chart) that for high humor levels, male viewers evaluate the product in the advertisement with lower scores; i.e., better, than female viewers. For low humor levels (right chart) this pattern is reversed. Here, female viewers evaluate the product better than male viewers.

PROBLEM 12: MULTIPLE REGRESSION

Background

Multiple regression analysis allows researchers to predict the values of one dependent or outcome variable from one or more independent or predictor variables. In the most basic version of regression analysis, all variables are assumed to be measured on an interval/ratio level. Regression analysis can also be used to examine the linear relationship between a set of independent variables and a dependent variable (see Chapter 8).

Assumptions

Multiple regression analysis assumes the following: 1) the dependent variable's values are normally distributed around the regression line or line of best fit in the population, this assumption is usually replaced by the assumption of normally distributed regression residuals; 2) there is a linear relationship between the dependent variable and the set of independent variables; 3) the variability of residuals is the same through the range of values for the independent variables; and 4) the residuals are not correlated among each other. We will not demonstrate how to test all of these assumptions, however.

Example: Communicative Action and Perceived Legitimacy

A researcher is interested in citizen participation in a policy-making process on smoking in public places. Specifically, this researcher wants to determine the effects of communicative action variables (comprehensibility, appropriateness, and speech conditions) on individuals' perceptions of policy legitimacy. The researcher states the following hypotheses: H_1a: With increasing perceived comprehensibility, perceived legitimacy increases; H_1b: With increasing perceived appropriateness, perceived legitimacy increases; H_1c: With increasing perceived speech conditions, perceived legitimacy increases; H_0: There is no relationship between perceived legitimacy, perceived comprehensibility, perceived appropriateness, and perceived speech conditions. See Chang and Jacobson (2010).

STEP-BY-STEP (BASIC ANALYSIS)

WINDOWS

1. Select **Packages, Install Packages, CRAN mirror (nearest you),** and **yhat.**
2. Select **Load Packages, yhat.**
3. Click **Data, New data set.**
4. Enter PROB12 as the name for the new data set.
5. Enter the data in Figure D31 into the Data Editor.

MAC

1. In the main R Console, select **Packages & Data,** and **Package Installer.**
2. In the dropdown menu select CRAN Binaries. Scroll down to the **yhat** package. It is very important to tick the checkbox to **Install dependencies,** and then **Install Selected.** Go to **Packages & Data,** and **Package Manager.** Scroll down until you see yhat. The left column ("status") will let you know whether it is loaded or not loaded. If yhat is "not loaded" tick the box to change its status to "loaded." You may need to hit the refresh button at the top right-hand corner to update the status. Once yhat has loaded you can minimize the Package manager.
3. In your text editor of choice (TextEdit, Microsoft Word, or OpenOffice Writer) enter the data as pictured in Figure D31, and save as PROB12.txt.
4. Click on **Data, Import data, from text file, clipboard, or URL.** Once the data set is imported, check the Message Window for any errors.
5. Enter PROB12 as name of the data set and check **Variable names in file.**

Figure D31. Problem 12 data

R Data Editor

	LEGIT	SPEECH	COMP	APPR
1	9.86	6.47	5.79	8.54
2	6.1	9.34	4.12	6.51
3	6.38	7.54	9.41	4.58
4	8.14	9.71	8.56	5.88
5	3.32	9.87	3.82	1.82
6	8.25	4.58	9.09	6.1
7	4.26	7.71	4.91	4.48
8	3.03	9.16	1.31	2.06
9	5.15	8.03	8.07	4.54
10	7.61	7.71	6.43	4.57
11	9.28	9.92	8.78	5.64
12	5.07	7.37	7.02	3.24
13	9.33	8.55	9.03	4.57
14	6.94	9.8	7.64	4.29
15	3.58	6.91	8.32	2.44

BOTH WINDOWS/MAC

6. Click **Data, Active data set, Save active set.**
7. Select **Statistics, Fit Models,** and **Linear Regression.**
8. Enter **PROB12.MODEL** as the name of the model.
9. Select LEGIT as the Response Variable, and SPEECH, COMP and APPR as Explanatory Variables. Click **OK.**
10. In the Script Window, enter and highlight the following: regr(PROB12.MODEL).
11. Click **Submit.**

Output (basic analysis)

Figure D32 provides the output of the multiple regression, including residuals statistics, regression coefficients, *R* squared, adjusted *R* squared, *F* statistic, degrees of freedom, and *p*-value for the model. The command "regr" produces beta-weights, which are standardized coefficients that provide information about the relative influence of each independent variable on the dependent variable.

Figure D32. Multiple regression output

```
Call:
lm(formula = LEGIT ~ APPR + COMP + SPEECH, data = PROB12)

Residuals:
     Min       1Q   Median       3Q      Max
-1.63132 -0.92640  0.09939  0.64721  2.02376

Coefficients:
            Estimate Std. Error t value Pr(>|t|)
(Intercept)  -2.6029     2.5354  -1.027 0.326653
APPR          0.9394     0.1926   4.878 0.000488 ***
COMP          0.3777     0.1459   2.589 0.025169 *
SPEECH        0.2579     0.2273   1.135 0.280611
---
Signif. codes:  0 '***' 0.001 '**' 0.01 '*' 0.05 '.' 0.1 ' ' 1

Residual standard error: 1.214 on 11 degrees of freedom
Multiple R-squared: 0.7805, Adjusted R-squared: 0.7206
F-statistic: 13.04 on 3 and 11 DF,  p-value: 0.0006079

$Beta_Weights
     APPR      COMP    SPEECH
0.7328768 0.3922717 0.1706390
```

Output Interpretation

For a basic analysis, only the table "coefficients" and the "R-squared" information are important. The table "coefficients" tells the researcher that both independent variables predict the dependent variable well, both perceived appropriateness ($\beta = 0.73$, $t = 4.88$, $df = 11$, $p > 0.001$ and perceived comprehensibility ($\beta = 0.39$, $t = 2.59$, $df = 11$, $p > 0.05$) are significant predictors of perceived legitimacy. Perceived speech conditions was not a significant predictor of perceived legitimacy ($\beta = 0.17$, $t = 1.13$, $df = 11$, $p > 0.05$). As the beta-coefficients and coefficient estimates indicate, with increasing perceived appropriateness and increasing perceived comprehensibility, perceived legitimacy increases. Comparing the two predictors, perceived appropriateness is the more important predictor. The regression model explains a high amount of variance in the dependent variable. With the three predictors 72.06% (adjusted *R*-squared) of variance in perceived legitimacy can be explained. At this point, we should check the adequacy of the model by conducting a basic analysis of the model's residuals.

STEP-BY-STEP (MODEL ANALYSIS)

1. Click **Models, Graphs, Basic diagnostic plots.**
2. Click **Statistics, Summaries, Shapiro-Wilk test of normality.**
3. Select **LEGIT** as the variable.

Output Interpretation (model analysis)

Although running the diagnostic plot provides four outputs, Figure D33 focuses your attention on the "Residuals versus Fitted" and the "Normal Q-Q" plots. The graph on the left shows residuals on the *y*-axis against fitted values on the *x*-axis. In this type of plot you do not want to see structure or pattern. Moreover, it is problematic if the scatter cloud increases as the fitted values increase. The graph on the right shows a Normal Q-Q plot, which should be a straight line if the errors are normally distributed. Since there appears to be some deviation from normality, we are going to check whether this is a problem by using the output of a Shapiro-Wilk normality test.

Figure D33. Multiple regression—Model diagnostic plot

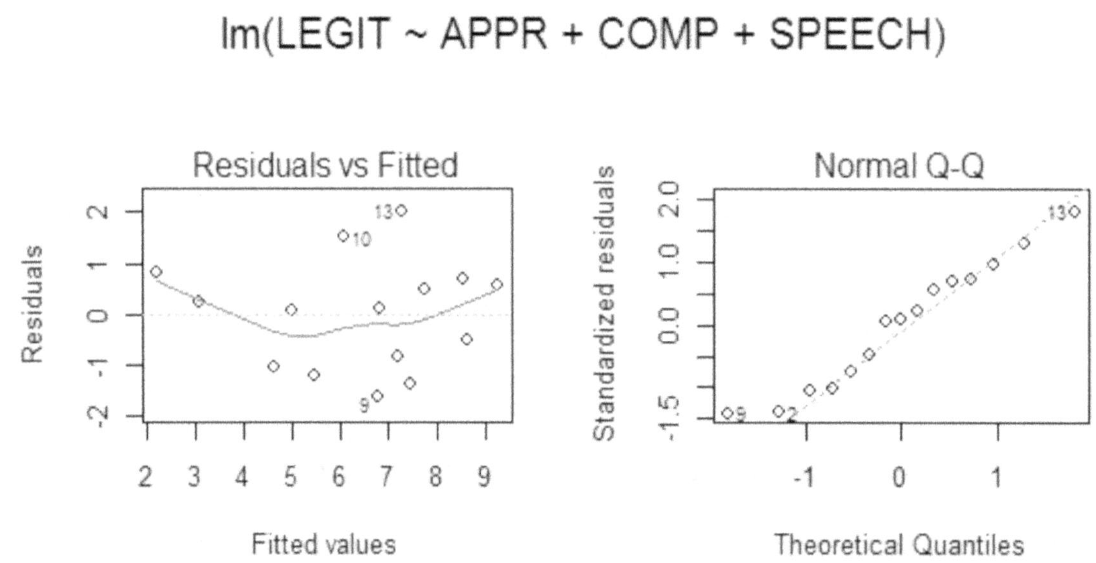

Figure D34. Shapiro-Wilk test of normality

```
        Shapiro-Wilk normality test

data:  PROB12$LEGIT
W = 0.9428, p-value = 0.419
```

The Shapiro-Wilk test of normality in Figure D34 indicates a non-significant *p*-value which is a desired result as it shows that our residual distribution does not significantly deviate from a normal distribution. So we proceed with our interpretation of Figure D33.

An advanced model analysis would also check the influence of individual values on the model coefficients and test whether the assumptions of homoscedastic (equal variance) and uncorrelated residuals are violated. An advanced model analysis is not provided here as this analysis requires more background than discussed in this introductory textbook.

FINAL NOTE

In this appendix we have provided a beginning overview of using the software package R and the graphical user interface R-Commander. We have overviewed several elementary functions, including generating descriptive data, as well as those more advanced functions related to inferential statistics. There are many, many more applications that one can use in R and R-Commander. We encourage you to read the online available documentation for R and R-Commander, and explore the additional statistical analyses available in this powerful and free software package for statistical analyses.

REFERENCES

Boyle, M, Schmierbach, M, Armstrong, C, et al. (2006). Expressive responses to news stories about extremist groups: A framing experiment. *Journal of Communication, 56*(2), 271–288.

Chang, C. (2006). Beating the news blues: Mood repair through exposure to advertising. *Journal of Communication, 56*(1), 198–217.

Chang, L., & Jacobson, T. (2010). Measuring participation as communicative action: A case study of citizen involvement in and assessment of a city's smoking cessation policy-making process. *Journal of Communication, 60*(4), 660–679.

Coyne, S. M., & Whitehead, E. (2008). Indirect aggression in animated Disney films. *Journal of Communication, 58*(2), 382–395.

Dixon, T. L., Schell, T. L., Giles, H., & Drogos, K. L. (2008). The influence of race in police-civilian interactions: A content analysis of videotaped interactions taken during Cincinnati police traffic stops. *Journal of Communication, 58*(3), 530–549.

Eveland, W. P., & Hively, M. H. (2009). Political discussion frequency, network size, and "heterogeneity" of discussion as predictors of political knowledge and participation. *Journal of Communication, 59*(2), 205–224.

Eastin, M.S., Appiah, O., & Cicchirillo, V. (2009). Identification and the influence of cultural stereotyping on postvideogame play hostility. *Human Communication Research, 35,* 337–356.

Fox, J. (2011). Rcmdr: R Commander. R package version 1.6-3. http://www.r-project.org/package=Rcmdr

Kim, S. (2011). Ppcor: Partial and semi-partial (part) correlation. R package version 1.0. http://CRAN.R-project.org/package=ppcor

Knobloch, L.K. (2006). Relational uncertainty and message production within courtship: Features of date request messages. *Human Communication Research, 32*(2), 244–273.

Li, H., Kuo, C., & Russell, M.G. (1999). The impact of perceived channel utilities, shopping orientations, and demographics on the consumer's online buying behavior. *Journal of Computer-Mediated Communication, 5*(2).

Nabi, R L., & Clark, S. (2008). Exploring the limits of social cognitive theory: Why negatively reinforced behaviors on TV may be modeled anyway. *Journal of Communication, 58*(3), 407–427.

Nimon, K. & Roberts, J.K. (2009). yhat: Interpreting Regression Effects. R package version 1.0-3. http://CRAN.R-project.org/package=yhat

Perry, S.D., Jenzowsky, S.A., King, C.M., Yi, H., Hester, J.B., & Gartenschlaeger. (1997). Using humorous programs as a vehicle for humorous commercials. *Journal of Communication, 47*(1), 20–39.

Taylor, M., Jowi, D., Schreirer, H., & Bertelsen, D. (2011). Students' perceptions of email interaction during student-professor advising sessions: The pursuit of interpersonal goals. *Journal of Computer-Mediated Communication, 16*(2), 307–330.

R Core Development Team. (2011). R: A language for statistical computing. R Foundation for Statistical Computing, Vienna, Austria. http://www.r-project.org

INDEX

www.ingramcontent.com/pod-product-compliance
Lightning Source LLC
LaVergne TN
LVHW080644011026
849160LV00014B/1313
* 9 7 8 1 4 6 5 2 1 2 2 3 8 *